Mike Sawicki

Trainee Guide

Concrete Finishing

Level 1

11-2 23107
11-9 23108

National Center for Construction Education and Research

Wheels of Learning
Standardized Craft Training

Prentice Hall
Upper Saddle River, New Jersey Columbus, Ohio

Prentice-Hall, Inc.
Upper Saddle River, New Jersey 07458

This information is general in nature and intended for training purposes only. Actual performance of activities described in this manual requires compliance with all applicable operations procedures under the direction of qualified personnel. References in this manual to patented or proprietary devices do not constitute a recommendation for their use.

Printed in the United States of America

10 9 8 7

ISBN: 0-13-010246-6

Prentice-Hall International (UK) Limited, *London*
Prentice-Hall of Australia Pty. Limited, *Sydney*
Prentice-Hall of Canada, Inc., *Toronto*
Prentice-Hall Hispanoamericana, S. A., *Mexico*
Prentice-Hall of India Private Limited, *New Delhi*
Prentice-Hall of Japan, Inc., *Tokyo*
Pearson Education Asia, *Singapore*
Editora Prentice-Hall do Brasil, Ltda., *Rio de Janeiro*

Preface

This volume is one of many in the *Wheels of Learning* craft training program. This program, covering more than 20 standardized craft areas, including all major construction skills, was developed over a period of years by industry and education specialists. Sixteen of the largest construction and maintenance firms in the U.S. committed financial and human resources to the teams that wrote the curricula and planned the national accredited training process. These materials are industry-proven and consist of competency-based textbooks and instructor guides.

The *Wheels of Learning* was developed by the National Center for Construction Education and Research in response to the training needs of the construction and maintenance industries. The NCCER is a nonprofit educational entity affiliated with the University of Florida and supported by the following industry and craft associations:

Partnering Associations

- ABC Texas Gulf Coast Chapter
- American Fire Sprinkler Association
- American Society for Training and Development
- American Vocational Association
- American Welding Society
- Associated Builders and Contractors, Inc.
- Associated General Contractors of America
- Carolinas AGC, Inc.
- Carolinas Electrical Contractors Association
- Construction Industry Institute
- Design-Build Institute of America
- Merit Contractors Association of Canada
- Metal Building Manufacturers Association
- National Association of Minority Contractors
- National Association of Women in Construction
- National Insulation Association
- National Ready Mixed Concrete Association
- National Utility Contractors Association
- National Vocational Technical Honor Society
- Painting and Decorating Contractors of America
- Portland Cement Association
- Steel Erectors Association of America
- University of Florida
- Women Construction Owners and Executives, USA

Some of the features of the *Wheels of Learning* program include:

- A proven record of success over many years of use by industry companies.
- National standardization providing "portability" of learned job skills and educational credits that will be of tremendous value to trainees.
- Recognition: upon successful completion of training with an accredited sponsor, trainees receive an industry-recognized certificate and transcript from the NCCER.
- Each level meets or exceeds Bureau of Apprenticeship and Training (BAT) requirements for related classroom training (CFR 29:29).
- Well illustrated, up-to-date, and practical information. All standardized manuals are reviewed annually in a continuous improvement process.

Acknowledgments

This manual would not exist were it not for the dedication and unselfish energy of those volunteers who served on the Technical Review Committee. A sincere thanks is extended to:

Gene Boeke Jr.
Rick Bohan
Ted Bowman
Bob Calderone
Anne Ellis
James (Jamie) A. Farny
Gus Lorber
Norbert Lovata
John Nehasil
Dennis Phillips
Michael J. Schneider
Gustave Rener
Dan Wolf

Introduction to Concrete Construction and Finishing

Module 23101

NATIONAL
CENTER FOR
CONSTRUCTION
EDUCATION AND
RESEARCH

INTRODUCTION TO CONCRETE CONSTRUCTION AND FINISHING

OBJECTIVES

Upon completion of this module, the trainee will be able to:

1. Define terms associated with concrete construction.
2. Identify the composition and characteristics of concrete.
3. Identify the uses of concrete as a building material.
4. Identify the effect of craftsmanship on finished concrete.
5. Explain the concrete construction process.
6. Identify site operation work requirements.
7. Explain the career potentials in concrete construction and finishing.

Prerequisites

Successful completion of the following Task Modules is recommended before beginning study of this Task Module: Core Curricula.

Required Trainee Materials

1. Trainee Task Module
2. Paper and pencil
3. Appropriate Personal Protective Equipment

COURSE MAP

This course map shows all of the task modules in the first level of the Concrete Finishing curricula. The suggested training order begins at the bottom and proceeds up. Skill levels increase as a trainee advances on the course map. The training order may be adjusted by the local Training Program Sponsor.

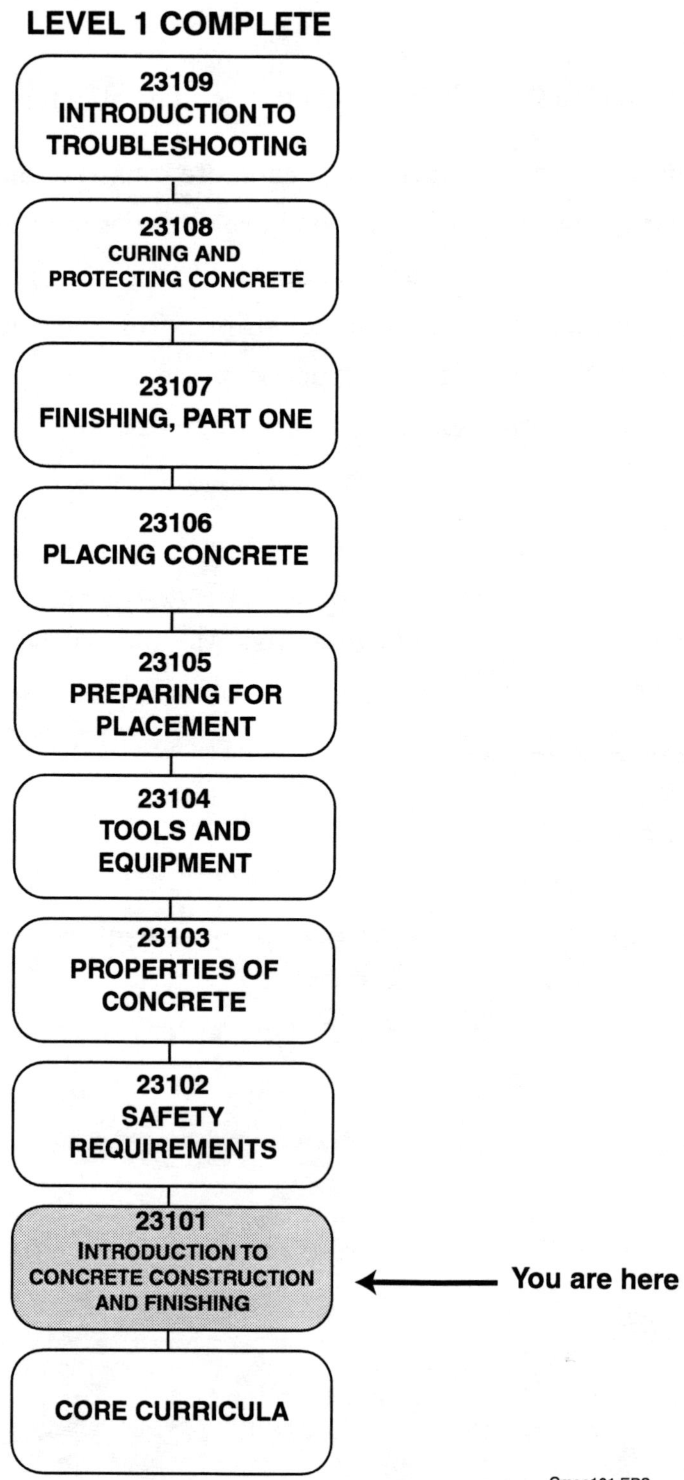

Cmap101.EPS

TABLE OF CONTENTS

Trade Terms Introduced In This Module

Admixture: A chemical or mineral other than water, cement, or aggregates added to concrete immediately before or during mixing to change its setting or curing time, or to reduce water, or to change overall properties of concrete.

Aggregate: A granular material, such as sand, gravel, crushed stone, or iron blast-furnace slag, which is used with portland cement to form concrete or mortar. This can also include recycled crushed concrete.

Air entrainment: The inclusion of air in the form of microscopic bubbles during the mixing of concrete. This gives frost resistance to the concrete and improves its workability.

Blast-furnace slag: The glassy material that is formed as a by-product of iron manufacture. It is blended in a pulverized state with portland cement to form blended hydraulic cements. It has various forms and can be used as an aggregate or a pozzolan in cementitious material. It consists essentially of silica, or aluminosilica glass containing calcium and other elements.

Cement: A powder ground from lime, iron, and other minerals heated in a kiln, that chemically reacts with water and serves as a binder of the ingredients in mortar and concrete.

Clinker: The partially fused by-product that is removed from a cement kiln after burning.

Compression: The force or load on a concrete member that acts to crush it.

Curing: Keeping fresh concrete at proper conditions of moisture and temperature to allow the cement to hydrate.

Expansive cement: A modified portland cement that when mixed with water, forms a paste that increases in volume during the early hardening period to compensate for shrinkage in the cement paste and to minimize cracking.

Fly ash: The fine residue from the burning of powdered or ground coal. It is carried from the fire through the boiler by flue gases.

Hydration: The chemical reaction that occurs between water and cement or other substances to form hardened concrete.

Paste: A mixture of cement and water without aggregates.

Plasticizing agent: A chemical compound that reduces internal friction in fresh concrete mixes to make placing easier or to reduce its requirement for water.

Portland cement: The product obtained by pulverizing clinker. It consists of hydraulic calcium silicates, with gypsum added during grinding. When mixed with water, portland cement forms the binder in portland cement concrete.

Post-tensioned: A method of reinforcing concrete with embedded cables or tendons tensioned after the concrete has hardened.

Pozzolan: A siliceous or siliceous and aluminous material that will, in finely divided form in the presence of moisture, chemically react with calcium hydroxide at ordinary temperatures to form compounds possessing cementitious properties. Pozzolans include pumice, fly ash, and blast-furnace slag.

Precast: A concrete member that is cast and cured away from the construction site.

Prestressed: A method of reinforcing concrete with embedded cables or tendons tensioned before the concrete hardens.

Screed: A manual or motorized straightedge tool used to strike off the surface of placed concrete to a predetermined grade.

Set time: The elapsed time between the adding of water and the hardening of concrete. Set time is affected by materials, mix proportions, admixtures, curing temperature, and weather.

Silica fume: A pozzolanic material that is a by-product of the manufacture of ferrosilicon. It is an extremely fine powder. Admixtures made with silica fume are often in liquid form.

Slip forming: A method of forming whereby the forms are moved ahead as the concrete is placed. This allows a continuous placement of concrete.

Tension: The load on a concrete member that acts to pull the member apart.

Tilt-up: A type of construction whereby the wall sections are formed, placed, and finished flat on the ground, and then tilted up and permanently fastened.

Water-cement ratio: The weight ratio of the amount of free water to the amount of cement in a paste, mortar, or concrete.

Workability: The property of freshly mixed concrete, mortar, or paste that determines the ease with which it can be mixed, placed, compacted, and finished.

1.0.0 INTRODUCTION

Concrete is one of the most widely used construction materials. It is strong and durable and can be readily formed into many shapes and sizes. It is also economical compared to other materials because the raw materials from which concrete is made are usually readily available.

People who work with concrete should know what concrete is made of and how it behaves. Because concrete is manufactured at or near the job site, the concrete worker needs to understand the safety requirements for mixing and placing concrete. They should also understand that the success of concrete construction depends largely on the worker's skills, knowledge, and craftsmanship.

This introductory module provides basic information that a concrete construction and finishing trainee should know. It answers questions about concrete and what it is used for, and describes concrete use in the construction process. It also discusses the roles of the craftworker in obtaining quality results. Those roles require a worker who meets the rigors and constraints of manufacturing a product, often under harsh site conditions. For those workers who can meet the requirements, there are interesting, challenging, and rewarding career opportunities available.

2.0.0 WHAT IS CONCRETE?

Concrete is a combination of **cement**, water, and **aggregates** mixed together. The most common type of cement is **portland cement** manufactured from limestone (calcium oxide) and other mineral compounds. The aggregate is a combination of fine and coarse stone particles such as sand, crushed rock, or gravel. The cement, when mixed with water, starts a chemical reaction called **hydration**, which causes the mixture to harden into concrete. *Figure 1* shows the proportions by volume of ingredients for typical air-entrained concrete. In other mixes, cement may range from 7 to 15 percent and aggregates from 60 to 75 percent. Changing the proportion of the ingredients changes the physical characteristics of the hardened concrete.

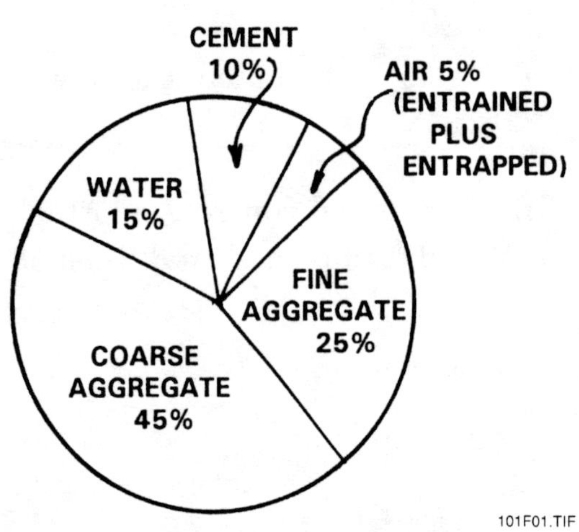

Figure 1. Composition Of Air-Entrained Concrete

2.1.0 CHARACTERISTICS OF CONCRETE

Concrete has characteristics that make it the most widely used construction material in the world. These characteristics include the following:

- *Availability of raw materials:* Limestone, clay, sand, gravel, and rock are available worldwide. Competing materials such as steel, aluminum, lumber, and asphalt are not always as readily available.
- *Versatility:* Concrete forms or molds into almost any shape or size. It does not require heat or elaborate casting facilities. This characteristic provides a tremendous flexibility in construction applications and partially accounts for the popularity of concrete as a building material.
- *Strength:* Concrete also accommodates a variety of strength applications. With its ability to be formed into an infinite variety of shape and sizes, concrete can be designed to meet a variety of strength needs. Strength changes with changes in the amount of cement used to make the **paste** that is mixed with aggregate. By adding steel reinforcement, concrete structural elements can be designed to carry almost any load.
- *Hardening in the presence of water:* Concrete hardens (sets up) when placed in water, so it can be used in wet locations, below ground level, and underwater. In fact, the wet or submerged conditions will help the concrete set properly and increase its ultimate strength.
- *Durability:* Concrete is extremely durable. It can last for hundreds of years. Forces that destroy other building materials do not affect concrete. It withstands heat and fire, which is highly desirable in many applications, such as housing. It resists corrosion and decay. In the same way, the stone-like nature of concrete makes it resistant to attack by vermin, rodents, and termites.

2.2.0 HISTORY

Cement has its roots in early Egypt. Records that show a cement-like material made from burned gypsum was used in the construction of the pyramids. The Greeks and Romans developed methods of burning limestone to produce quicklime that was used to make mortar. A superior mortar was produced by adding fine volcanic material called tuff. The Romans used this cement to make a form of concrete. It was used in many Roman structures, such as the Colosseum, which are still standing today.

During the Middle Ages, the art of making good concrete was almost lost. In the mid 1700s, masons in England and Europe experimented with the cementing qualities of gypsum, limestone, and other materials. Nothing significant occurred until 1824 when a brick mason in England obtained a patent for a material he called portland cement. It was called portland because concrete made with it resembled limestone quarried near Portland, England. By 1850, the making of portland cement was well established in England, Germany, and Belgium.

Shortly after the Civil War, portland cement was shipped from Europe to the United States. Beginning in 1871, United States businesses began producing portland cement on their own. As the United States expanded and developed, the need for concrete for canal locks and dams, railroads, bridges, buildings, and pavements steadily increased. Technical advances since the 1930s included the use of steel rods to reinforce concrete, the development of more fluid concretes for forming more shapes, the development of quality control methods for the ready mix industry, and the development of fire standards, **air entrainment**, and special cements. The increasing need for cement coupled with the technical advances provided dynamic growth in the concrete construction industry.

Concrete continues to advance in its applications. The past 50 years have seen dramatic advances that include the following:

- Aesthetic expression in textures, colors, shapes, and sizes for interior and exterior uses
- **Precast** and **prestressed** structural elements
- **Tilt-up** construction
- **Slip forming**
- Pumps designed to transport and place concrete for buildings
- Development of white portland cement and colored concrete
- **Expansive cement**
- Lightweight aggregates and lightweight concrete
- Epoxy-coated reinforcing steel to reduce corrosion
- Technology to support high-tolerance superflat floors

As new technology becomes available, materials manufacturers, equipment suppliers, and contractors continue to apply this knowledge toward the improvement of the concrete industry.

2.3.0 COMPOSITION

As defined earlier, concrete is composed of portland cement, water, and aggregate. Frequently, concrete also contains a chemical or mineral **admixture**. The cement and water form a paste that when mixed with the aggregate, binds the aggregate particles together. When the paste hardens, the mixture becomes a rock-like mass. This hardening is the result of a chemical process called hydration. It can take place in air or under water. Concrete does not harden by drying. It needs moisture to hydrate and harden.

Portland cement consists primarily of compounds of lime, silica, alumina, and iron. It is manufactured from selected materials using closely controlled processes, as shown in *Figure 2*. The process starts with the quarrying of limestone, shale, and clay. The quarried materials are crushed and stored in bins. The crushed materials are then proportioned, mixed, and ground into fine particles in a roller mill. The blended material passes to a storage

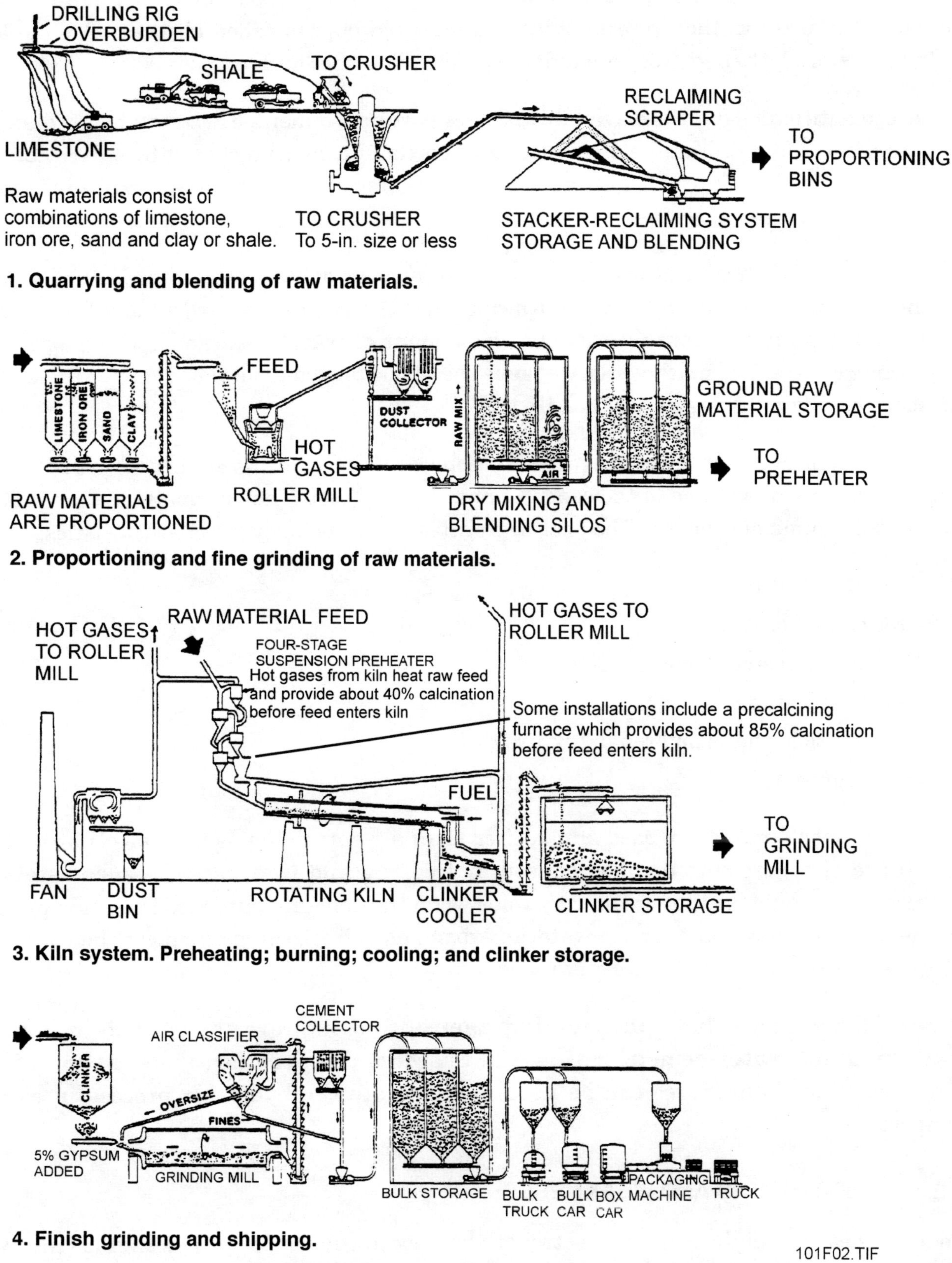

Figure 2. Manufacturing Process For Portland Cement

area, then through a preheater and a rotary kiln. After firing at extremely high temperatures, the particles fuse together to form small marble-sized objects called **clinker**. The clinker is cooled and stored, then ground into a fine powder called portland cement.

Other cementitious agents are available besides portland cement. These can be used alone or with portland cement. Materials blended with portland cement include **blast-furnace slag**, **fly ash**, and **silica fume**. These materials are generally known as **pozzolans** and add strength to concrete.

The water used with cement can be any natural, noncarbonated water that is drinkable and has no pronounced taste or odor. Its principal purpose is to cause hydration of the cement. There is a definite relationship between the amount of water used and the strength of the resultant concrete. Too much water weakens the concrete. Too little water makes the concrete too stiff, thereby reducing its **workability**.

The most commonly used aggregates for concrete are sand, gravel, and crushed rock. Aggregates provide volume to the concrete mix at a low cost. They normally make up 60% to 75% of the volume of concrete. The best aggregates have the following characteristics:

- Hard and durable
- Free of dust or dirt
- High compressive strength
- Chemically stable
- Good particle shape and rough texture
- Proper gradation

One of the more frequently used admixtures is an air-entraining agent which improves the resistance of concrete to damage caused by freezing and thawing. The agent stabilizes microscopic air bubbles that create tiny voids when the concrete hardens. The voids produced by the bubbles allow space for the water to expand as it freezes. This reduces the destructive effect of freezing water on the concrete.

Other admixtures include **plasticizing agents** that improve the workability of fresh concrete at low **water-cement ratios**, and retarders that lengthen the **set time** of fresh concrete. Other admixtures can be used to accelerate the hydration process in extreme temperatures.

2.4.0 STRENGTHS AND WEAKNESSES

Concrete has very high compressive strength but low tensile or flexural strength. This means that it is hard to crush but easy to pull apart. Therefore, you need to be careful how concrete is used. The objective is to capitalize on its strengths and compensate for its weaknesses.

The tensile strength of concrete structural elements can be increased by using reinforcing steel. Structural elements that are subject to tensile stresses include beams, columns, slabs, decks, and walls. For structural members in **compression**, very little structural reinforcing may be needed. These members often include footings, bridge piers, and pavements. Flexural strength is resistance to bending, and is a combination of compressive and tensile strength.

Figure 3 illustrates the significant difference between the strength of concrete in compression and **tension**. In this example, the compressive strength is 4,000 pounds per square inch and is 10 times the tensile strength. This means that a 2 × 2 × 2 inch cube of concrete can support a weight of 16,000 lbs. However, that same cube of concrete can be pulled apart by a force of just 1,600 lbs.

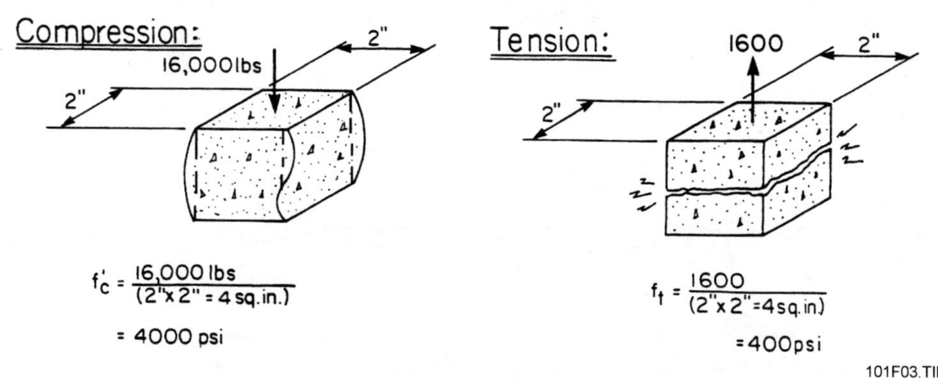

Figure 3. Concrete Strength In Compression And Tension

Figure 4 illustrates what happens to a plain concrete beam under load when the tensile strength of the concrete is exceeded. There will be some bending up to the point where the tensile strength is exceeded. As the beam bends, the bottom begins to crack due to the tensile force pulling the ends of the beam in opposite directions. The beam breaks because its flexural strength has been exceeded.

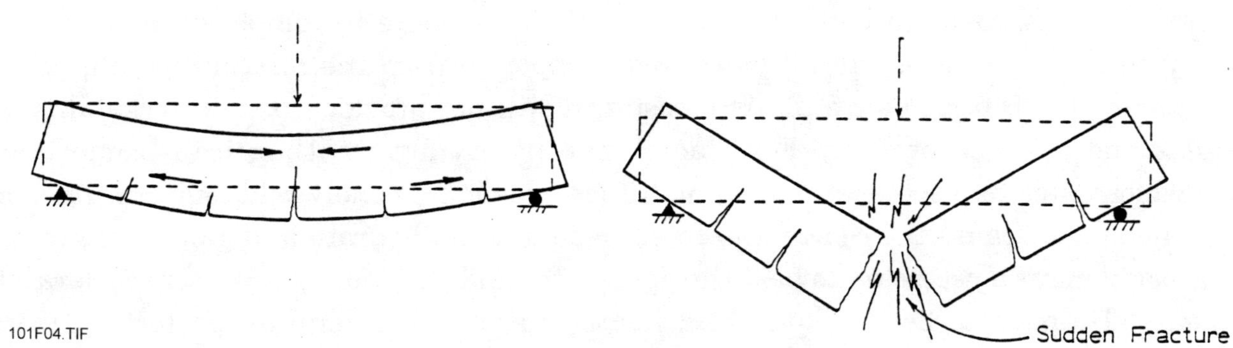

Figure 4. Concrete Beam Failure In Flexure

In *Figure 5,* the same beam has reinforcing steel placed at its bottom. With reinforcement, the beam can carry a tensile or flexural load much greater than plain concrete. Structural designers try to set the amount of reinforcement so that the compressive and tensile stresses are equal. Because the tensile and flexural strength of steel is 10 to 20 times the compressive strength of concrete, it does not require much steel to achieve equal stress throughout the cross-section of the beam.

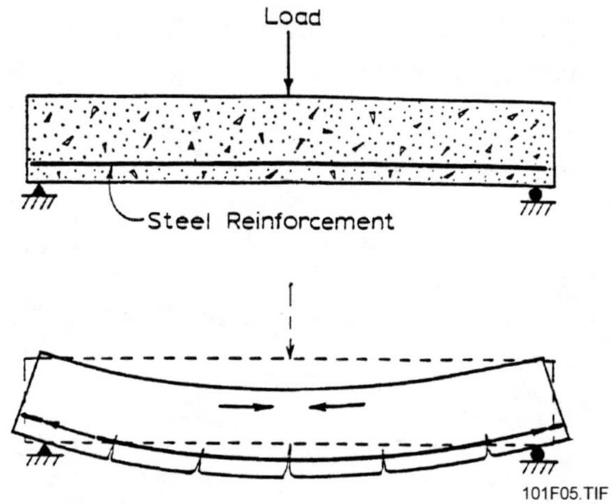

Figure 5. Reinforced Concrete Beam In Flexure

Concrete gains strength over time, so the compressive strength of concrete is usually specified as the strength of the concrete 28 days after it is placed. *Figure 6* provides typical age and strength relationships for plain concrete without air entraining at various water-cement ratios. Notice that the less water (0.4) there is in the concrete, the greater the strength. The concrete gets stronger as each day passes, until it reaches its design strength at 28 days. After 28 days the concrete continues to gain strength, but the gain tapers to a very slow rate.

Although concrete hardens in both air and water, it is sensitive to temperature extremes. Extremely high temperatures cause the water to evaporate from the mixture more quickly, thereby making the concrete too stiff. As a result, it will not provide enough workability to consolidate and to finish properly. For placements in very hot weather, ice is sometimes substituted for water, or a set-retarding admixture is added. Extremely cold temperatures, on the other hand, may cause the concrete to freeze before it can hydrate and gain strength. In cold weather, it may be necessary to heat the aggregates and water or to use a set accelerator. Forms may also need to be insulated and heated to provide adequate protection from freezing. Freshly placed concrete cannot be allowed to freeze, as it will never reach its specified strength.

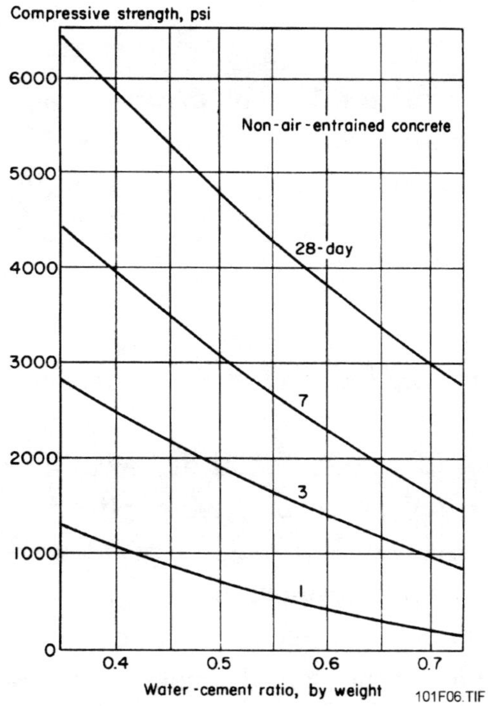

Figure 6. Example Strength Gain Curves For Concrete

3.0.0 USES OF CONCRETE

Concrete used in construction can take several forms — plain, reinforced, prestressed, **post-tensioned**, and precast. Plain concrete has no reinforcement of any type. Reinforced concrete has some type of reinforcement to increase its tensile strength. Prestressed concrete uses a technique that places the concrete under compression so that it can resist greater amounts of tension.

Prestressed concrete structural elements include high-strength steel strands that are placed under tension and stretched. The concrete is then placed and the strands are embedded in the concrete. After the concrete hardens and **cures**, the loads on the strands are released. The strands try to contract, but cannot, thus imposing a compressive load to resist the tensile loads on the concrete member. Post-tensioning works in a similar fashion, but the compressive force is applied through tension cables located in hollow tubes cast in the structural member. After the concrete cures, the cables are tensioned and secured.

Precast concrete is a term applied to concrete items that are cast and cured at a location other than their final placement in the structure. Precast items can be plain, reinforced, and prestressed. They can be almost any size. Prestressed concrete bridge girders 6 feet high and up to 200 feet long are routinely fabricated in precast facilities, then transported to the construction site to be placed. Precast and prestressed structural elements offer precise tolerances and better quality control, and represent technological advances in concrete engineering.

3.1.0 TYPES OF CONSTRUCTION

Different types of construction require different types of concrete. Plain and reinforced concrete, for example, are normally used in slab-on-grade construction. Reinforced, prestressed, post-tensioned, and precast concrete are usually associated with elevated construction.

3.1.1 Slabs On Grade

Construction of slabs on grade includes flatwork such as slabs and concrete paving for roads, airfields, parking lots, driveways, and sidewalks. Plain or reinforced concrete may be used in these applications. Typically, sidewalks, residential driveways, and floors for houses will be approximately 4 inches thick. Normally, none of these applications would require reinforcing. Floors for industrial buildings, however, are usually greater than 6 inches thick and will likely be reinforced. Highway pavements are usually 9 or more inches thick and may or may not be reinforced. Airfield pavements will range from 11 inches to 18 inches thick and may or may not be reinforced. Two examples of horizontal construction are shown in *Figures 7* and *8*.

Figure 7 shows a reinforced concrete slab being placed. The concrete has been transported to the site in a ready-mix truck. Placing is being made on a prepared base of compacted gravel. The concrete workers are moving the concrete into place with concrete hoes, or come-alongs.

101F07.TIF

Figure 7. Placing A Reinforced Concrete Slab

Figure 8 shows a section of concrete highway pavement being placed. One lane is being constructed at a time. In the foreground, dump trucks are depositing concrete. Behind them, a concrete paving machine is being used to spread and consolidate the concrete. Following behind it is another large machine that is leveling and screeding the concrete. Floating and straightedge work will be done by concrete workers after the screeding.

101F08.TIF

Figure 8. Placing Concrete Highway Pavement

3.1.2 Elevated Structural Elements

Elevated concrete construction includes columns, beams, piers, caps, and walls, as well as elevated structural decks. Most vertical concrete construction is reinforced. It may be prestressed, post-tensioned, or precast. *Figure 9* shows columns formed from fiber tubes that are braced for the placement of concrete. When finished, these columns will support a roof or a floor slab. Although the steel bars are not visible, the columns are reinforced. The reinforcing steel is placed inside the fiber tubes and tied into the footings on the ground.

101F09.TIF

Figure 9. Columns Formed from Fiber Tubes

Figure 10 shows a high-rise apartment building under construction. The walls of reinforced concrete are formed by manufactured forms that are designed to slip upward as the concrete is placed and cured. Slip forms such as these eliminate the need for the entire height of the structure to be formed at one time before concrete placement can start. For massive structural elements, slip forming solves the problem of supporting a very large quantity of fresh concrete until it hardens.

Figure 10. Slip Forming Exterior Concrete Walls

3.1.3 Pipes And Culverts

Concrete is used to make many types of drainage structures. These structures are either precast or cast in place. Typically, drainage pipes are manufactured in precast yards and come in a variety of sizes and shapes. *Figure 11* shows a common design of pipe with a cut-away section exposing the reinforcing steel. Culverts are used to channel water under a facility such as a highway. Concrete culverts can be either cast in place or precast. Most cast-in-place culverts are usually called box culverts because of their rectangular design.

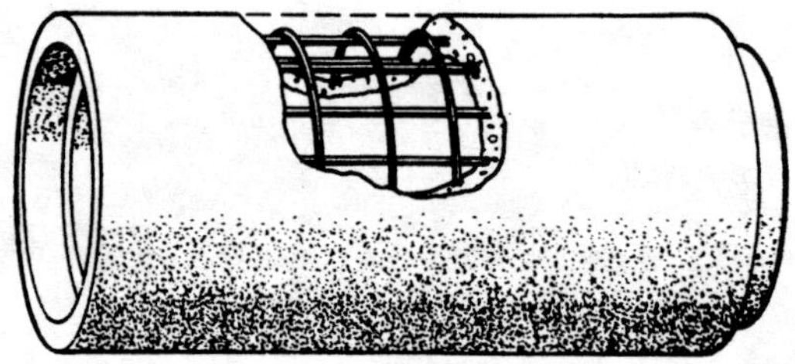

Figure 11. Reinforced Concrete Pipe

3.1.4 Other Structures

Concrete is used in many other large and small applications. Large applications include locks, dams, bridges, docks, tunnels, and paved channels. Traffic barriers, curbs and gutters, retaining walls, noise walls, light poles, railroad cross ties, and storm drains are a few of the smaller applications in the transportation industry. *Figures 12* through *15* show examples of very different uses of concrete in the construction industry.

Figure 12 shows two flights of concrete steps that are divided to accommodate the terraced slope. These steps were built using wood forms designed to fit the space. Because concrete is so versatile, it is easily made to fit almost any space.

101F12.TIF

Figure 12. Residential Concrete Steps

Figure 13 shows a building technique called tilt-up construction. The walls are formed, cast, and cured on the ground. Then, they are tilted and fastened into their final position. In *Figure 13*, a mobile truck crane is lifting tilt-up panels into position. Tilt-up construction is used for commercial and industrial buildings of one or two stories. Windows and other openings are framed out with metal or wood forms before the concrete is placed into the main form.

101F13.TIF

Figure 13. Tilt-Up Construction

Figure 14 shows the Sunshine Skyway Bridge over Tampa Bay. Located on Florida's west coast, the bridge was designed to accommodate high traffic volumes and provide for passage of ocean-going ships underneath. The concrete mix incorporated fly ash as a strength-enhancing admixture. This structure is an excellent example of innovation and technical capability to build large concrete structures in a harsh marine environment.

101F14.TIF

Figure 14. Sunshine Skyway Bridge

3.2.0 THE CONSTRUCTION PROCESS

In general, the construction process for concrete construction follows the same steps for both elevated and on-grade construction. The process for flatwork is discussed in the following sections.

3.2.1 Layout

The layout of any structure begins with the building plans. Architectural plans have been used for centuries as a means of transferring the ideas and designs of an architect or engineer into something that could be readily seen and interpreted by others. Construction drawings describe the size, shape, location, and specifications of the elements of a structure. These drawings use standard symbols and notes that are recognized by tradespeople in the construction industry. The trainee module *Introduction to Blueprints* in the NCCER Core Curricula provides a basic overview of plan and drawing interpretation and use. Plan dimensions are given in common measure. Over the next few years, the construction industry will convert to metric measures. *Appendix A* lists units of metric measure and conversion factors.

The site or plot plan in *Figure 15* shows the essential information needed to locate a building on a lot, locate underground utilities, and set elevations for the foundation. The persons responsible for establishing the location of the building on the site will set the necessary references and stakes for layout of the building. After excavation, these points are used to establish the line and grade for the concrete forms. Typically, a tape measure, string line, and level are needed to lay out the forms. For larger jobs, surveying equipment and other instruments are used.

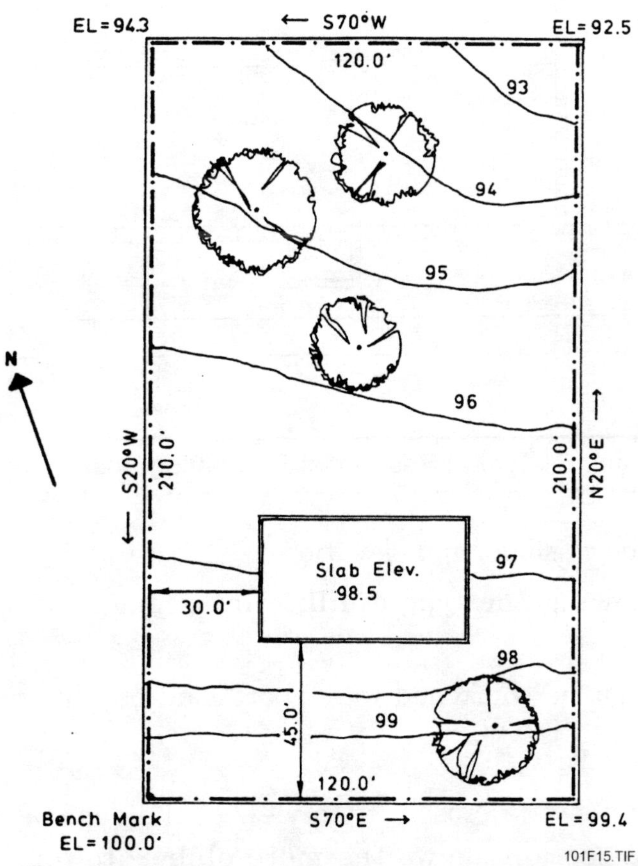

Figure 15. Typical Residential Site Plan

Figure 16 is a typical drawing for the layout of a residential building foundation. Foundation drawings include plan views, sections, and details. They should include symbols, dimensions, and notations to reveal information about the footing, the foundation wall, and the foundation surface or flooring. In most cases, footings, flooring, and foundation flooring require some degree of finishing. A concrete floor slab is often part of a building's foundation system. For floor slabs, the plans should show the following:

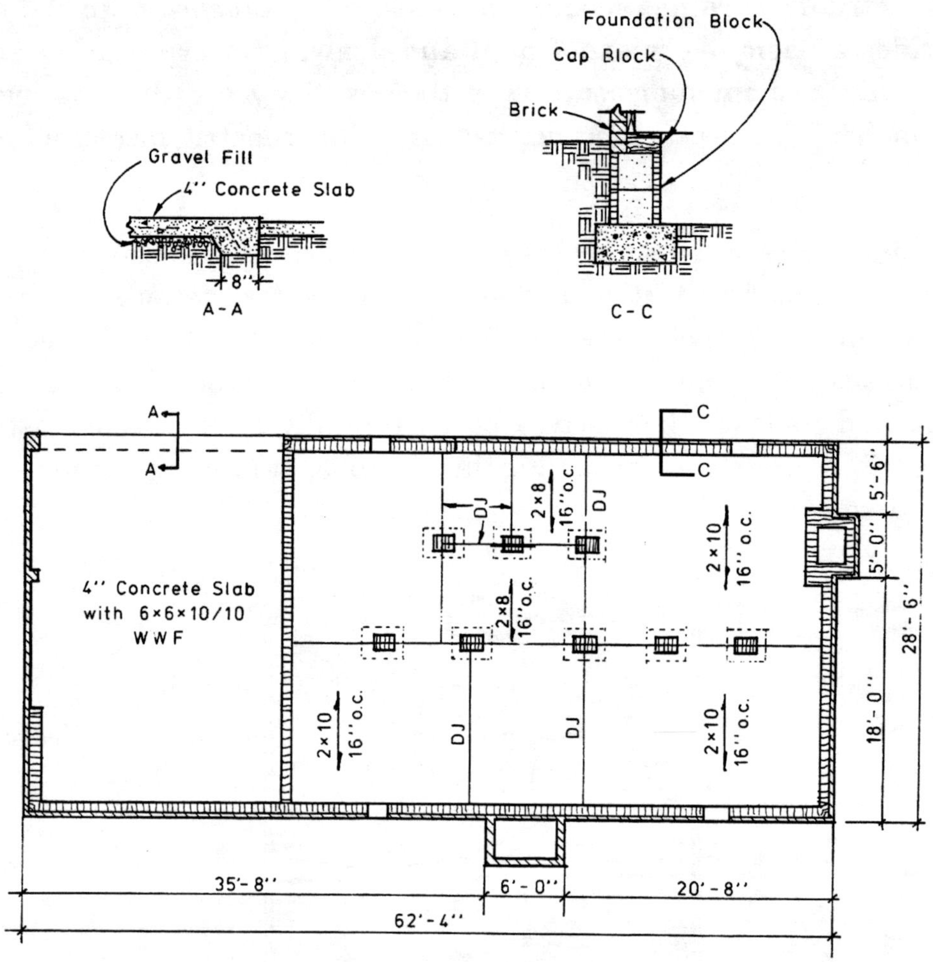

Figure 16. Typical Residential Foundation Plan

- The slab's position, thickness, slope, and elevation
- Sections and details showing the type of fill, reinforcement if specified, and vapor retarders
- Areas where the slab must be thickened to support load-bearing walls or depressed to receive finished flooring

In *Figure 16*, the A-A section shows a cross-section view of the edge of the garage slab, and the footing detail. The C-C section shows the more elaborate foundation and footing supporting the house itself.

3.2.2 Forming

Forming is a very important part of the concrete construction process because the concrete will take the shape of the constructed form. Once the concrete has set and the form has been stripped away, the shape of the concrete can be modified only with great difficulty.

Forming includes the forms, supports, hardware, and bracing to contain the fresh concrete. Formwork must stay in place until the concrete is strong enough to carry its own weight and any construction loads. Forms usually fall into one of three types:

- *Job-built forms* – The components of this type of form are assembled and fabricated piece by piece on the job site and are normally used only once. *Figure 17* provides an example of a job-built wall form. Most job-built forms are made of 2 × 4 lumber and plywood sheets. They should be straight and free from warping, and have sufficient strength and support to resist concrete pressure without bulging.

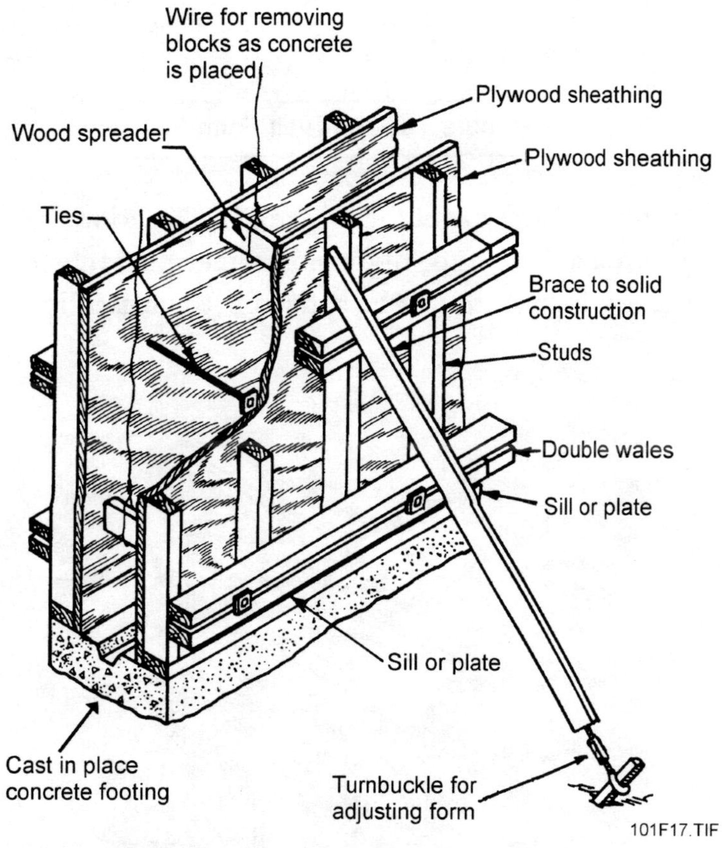

Figure 17. Job-Built Wall Form

Figure 18 shows another type of job-built form for residential and commercial construction. This type of forming does not need as much vertical bracing because most of the weight of the concrete is supported by the ground.

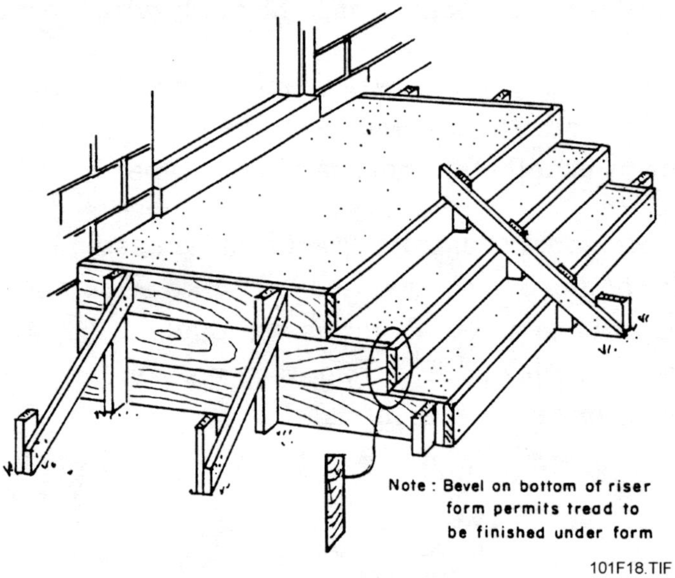

Figure 18. Job-Built Form For Steps

- *Prefabricated forms* – This type of form is made in sections and assembled on the job site so that it can be reused. This type of form is sometimes referred to as a gang form. *Figure 19* shows a wall gang form that can be reused both vertically and horizontally.

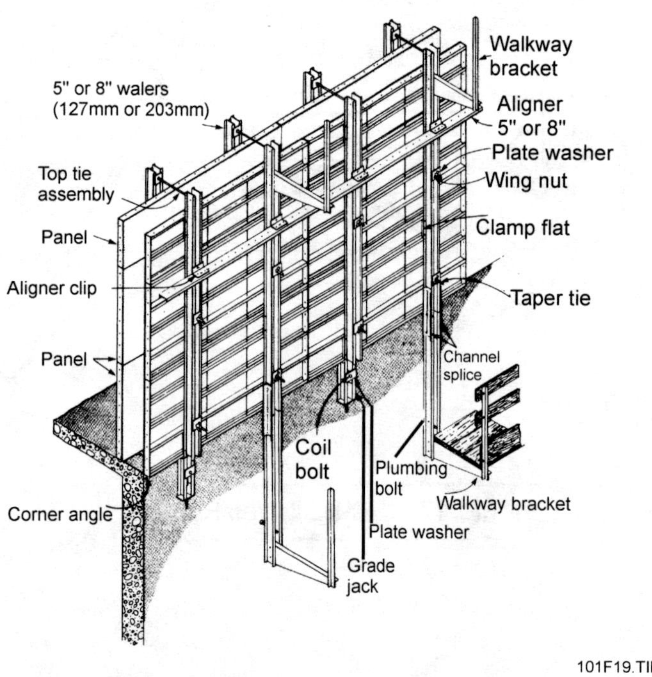

Figure 19. Prefabricated Wall Form

- *Manufactured forms* – This type of metal, plastic, or cardboard form is purchased from a form manufacturer and is normally reused many times. *Figure 20* shows a manufactured column form made of steel. The picture on the left shows an assembled column form with a fluted top. The picture on the right shows a finished concrete column after the metal form has been stripped away.

Figure 20. Manufactured Form For Concrete Columns

Safety for concrete workers is a primary consideration. The design and bracing of the formwork are important. Because fresh concrete is a very heavy material weighing over two times as much as water, it creates a fluid pressure against the side of the forms. The higher the form is filled, the greater the pressure. If the forms are not designed or assembled correctly, the forms can collapse and cause injury to workers in the vicinity. The forms should always be inspected for correct construction and assembly before placing concrete.

3.2.3 Placing

Concrete may be mixed on the job site or at a central batching plant and then delivered to the job site by truck. Once mixed or delivered, the concrete is moved to the placement location and placed in the forms. For concrete that is delivered by truck, it is best if the concrete can be placed directly into the forms. If not, the concrete is placed into wheelbarrows, buggies, or buckets for transport to its final position.

In general, concrete should be placed in horizontal layers of uniform thickness, each layer being thoroughly consolidated before the next is placed. The rate of placing should be rapid enough so that the layer of concrete has not yet set when a new layer is placed upon it. This will avoid flow lines, seams, and planes of weakness (cold joints) that result when freshly mixed concrete is placed on hardened concrete. Layers should be about 6 to 20 inches thick for reinforced members, 15 to 20 inches thick for mass work, the thickness depending on the width between the forms and the amount of reinforcement.

Consolidation is the process of compacting fresh concrete to mold it within the forms and around embedded items and reinforcement. Consolidation eliminates voids, pockets of aggregates, and honeycombs in vertical construction. It also helps blend the placed layers of concrete together. Consolidation is normally accomplished by an internal vibrating action that makes the concrete behave temporarily as a liquid, thereby filling in voids and eliminating air pockets. *Figure 21* shows a worker using a hand held vibrating machine to consolidate freshly placed concrete.

101F21.TIF

Figure 21. Consolidating With An Internal Vibrator

The most critical time during concrete construction is when the concrete is being placed in the form. Care must be taken not to bump, damage, or otherwise disturb the forms or bracing. The concrete craftworker must be especially alert to noise or movement of the forms that might indicate that a collapse is occurring or is going to occur.

3.2.4 Finishing

Finishing horizontal concrete surfaces involves screeding, bullfloating, edging, jointing, floating, troweling, and brooming. Finishing begins after concrete is placed, and continues at different stages of setting. The finishing of elevated concrete elements that have been formed occurs when the forms are removed. Exposed surfaces are finished as the fresh concrete hardens. The finishing effort for the top of a wall is small compared to the effort required to finish a slab.

To finish a slab, the concrete workers strike the surface of the fresh concrete with a **screed**. This levels and consolidates the surface. Next, the screeded surface is floated with a bullfloat or a highway straightedge in order to fill minor low spots and cut down high spots. Floating also embeds the coarse aggregate particles into the surface. After floating, the surface may also be troweled to produce a smooth durable surface texture.

Screeding or strikeoff is the process of cutting off excess concrete to bring the top surface of a slab to proper grade. This can be done manually with a straightedge or with a mechanized screed that is either pulled manually or motorized. *Figure 22* shows a screed striking the concrete surface for a floor. The concrete workers are spreading the concrete in front of the screed as it slowly advances. Note the change in texture behind the screed.

101F22.TIF

Figure 22. Mechanized Screed

Figure 23 depicts a floor slab being bullfloated to remove high points, fill in low spots, and embed the coarse aggregate. This should be done immediately after strikeoff. Note the difference in surface texture between floated and non-floated areas.

Figure 23. Using A Long-Handled Bullfloat

On some jobs, the highway straightedge ("bumpcutter") is used instead of a bullfloat. The highway straightedge, as shown in *Figure 24*, has a 10- to 12-foot blade with a rectangular cross section and a long pole handle. The highway straightedge is used to finish slabs to superflat tolerances.

Figure 24. Using A Highway Straightedge

Edging is required along all edge forms and for isolation and construction joints in floors and outdoor slabs such as walks, drives, and patios. Edging densifies and compacts the concrete next to the form, where floating and troweling are less effective, making it more durable and less vulnerable to spalling and chipping.

After the concrete has been hand edged and hand jointed, it should be floated with a wood or metal hand float or with a power float like that shown in *Figure 25*.

101F25.TIF

Figure 25. Using A Power Float

The purpose of floating is threefold:

- To embed aggregate particles just beneath the surface
- To remove slight imperfections, humps, and voids
- To compact the mortar at the surface in preparation for additional finishing operations

Concrete should not be overworked as this may bring an excess of water and fine material to the surface and could result in additional surface defects.

Floating produces a relatively even texture that has good slip resistance. It is often used as a final finish, especially for exterior slabs. When this will be the final finish, floating may be done a second time to partially hardened concrete. If superflat tolerances are required, a highway straightedge will be used for the second float.

Where a smooth, hard, dense surface is desired, floating should be followed by steel troweling. *Figure 26* shows the use of a hand trowel to produce the desired pattern and surface effect. Concrete craftworkers often lean on a hand float to support their weight as they trowel. Note the differences in the surface textures. Troweling should not be done on a surface that has not been floated.

101F26.TIF

Figure 26. Hand Troweling A Floor Slab

The first troweling may produce the desired surface free of defects. However, surface smoothness, density, and wear resistance can all be improved by additional troweling.

3.2.5 Joints And Edging

Concrete shrinks as it hardens, expands as the temperature rises, and contracts as the temperature falls. This movement can create cracks in concrete slabs. Jointing greatly reduces unsightly random cracking in concrete slabs. There are three basic types of joints.

- Contraction or control joints minimize uncontrolled shrinkage cracks when the concrete contracts.
- Isolation joints isolate the slab from other structural members and minimize cracking that is caused by structural members expanding and contracting at different rates.
- Construction joints are located where the slab placement is interrupted or ended.

Figure 27 shows the control joints, isolation joints, and construction joints normally found in a floor slab on grade. Isolation joints allow the slab to move up or down in relation to walls, footings, or columns. Control (contraction) joints allow slabs to shrink or contract without cracking between the joints. Construction joints are stopping points between placements. These joints use keys, dowels, or other connectors for load transfer purposes.

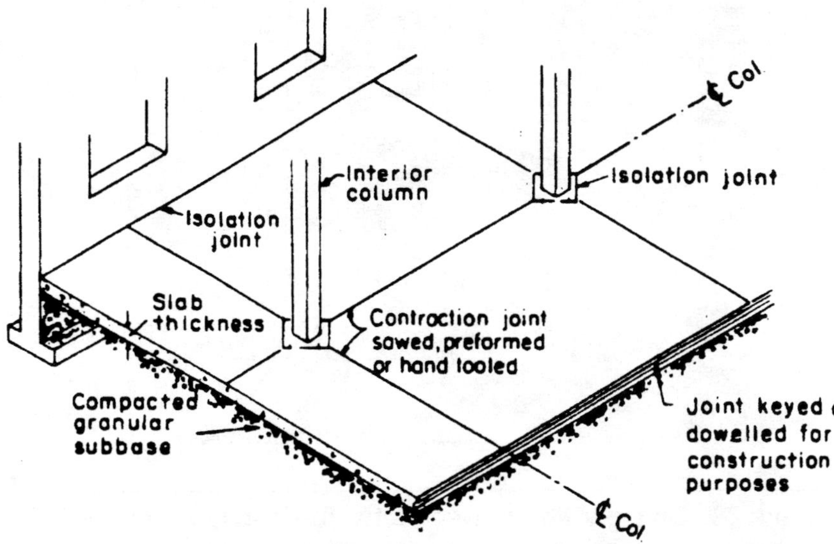

Figure 27. Joints In A Concrete Slab

Joints can be formed, sawed, or hand tooled. Control joints are normally sawed with a diamond blade saw after the concrete has hardened, or hand tooled with a joint tool in the fresh concrete. Construction joints are normally formed to act as isolation or control joints.

Edging is most commonly done on sidewalks, driveways, and steps. It leaves a rounded edge that is less likely to chip than a square edge. It is usually done with an edging tool on fresh concrete but can also be done with a grinding machine after the concrete hardens. Grinding produces a beveled edge instead of a rounded edge.

3.2.6 Curing

Floors and other flatwork have the greatest amount of exposed surfaces from which moisture can evaporate. Curing is the process of maintaining a favorable moisture content and temperature in concrete so that it achieves the desired strength and durability. Curing is important for all concrete and should start immediately after finishing is completed. There are three methods of controlling the evaporation of moisture from the concrete.

- Replacing evaporated moisture by keeping the concrete surface wet. This is accomplished by covering the surface with water, spraying it with water, or covering it with a material such as burlap or synthetic materials that are kept continuously wet. These methods afford some cooling through evaporation, which is beneficial in hot weather. *Figure 28* shows a floor sprayed with water to keep the concrete moist.

101F28.TIF

Figure 28. Curing With Sprayed Water To Keep Concrete Moist

- Sealing the surface to keep moisture from escaping. Synthetic materials, waterproof paper, and membrane-forming curing compounds are the most commonly used methods of sealing the surface. *Figures 29* and *30* show the application of synthetic material and curing compounds to control moisture.

101F29.TIF

Figure 29. Sealing A Surface With Synthetic Material

Figure 30. Spraying A Membrane Curing Compound On A Slab

- Accelerating strength gain by supplying heat and additional moisture to the concrete. This is usually accomplished with live steam, heating coils, or electrically heated forms or pads.

The method or combination of methods chosen depends on factors such as availability of curing materials, size and shape of concrete, production facilities, aesthetic appearance, and economics.

3.2.7 Form Removal

It is advantageous to leave forms in place as long as possible to continue the curing process. Forms can usually be removed when the concrete has reached its design strength. Forms for slabs on grade can be stripped off a day after placement. Wall and column forms can usually be stripped well before beams and floor slabs not on grade. However, early removal of forms can cause collapse of the structure.

The designer should specify the minimum strength requirements for various members for form removal. The age-strength relationship should be determined from a representative sample of concrete that is used in the structure and field-cured under job conditions. Strengths are affected by the materials used, temperatures, and other conditions. The time required for form removal, therefore, will vary from job to job.

4.0.0 IMPORTANCE OF CRAFTWORKERS

The success of concrete construction depends on the concrete performing as it was designed. Because concrete is a wholly manufactured product, its performance depends on the skills of the workers who mix, place, finish, and cure the concrete.

4.1.0 MIXING

The concrete craftworker must understand the factors that influence the quality, economy, speed of placement, and ease of finishing the concrete mix. Each job is different and adjustments may need to be made to the mix or placement site. These adjustments need to be made to handle factors such as these:

- Where the concrete is deposited. If the base is dry and porous, concrete deposited directly on grade may dry out too quickly. To prevent this from happening, the base may need to be moistened with water so that the water in the concrete is not absorbed too quickly. Or, the mix may need to be adjusted to provide a more fluid concrete. The concrete supervisor is the one who makes the adjustments on the ground or requests that the adjustments be made to the mix.
- How the concrete is transported. How the fresh concrete is transported from the mixer to the placement site is an important consideration. A very stiff mix cannot be easily chuted, conveyed, pumped, or placed by bucket. A very fluid mix is not easily wheelbarrowed or buggied.
- Weather conditions. Extremes in weather must be considered. If the weather is too hot, chilled water and ice may need to be added to meet temperature requirements. If it is too cold, the aggregates and mixing water may need to be heated. Admixtures to retard or accelerate the hydration process may be necessary.

Throughout the placement process craftworkers are necessary to plan and execute the work efficiently, effectively, and safely.

4.2.0 PLACING

Placement must be well planned and organized. Teamwork is a main consideration for large jobs. Once the concrete has been mixed and delivered, the placement becomes time-sensitive and delays cannot be tolerated. Any delays during this operation impair quality and cost of the finished product.

When the concrete craftworker is knowledgeable about the equipment and techniques used for placing, this part of the construction operation is likely to run smoothly. Therefore, well-trained and skilled workers must be ready to perform this work.

4.3.0 FINISHING

After placing, the concrete must be finished to meet the specifications and job requirements. Finishing represents a major activity in the construction process that requires knowledge and skill. Since many of the tasks in the finishing process are manual, proper execution will result in quality appearance of the concrete and in long-term durability.

Highway straightedging, bullfloating, edging, jointing, and troweling are all hand-guided tool operations. Each of these tasks requires a high level of skill by the concrete craftworker. Without this skill the concrete would not be finished adequately and its utility, durability, appearance, and quality would be compromised.

4.4.0 CURING

Final concrete strength and durability are a result of the curing process. Improper curing will significantly reduce both integrity and strength of the structure. Curing keeps the fresh concrete at temperature and moisture levels that support hydration.

Curing is started as soon as finishing is complete. Concrete craftworkers implement the proper curing method and perform the curing activities. Their diligence and skill will insure a quality end product. Inattention will result in a poor quality product.

5.0.0 CAREERS IN CONCRETE CONSTRUCTION AND FINISHING

Most work involving concrete is centered in the construction industry. This involves not only residential, industrial, and commercial buildings, but other structures such as networks of roads, bridges, airports, and marine terminals. Other areas include dams, water and sewer systems, and utilities.

Construction is the second largest industry in the United States, employing over five million workers. As the population of the country increases, so will the need for new construction. Also, more American companies are fulfilling the needs for construction overseas. With this increased demand for construction comes a stronger need for trained craftworkers.

Today, concrete is one of the most widely used construction materials because it is economical and readily available in most areas. It is an essential part of most vertical structures and provides one alternative for construction in many other types of structures. The production, placing, finishing, and other activities required to complete concrete work are largely dependent upon the skilled craftworker who knows the trade.

The concrete finishing trade recognizes various levels of skill or classes of experience. These include:

- Trainee or apprentice
- Journeyperson
- Foreperson
- Superintendent
- Contractor

Generally, with increased skills and an increase in job experience, the financial rewards become greater. But, as the craftworkers' careers advance, their need for more knowledge increases along with their obligations, duties, and responsibilities.

Most untrained or unskilled workers start out as a laborer on a construction project. Once the person has decided on a particular trade, they must seek training and experience in that area. At this point they usually enter an apprenticeship program where they have an opportunity to work with craftworkers specializing in the trade. Training and apprenticeships will vary among the trades. Also, each state within the United States sets its own requirements for recognizing and certifying trained and skilled craftworkers.

Once an apprenticeship has been completed, the trainee is recognized as a fully trained finisher or journeyperson. If journeypersons show a great deal of skill, leadership qualities, and a desire to excel in their trade, they may be chosen to supervise a group of workers (foreperson) or to be in charge of all the field work for the contractor (superintendent). With a great deal of experience and added study, craftworkers may become contractors in charge of the whole construction job: scheduling workers and work, preparing bids, inspecting work, and running their own businesses.

6.0.0 WORK REQUIREMENTS

Employers in the concrete construction and finishing industry look for conscientious, hard working craftworkers who are dependable and reliable and who take pride in themselves and their work. Other qualities that employers appreciate are a cooperative attitude, safety awareness, a considerate manner, initiative, and a willingness to assume responsibility.

All successful construction jobs require the efficient use of labor and materials. When planning construction jobs, the contractor must accurately estimate the time that it will take to do the job. They can be certain that their estimates are accurate only if they know that workers are skilled and dependable. Good craftworkers will have the knowledge, skills, abilities, and attitudes to make this happen.

6.1.0 KNOWLEDGE, SKILLS, AND ABILITY

Becoming a good concrete finisher takes more than placing concrete and smoothing it. A competent finisher is one who can be trusted to perform the required work and meet the project specifications.

Concrete craftworkers need to know how to place, finish, and cure the concrete. They need to know:

- What materials go into concrete, and how they react with each other
- How to safely assemble and take apart forms
- How to install reinforcing steel
- How to transport concrete by wheelbarrow, buggy, conveyor, bucket, or pump
- How to use hand and power screeds, compactors, floats, trowels, jointers, edgers and finishing tools, and other equipment on the job
- Curing methods and techniques
- The safety precautions necessary to protect themselves and other workers on the job

Concrete craftworkers need to be able to apply their trade skills to what they face each day on the job. The best way to do the work at a particular job site will depend on the layout of the work, what is happening around the concrete site, and the environmental conditions surrounding the project. Mixing, transporting, and placing fresh concrete is hard physical work because of the weight of the material. Much of the work of concrete construction is outside where the worker is exposed to variations in temperature and weather. A concrete finisher must be able to work under these conditions and not be distracted by them.

In summary, concrete craftworkers must:

- Know their craft
- Have the skills to apply that knowledge to the work at hand
- Have the ability to do that work safely under all conditions

6.2.0 ATTITUDE TOWARD WORK

Attitude can build an invisible bridge, or build an invisible wall, between us and others. No one wants to hang around a grouch or count on someone who is not dependable. No one minds helping someone who can do something in return or working with a friendly, cooperative partner. On top of knowledge, skills, and ability, the concrete craftworker needs the right attitude. Your attitude comes from how you think and feel about your work and yourself. Remember, a positive attitude is contagious.

6.2.1 Dependability

A concrete craftworker must be dependable. Concrete construction is a closely timed operation. Once started, it cannot be stopped without waste of material and effort. Employers need workers who report to work on schedule. An undependable, absent worker will slow or stop concrete work and cost the project time and money.

6.2.2 Responsibility

Every craftworker must assume responsibility for doing the assigned work in the proper manner. Every craftworker must assume responsibility for working safely, using skill, knowledge, and ability. The concrete craftworker must be responsible enough to work without supervision, and to work until the task is finished. Employers are always in need of workers who are ambitious and want to become a foreperson or supervisor, responsible for teams, crews, and job operations. Being responsible for what others do may be your career goal. The path to that goal starts with being responsible for what you do.

Responsibility includes being proactive. This is the art of looking ahead. Being proactive calls for taking steps to be sure things will go as planned. Anticipating what you will need to do a job, and having it on hand, is being proactive. And when things do not go as planned, the responsible concrete worker will do what is needed to get them on track again.

6.2.3 Team Player

When forming, placing, and finishing take place, a large amount of work must be done in a short time. Planning and teamwork are needed to do so efficiently and safely. Supervisors will form teams of two or more workers to do specific tasks, then form different teams to do different tasks.

Concrete workers on a job site will find themselves teaming with different people at different times. Being a team player becomes important. Everyone must be willing to work together. Everyone must be willing to bring their best attitude to the team. Team members need to be able to depend on each other. Team priorities ("Finish the task") must be more important than individual priorities ("I want to do something else").

Concrete craftworkers are also part of a larger team, composed of all the workers in all the trades at that job site. Playing on this team calls for protecting the work of other craftworkers. This can mean keeping fresh concrete out of drain inserts in a slab. This can also mean checking the anchoring of electric cable raceways before placing fresh concrete around them. Team players need to look out for each other.

6.2.4 Pride

Pride in what you do comes from doing high-quality work in a timely manner, and knowing you are doing your best. Being proud of what you do can overflow into other areas. Proud craftworkers take pride in their personal appearance. Proud craftworkers take pride in their tools. They have a complete set of well-kept tools and other special equipment they need to do their jobs.

Work so that you can be proud of what you do, and how you do it. Being proud of what you do is an important part of being proud of who you are.

6.3.0 SAFETY

Concrete workers operate in a high-risk environment, in terms of safety. All around them are stacks of heavy materials, trucks, ramps, and loads going by on cranes. Forms can collapse, other workers can drop things, and runways can slide out from under wheelbarrows. Placement sites have many possibilities for accidents.

Concrete craftworkers must think and practice safety at all times. Their work must be planned so that it is first safe and second efficient. Forms and their supports must be able to handle the load of the fresh concrete without collapsing. Cranes and slings must be checked before use. Overhead paths must be cleared for buckets. Tools must be checked and used properly.

All craftworkers must wear hard hats and other protective gear. Concrete is caustic and can irritate skin. Craftworkers need to keep track of the rest of their crew and of other crews around them. They need to be on the lookout for movement of forms and unfamiliar noises. Craftworkers need to have the knowledge, skill, and ability to:

- Recognize an unsafe situation
- Alert fellow crew members to workplace dangers
- Take evasive or corrective action

6.4.0 QUALITY

The latest ideas about quality in work are not new to concrete workers at all. Those who work in concrete construction and finishing have been concerned about quality for over 200 years. The quality of concrete depends on many factors. The craftworker may have little control over the mix design and the quality of the raw materials; however, the craftworker does control placement, finishing, and curing.

As discussed above, the quality of the finished concrete product depends directly on the knowledge, skill, and ability of the concrete craftworker. Quality work will be easily recognized. Poor work will be even more easily recognized. Given the durability of concrete, the quality of the work will be a monument to the skill of the craftworker for a very long time. The skilled, proud craftworker always strives for the highest quality that can be achieved.

SUMMARY

Concrete has many properties that make it the most commonly used construction material in the world. It is versatile and durable; its raw materials are easily available.

Although concrete has been in use for more than 2,000 years, its use was limited until portland cement was developed around 1824. Since then, concrete has grown in use throughout the world. It is used for many types of construction, such as roads, bridges, drainage and navigation structures, buildings, sidewalks, and parking lots. Its uses continue to grow.

Concrete is strong in compression but weak in tension or flexure. If large tensile or flexural forces are developed, it must be reinforced. Concrete gains strength over time; it needs proper curing conditions to develop its designed compressive strength. The construction process for concrete work involves erecting the concrete forms, mixing the concrete, placing the concrete, finishing, and curing it.

The success of concrete construction depends on the concrete performing as it was designed. Performance and quality depend on the skills of the concrete craftworkers. Their role is vital to successful performance. Concrete craftworkers work in a unique and demanding environment. To meet the job conditions and challenges, employers want workers who are dependable and reliable and who take pride in themselves and their work. These workers must have a positive attitude, and the initiative to acquire the education, skills, and experience needed to make themselves a craftworker of the highest professional status.

References

For advanced study of the topics covered in this task module, the following reference materials are suggested:

Concrete Construction and Estimating, Craig Avery, Craftsman Book Company, Carlsbad, CA, 1991.

Concrete Manual, International Conference of Building Officials, Whittier, CA, 1992.

Concrete Parking Lots: Eight Steps to Quality Construction, Videotape, Portland Cement Association, Skokie, IL.

Concrete Technology, Third Edition, George R. White, Delmar Publishers, Inc., Albany, NY, 1991.

Design and Control of Concrete Mixtures, Thirteenth Edition, Portland Cement Association, Skokie, IL, 1988.

Finishing Concrete Flatwork, Videotape 32:00 minutes, Portland Cement Association, Skokie, IL, 1991.

Guide for Concrete Floor and Slab Construction, ACI 302.1R-96, American Concrete Institute, Farmington Hills, MI, 1997.

The Contractor's Guide to Quality Construction, American Concrete Institute, Farmington Hills, MI, 1997.

ACKNOWLEDGMENTS

Figures 1, 9, 17, 19, 20, and 27-29 courtesy of American Concrete Institute
Figures 2, 6, 12, and 18 courtesy of Portland Cement Association
Figure 3-5, and 11 courtesy of Roy Jorgensen Associates, Inc.
Figure 7, 21-26, and 30 courtesy of Baker Concrete Construction
Figure 8 courtesy of International Road Federation
Figure 10 courtesy of PERI Formwork Engineers
Figure 13 courtesy of Delmar Publications
Figure 14 courtesy of Federal Highway Administration
Figure 15 National Association of Homebuilders
Figure 16 courtesy of Associated General Contractors

SI* (MODERN METRIC) CONVERSION FACTORS

APPROXIMATE CONVERSIONS TO SI UNITS

Symbol	When You Know	Multiply By	To Find	Symbol
		LENGTH		
in	inches	25.4	millimeters	mm
ft	feet	0.305	meters	m
yd	yards	0.914	meters	m
mi	miles	1.61	kilometers	km
		AREA		
in^2	square inches	645.2	square millimeters	mm^2
ft^2	square feet	0.093	square meters	m^2
yd^2	square yards	0.836	square meters	m^2
ac	acres	0.405	hectares	ha
mi^2	square miles	2.59	square kilometers	km^2
		VOLUME		
fl oz	fluid ounces	29.57	milliliters	mL
gal	gallons	3.785	liters	L
ft^3	cubic feet	0.028	cubic meters	m^3
yd^3	cubic yards	0.765	cubic meters	m^3
NOTE: Volumes greater than 1000 l should be shown in m^3.				
		MASS		
oz	ounces	28.35	grams	g
lb	pounds	0.454	kilograms	kg
T	short tons (2000 lb)	0.907	megagrams (or "metric ton")	Mg (or "t")
		TEMPERATURE (exact)		
°F	Fahrenheit temperature	5(F-32)/9 or (F-32)/1.8	Celsius temperature	°C
		FORCE AND PRESSURE OR STRESS		
lbf	poundforce	4.45	newtons	N
lbf/in^2	poundforce per square inch	6.89	kilopascals	kPa

APPROXIMATE CONVERSIONS FROM SI UNITS

Symbol	When You Know	Multiply By	To Find	Symbol
		LENGTH		
mm	millimeters	0.039	inches	in
m	meters	3.28	feet	ft
m	meters	1.09	yards	yd
km	kilometers	0.621	miles	mi
		AREA		
mm^2	square millimeters	0.0016	square inches	in^2
m^2	square meters	10.764	square feet	ft^2
m^2	square meters	1.195	square yards	yd^2
ha	hectares	2.47	acres	ac
km^2	square kilometers	0.386	square miles	mi^2
		VOLUME		
mL	milliliters	0.034	fluid ounces	fl oz
L	liters	0.264	gallons	gal
m^3	cubic meters	35.71	cubic feet	ft^3
m^3	cubic meters	1.307	cubic yards	yd^3
		MASS		
g	grams	0.035	ounces	oz
kg	kilograms	2.202	pounds	lb
Mg (or "t")	megagrams (or "metric ton")	1.103	short tons (2000 lb)	T
		TEMPERATURE (exact)		
°C	Celsius temperature	1.8C + 32	Fahrenheit temperature	°F
		FORCE AND PRESSURE OR STRESS		
N	newtons	0.225	poundforce	lbf
kPa	kilopascals	0.145	poundforce per square inch	lbf/in^2

* SI is the symbol for the International System of Units. Appropriate rounding should be made to comply with Section ASTM E380.

REVIEW/PRACTICE QUESTIONS

Choose the correct answer for each of the following questions.

1. A mixture of cement, water, and sand is called _____.
   a. portland cement
   b. concrete
   c. hydraulic cement
   d. limestone

2. Concrete is strongest in _____.
   a. tension
   b. shear
   c. deflection
   d. compression

3. Portland cement was so named because it _____.
   a. resembled rocks found in Portland, Oregon
   b. was the name of the cement's developer
   c. sounded more official than concrete
   d. resembled limestone quarried at Portland, England

4. Concrete weighs _____ as water.
   a. the same as water
   b. 1.5 times as much
   c. over 2 times as much
   d. 18 times as much

5. The concrete finisher influences the quality of the finished concrete by controlling the _____.
   a. mix design
   b. raw materials
   c. testing of samples
   d. placement, finishing, and curing

6. In what year did production of portland cement begin in the United States?
   a. 1824
   b. 1845
   c. 1871
   d. 1895

7. A common concrete mix admixture is _____.

   a. calcium hydroxide
   b. plasticizing agent
   c. air reducing agent
   d. slip forming

8. Design strength of concrete is specified as the strength _____ days after hardening.

   a. 3
   b. 7
   c. 10
   d. 28

9. Tensile strength of a concrete member is increased by _____.

   a. adding more cement
   b. reducing the water in the mix
   c. using reinforcing steel
   d. changing the aggregate size

10. Concrete hardens through a process called _____.

   a. hydration
   b. drying
   c. chemical stabilization
   d. evaporation

11. Prestressing places concrete in _____.

   a. tension
   b. compression
   c. shear
   d. deflection

12. Plain concrete contains _____.

   a. reinforcing
   b. no reinforcing
   c. only water and cement
   d. no air entrainment

13. Portland cement concrete pavements are usually more than _____ inches thick.

   a. 4
   b. 6
   c. 9
   d. 1

14. Most box culverts are _____.

    a. precast
    b. cast in place
    c. assembled on site
    d. used under highways

15. The highway straightedge is used to _____.

    a. screed highways
    b. consolidate placed concrete
    c. measure minimum strength
    d. make superflat surfaces

16. The primary reason for inspecting forms prior to concrete placement is _____.

    a. to fill in high and low spots
    b. to prevent leakage
    c. safety
    d. to insure that reinforcing is in place

17. The most critical time during concrete placement is _____.

    a. when forms are being erected
    b. when concrete is being placed in the forms
    c. after consolidation
    d. during finishing

18. Screeds are used to _____.

    a. vibrate the concrete
    b. support the weight of the finisher
    c. provide a smooth surface
    d. level and consolidate the surface

19. Bullfloats are used to _____.

    a. smooth surfaces after troweling
    b. fill in low spots and cut high spots
    c. embed coarse aggregate above the surface
    d. finish floors

20. When constructing a smooth finish industrial floor, troweling is the _____ step.

    a. first
    b. second
    c. third
    d. final

21. Expansion or isolation joints are used to _____.
    a. eliminate cracking
    b. separate slabs from structural members
    c. separate slabs placed at the same time
    d. protect the appearance of slabs

22. Curing maintains a _____ and temperature in newly placed concrete.
    a. high degree of smoothness
    b. weather protective surface
    c. satisfactory moisture content
    d. strength

23. Concrete forms for slabs-on-grade can be removed as soon as _____.
    a. the concrete is finished
    b. the concrete is edged
    c. two days after placing
    d. one day after placing

24. A concrete worker must think and practice _____ at all times.
    a. safety
    b. efficiency
    c. evasive action
    d. skills

25. In order to work effectively on larger concreting projects, you must _____.
    a. do extra work all the time
    b. know how to design concrete mixes
    c. be a team player
    d. know first aid

ANSWERS TO REVIEW/PRACTICE QUESTIONS

Answer	Section Reference
1. b	2.0.0
2. d	2.4.0
3. d	2.2.0
4. c	3.2.2
5. d	6.4.0
6. c	2.2.0
7. b	2.3.0
8. d	2.4.0
9. c	3.0.0
10. a	2.0.0
11. b	3.0.0
12. b	3.0.0
13. c	3.1.1
14. b	3.1.3
15. d	3.2.4
16. c	3.2.2
17. b	3.2.3
18. d	3.2.4
19. b	3.2.4
20. d	3.2.4
21. b	3.2.5
22. c	3.2.6
23. d	3.2.7
24. a	6.3.0
25. c	6.2.3

NCCER CRAFT TRAINING USER UPDATES

The NCCER makes every effort to keep these manuals up-to-date and free of technical errors. We appreciate your help in this process. If you have an idea for improving this manual, or if you find an error, a typographical mistake, or an inaccuracy in the NCCER's Craft Training Manuals, please write us, using this form or a photocopy. Be sure to include the exact module number, page number, a description of the problem, and the correction, if possible. Your input will be brought to the attention of the Technical Review Committee. Thank you for your assistance.

Instructors – If you found that additional materials were necessary in order to teach this module effectively, please let us know so that we may include them in the Equipment/Materials list in the Instructor's Guide.

Write: Curriculum and Revision Department
National Center for Construction Education and Research
P.O. Box 141104
Gainesville, FL 32614-1104

Fax: 352-334-0932

Craft ____________________ Module Name ____________________

Module Number ____________________ Page Number(s) ____________________

Description of Problem

____

____

____

____

(Optional) Correction of Problem

____

____

____

(Optional) Your Name and Address

____

____

____

notes

Safety Requirements

Module 23102

SAFETY REQUIREMENTS

NATIONAL
CENTER FOR
CONSTRUCTION
EDUCATION AND
RESEARCH

OBJECTIVES

Upon completion of this module, the trainee will be able to:

1. Describe and wear different types of safety gear for the work site.
2. State the guidelines for dressing appropriately for concrete work.
3. Describe how to safely handle concrete when forming, mixing, placing, curing, and finishing.
4. Describe safety precautions to follow when working in extreme heat and cold.
5. Describe safety precautions to follow when working with hazardous materials.
6. Describe proper procedures for safely handling and maintaining concrete construction tools safely.

Prerequisites

Successful completion of the following Task Modules is recommended before beginning study of this Task Module: Core Curricula; Concrete Finishing Level 1, Module 23101.

Required Trainee Materials

1. Trainee Task Module
2. Pencil and paper
3. Appropriate Personal Protective Equipment

COURSE MAP

This course map shows all of the task modules in the first level of the Concrete Finishing curricula. The suggested training order begins at the bottom and proceeds up. Skill levels increase as a trainee advances on the course map. The training order may be adjusted by the local Training Program Sponsor.

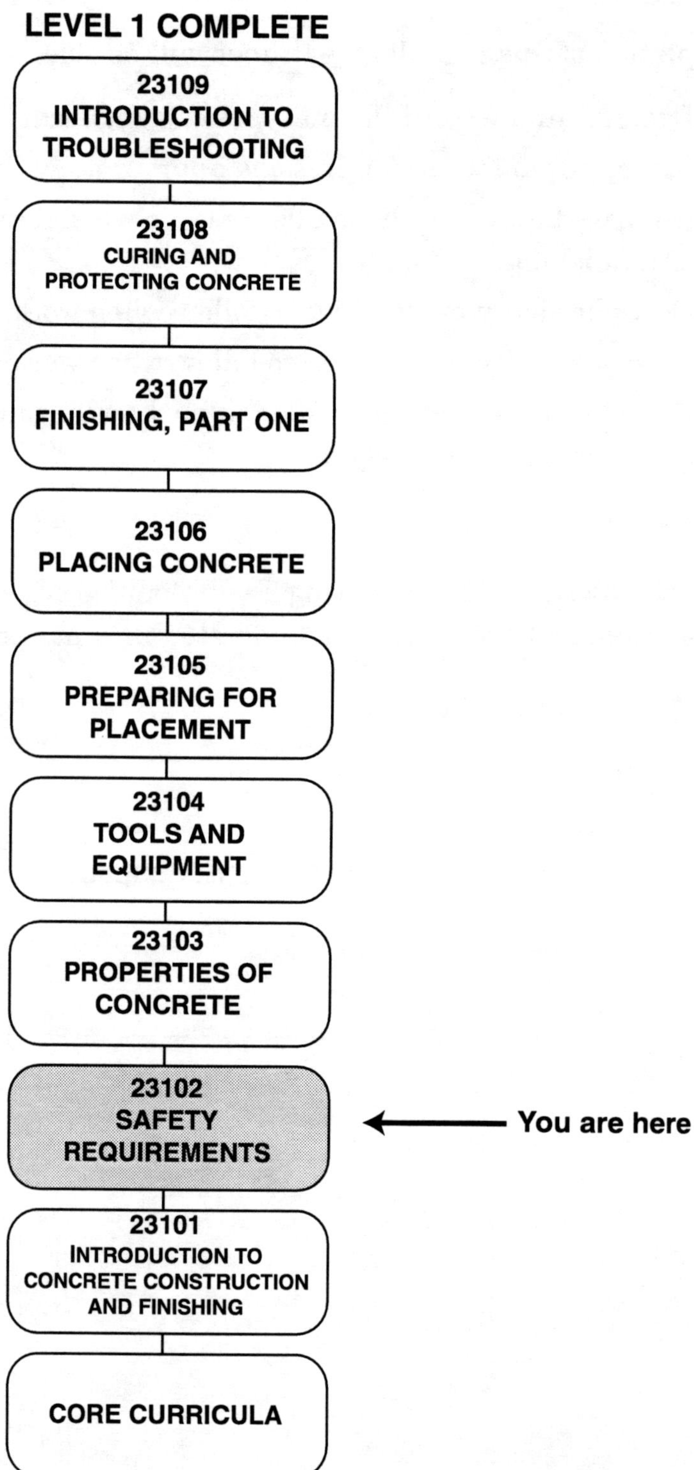

Cmap102.EPS

TABLE OF CONTENTS

TABLE OF CONTENTS (Continued)

Trade Terms Introduced In This Module

Accident: An unexpected, unintentional event that causes injury, death, or property damage.

Alkaline: Referring to a substance with a pH of more than 7; salts of alkali metals, mainly potassium and sodium. These chemicals can occur in materials that make up concrete and mortar.

Carelessness: Inattentiveness; failure to pay attention.

Caustic: A material that is characterized as capable of destroying the texture of anything or eating away its substance by chemical action.

First aid: The immediate care given to the victim (of an accident or sudden illness) at the work site until the services of a medical professional can be obtained.

Hazard: A source of danger creating the threat of an accident or incident.

Hygroscopic: The quality of absorbing moisture readily, especially from the skin. Wet concrete is hygroscopic.

Material Safety Data Sheet (MSDS): Printed information supplied by manufacturers of chemicals and other substances that explain the hazards of the material, proper handling procedures, and emergency actions.

Occupational Safety and Health Administration (OSHA): Part of the Department of Labor that oversees workplace health and safety issues.

Safety: Freedom from danger, risk, or injury.

Shoring: Props or posts of timber or other material in compression used for temporary support for excavations, formwork, or unsafe structures. Also, the process of erecting shores.

Toxic: Being harmful, poisonous, or containing a lethal chemical (toxin).

Unsafe act: The process of performing an action that is known to be unsafe or possibly hazardous to an individual or others. For example, unauthorized use of equipment is an unsafe act.

Unsafe condition: Hazardous situations that exist due to the disregard or lack of action on the part of a worker. For example, defective tools and equipment can present unsafe conditions.

Wind chill: The cooling effect of air moving over a body which takes into account both temperature and wind speed. Wind chill is expressed as a value of the amount of heat lost per unit area per unit of time.

1.0.0 INTRODUCTION

This module introduces **safety** and safe practices relating to concrete construction. It covers general safety, appropriate personal protective equipment, and personal protective actions that you will need to know when working with concrete and concrete equipment and tools. Being aware of safety practices and applying them will help to keep you and your fellow workers from being injured on the job.

1.1.0 WHAT IS SAFETY?

Basic safety is an attitude. It is a way of working on the job. The time spent learning and practicing safety procedures can literally save your life—and the lives of others. According to the National Safety Council, organized safety programs have saved more than three million lives so far, thereby preventing three million tragic deaths.

The ultimate responsibility for on-the-job safety rests with each and every worker. It is every worker's responsibility to learn:

- How to work around equipment and other people
- Requirements for safety gear and clothing for each job
- Safety rules and procedures associated with each job
- General work site safety and the **hazards** associated with each job
- Environmental safety as it affects each job

Failure to follow safety regulations and improper use of protective equipment can lead to an **accident** and possibly painful injury. Lost time injuries tend to be more severe and may even lead to the long-term reduction of a person's physical ability.

This module presents information about safety requirements and procedures, as well as safety tips for working with and maintaining concrete finishing tools. The procedures, practices, and rules presented in this module will help you identify and avoid hazardous conditions on the job site. By practicing safe procedures and keeping a safe attitude you will be doing your part to keep your workplace free from accidents and protecting yourself and your fellow workers from harm.

A skilled concrete craftworker must be constantly aware of job site hazards and the problems they might cause to others in the work area. The concrete finishing profession requires workers to use many potentially dangerous tools and to work in areas where an accident can occur as a result of even momentary **carelessness**.

1.2.0 WHY SAFETY IS A CONCERN

On-the-job safety is a concern not only because many companies require safe work records, but, more importantly, because safe working practices and conditions ensure a safe operation. This, in turn, helps to ensure the safety and well-being of all employees. Thinking before you act, staying alert, and being aware of where you are and what you're doing can help prevent personal injury. It can also help to ensure that you'll be able to continue working, and so continue to provide for your family.

Working responsibly and safely can also help prevent harm to co-workers. The inexperienced worker is arguably the most feared person on the construction site. Experience has shown that new workers cause many of the accidents on the job, although they themselves may not be injured. This is why it is important to encourage everyone, especially new personnel, to be safety conscious. The use of safe working practices also helps your employer because it means that everyone will be able to work, that the job is more likely to progress smoothly, and that it will finish on time and within budget.

1.3.0 BARRIERS TO SAFETY

Unfortunately, in today's environment there are some barriers to safety. These include the tools and equipment used on the job, the condition of the job site, and the attitude of some workers.

In order to work safely and securely, all workers must be in good physical and mental condition. They should be free from stress, drugs, and alcohol, and be focused on the job at hand. The tools and equipment used on the job must be well-maintained and in good working condition, and the operator must know how to use them appropriately. The job site should be clean and free of any hazardous materials or conditions.

1.4.0 COSTS OF UNSAFE PRACTICES

Unsafe working conditions and practices can result in:

- Personal injury or death
- Injury or death of workers
- Damage to equipment
- Damage to the work site
- Danger to the general public

According to the **Occupational Safety and Health Administration (OSHA)**, there were more than 6,000 work-related deaths in 1996. The construction industry ranks as the number two industry for most deaths, and falls are the leading cause of worker fatalities. Each year, on average, between 150 and 200 workers are killed and more than 100,000 are injured as a result of falls at construction sites.

Aside from the pain and suffering for the individuals involved and their families, injury and death affect others as well. For example, accidents can slow down or stop a job, thereby putting the entire operation in jeopardy, and possibly resulting in site-wide layoffs. A high accident rate can cause an employer's insurance rates to increase, thereby making him less competitive with other construction companies and less likely to secure future work.

Damage to equipment and the work site also affects more than just the one job. Damage to the equipment can either stop a job for a period of time until the equipment can be fixed, or increase the cost of a job because other machinery must be used. Damage to the work site can also increase the cost of the job, requiring repair and rework. This is why safety must be made a part of every job, and a part of everyone's responsibility.

2.0.0 SAFETY GEAR AND CLOTHING

One of the first steps to ensuring a safe operation is to wear appropriate safety equipment and clothing. Safety gear and clothing, when worn correctly, can help protect you from harm. The type of clothing and equipment required varies with job responsibilities and daily activities. It is the responsibility of all workers to see that they wear the appropriate clothing for the tasks they are performing. Your supervisor will provide a specific list of safety clothing and equipment required for your job.

2.1.0 TYPES OF PROTECTIVE GEAR

The general types of safety equipment used by concrete finishers are described in this section.

The hard hat, shown in *Figure 1*, is used to protect the worker's head from falling and flying materials or from bumps to the head. Construction sites are filled with activity from all directions. Therefore, it is important to wear a hard hat at all times, especially when working under ladders or scaffolding.

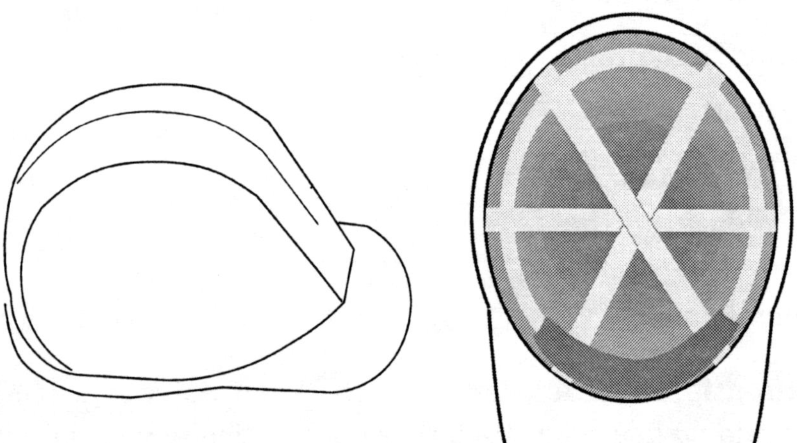

102F01.tif

Figure 1. Hard Hat

The best hard hats are made of rigid plastic and have an inner lacing or webbing (as shown in *Figure 1*). This can be adjusted, but it should not be removed because this is what enables the hard hat to absorb the shock of an external blow.

The safest way to wear the hard hat is slightly raised off the head. You should also inspect your helmet every time it is used. If there are any cracks, dents, or deep scratches in the shell, you must get a new hard hat. If any webbing straps are worn or torn, you should replace the webbing before the next use.

Goggles or a face shield, as shown in *Figure 2*, are used to protect the worker's eyes and face from flying chips of concrete or other debris that might be released into the air. They should be worn whenever there is any chance of an eye injury, such as when you are chipping or grinding. Goggles should also be worn when mixing concrete and whenever a truck is offloading concrete.

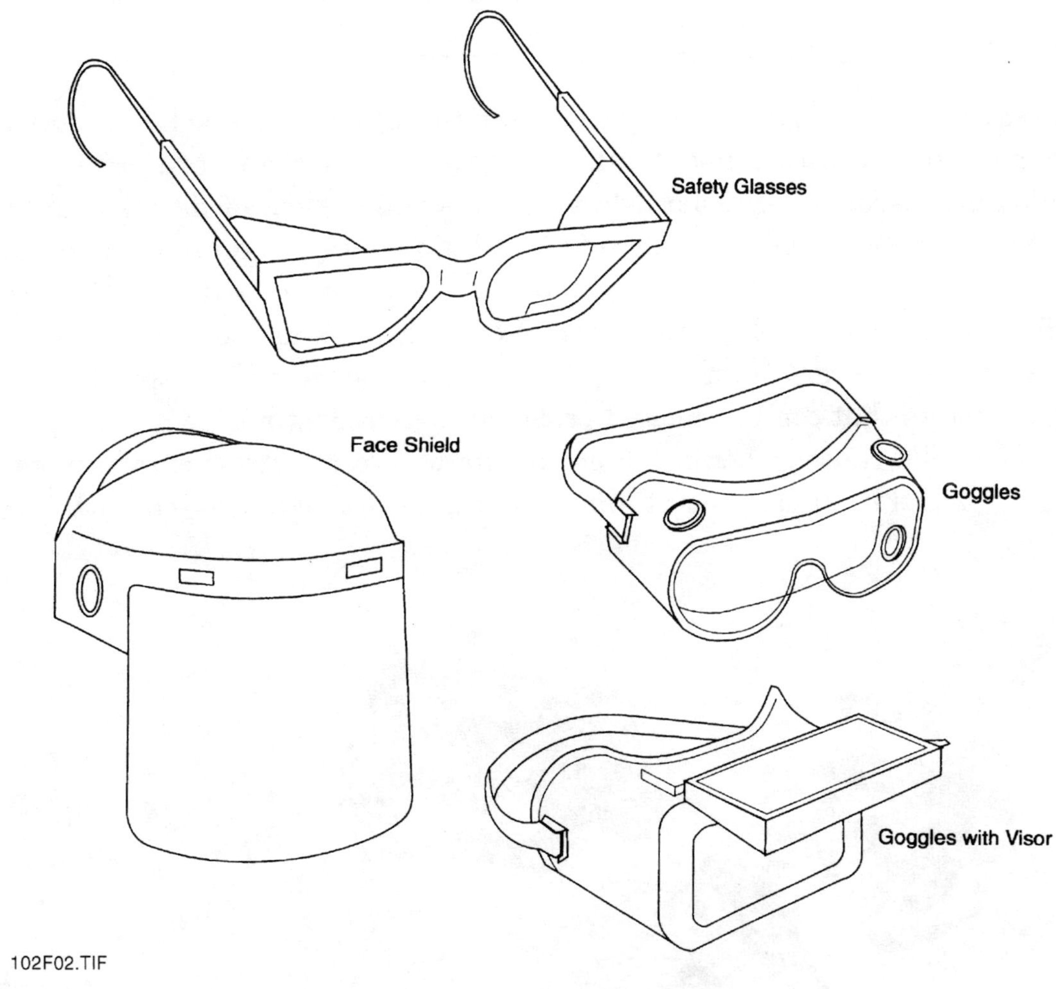

Figure 2. Safety Glasses And Face Shield

The two basic types of hearing protectors, shown in *Figure 3*, are used to protect the workers' hearing from harsh or loud noise. Construction sites are full of noise. Most construction companies follow OSHA rules in deciding when hearing protection must be used. Basically, protectors should be worn any time there is loud noise, such as while using a portable mixer or running a troweling machine. Protectors should be cleaned regularly with soap and water to prevent ear infection.

102F03.TIF

Figure 3. Hearing Protectors

Respirators and dust masks like the ones shown in *Figure 4* are used to protect the worker's lungs from dust and dirt. They should be worn whenever you are grinding, cutting, or roughening cured concrete. Federal law dictates which type of device should be used to protect the worker from different types of hazards. Make sure you use the appropriate one for your specific situation. You should also check the respiratory device for damage, and for a proper fit. Make sure you clean the filter after each day's use, or more often if necessary. There are many different types of respirators and dust masks. Respirators usually have filters or canisters that can be replaced or cleaned. Most dust masks are the disposable type and should not be reused or transferred to another worker. Your employer should have a program for distributing and maintaining breathing protection based on the type of work being done.

102F04.tif

Figure 4. Respirators And Masks

Rubber gloves (*Figure 5*) are used to protect the worker's hands from **caustic** or **toxic** materials. Fresh concrete, for example, can cause skin irritation and possible contact dermatitis. Concrete can also cause scrapes and cuts to your hands, as well as further injury to existing cuts. That is why it is important to wear rubber gloves whenever handling fresh or wet concrete. When gloves become worn, torn, or wet, they should be replaced and disposed of properly. Leather gloves can be worn whenever you are using a hand tool or performing a potentially dangerous operation in dry circumstances.

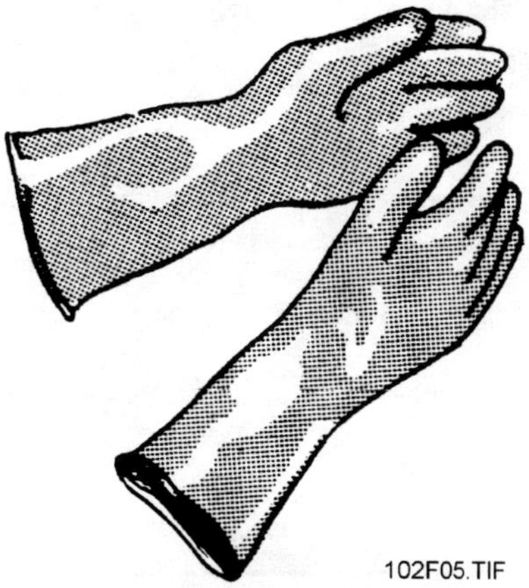

102F05.TIF

Figure 5. Rubber Gloves

The knee pads shown in *Figure 6* are used to cushion knees when craftworkers must kneel down on concrete, wood forms, or other hard material. They protect the knees against dampness and painful injuries caused by kneeling on small pebbles. When wearing knee pads, make sure they are securely strapped to your legs with no dangling straps or ties.

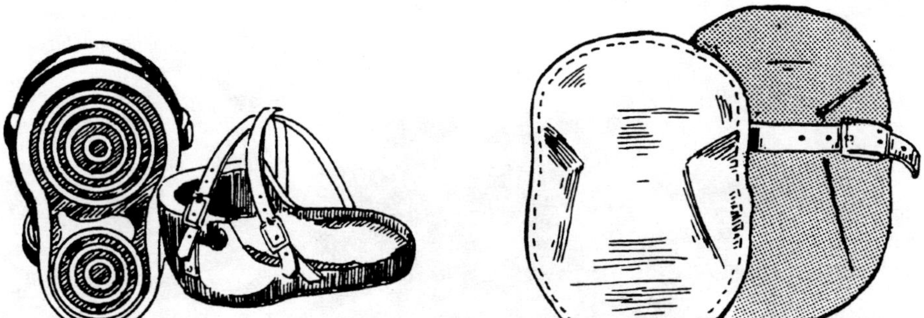

102F06.TIF

Figure 6. Knee Pads

Knee boards, shown in *Figure 7*, are used to distribute the worker's weight evenly when finishing or troweling. They can prevent you from sinking into the concrete or from forming a depression in the surface. They are necessary when hand floating and troweling large floor areas or wide sidewalks.

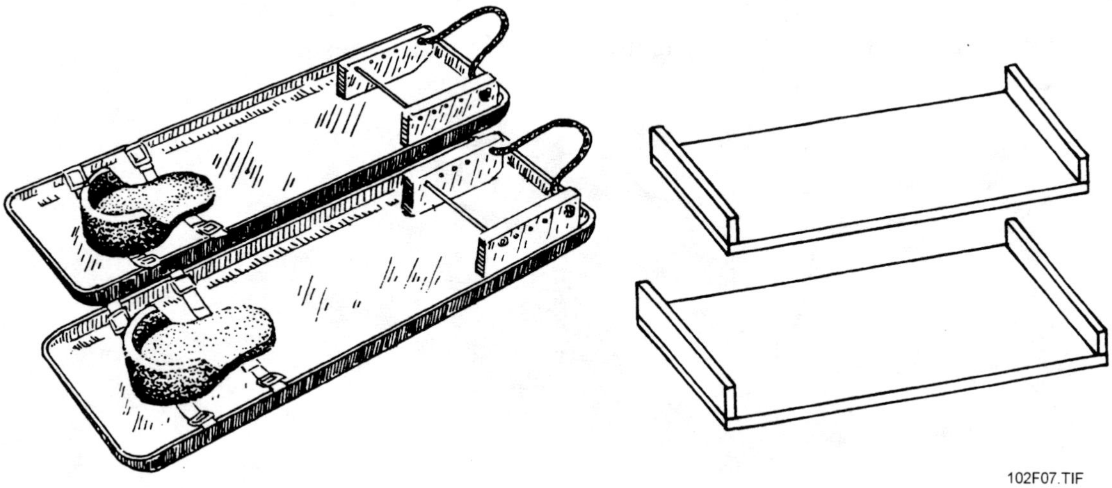

102F07.TIF

Figure 7. Knee Boards

Safety shoes or boots are used to protect the worker's feet from injury. The type of footwear you wear will be determined by the type of work you are doing. Typically, work boots like the ones shown in *Figure 8* are made of leather and have steel toes. Tennis shoes or soft walking shoes should never be worn while working. If it is necessary to stand in fresh concrete while it is being placed and screeded, you will need to wear rubber boots over your regular boots to protect the boots and your feet. Make sure the rubber boots are high enough to prevent concrete from flowing over the tops.

102F08.TIF

Figure 8. Safety Boots

2.2.0 WEARING SAFETY GEAR AND CLOTHING

In general, your employer is responsible for making sure that employees wear appropriate personal protective equipment whenever they are exposed to possible hazards to their safety. In turn, they are responsible for wearing the gear and clothing assigned.

You should wear the appropriate clothes and protective equipment for the specific work being performed. *Figure 9* is a picture of a concrete finisher wearing all the protective gear (except a safety harness).

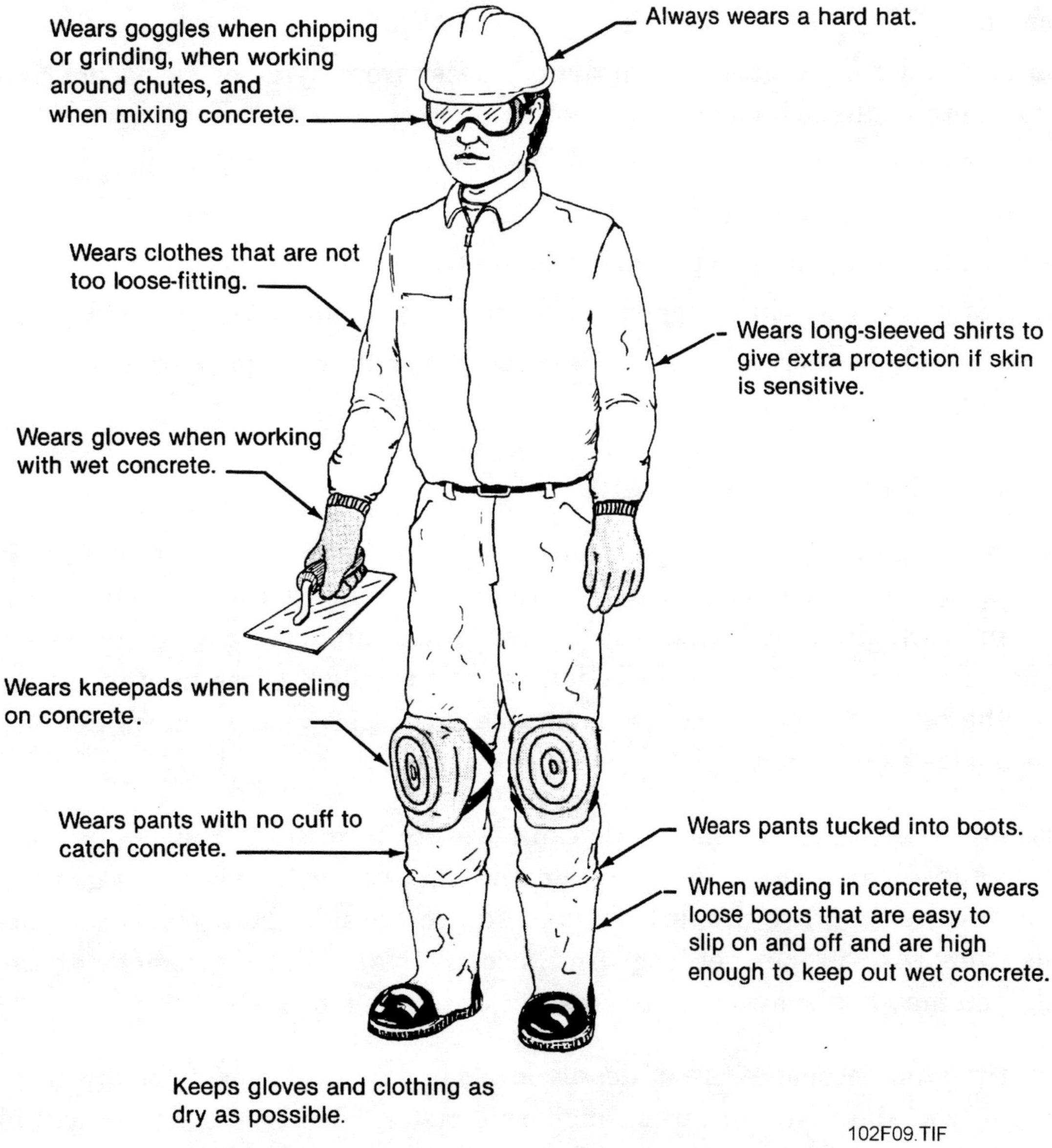

Figure 9. Properly Dressed Concrete Worker

Not all gear needs to be worn all the time. The safety harness, for example, is only needed for work on elevated surfaces where fall protection is not provided.

It is important to wear the necessary equipment and take the following safety precautions when performing concrete work:

- Remove all jewelry, including rings, bracelets, necklaces, and earrings. Jewelry can get caught on or in equipment, which could result in a lost finger, ear, or other appendage.
- Confine long hair in a ponytail or under your hard hat. Flying hair can obscure your view or get caught in machinery.
- Wear close-fitting clothing that is appropriate for the job. Clothing should be comfortable and should not interfere with the free movement of your body. Clothing or accessories that do not fit tightly, or that are too loose or torn, may get caught in tools, materials, or machinery.
- Wear face and eye protection when there is a risk from flying particles, debris, or hazards such as hot or caustic liquids.
- Wear hearing protection when using or working around noisy machinery.
- Wear respirators or dust masks as required.
- Wear a long-sleeved shirt to provide extra protection for your skin.
- Protect any exposed skin by applying skin cream, body lotion, or petroleum jelly.
- Wear hard sole leather safety shoes or boots. Never show up for work dressed in sneakers or gym shoes.

2.3.0 CARBON MONOXIDE METERS

In some instances, wearing the proper safety gear and clothing may not be enough to protect you from the hazards of the work site. For example, in cold weather, portable heaters may be set up at the work site in order to keep workers warm and create a proper environment for concreting. However, if proper ventilation is not provided, these heaters can use up the oxygen in the air and produce carbon monoxide. Also, gas-powered machinery will produce carbon monoxide as an exhaust.

Odorless and tasteless, carbon monoxide can seep into your lungs, causing nausea, vomiting, headaches, and—if not caught in time—death. One way to prevent this gas buildup is to make sure all heaters are properly vented to the outside. However, there still may be problems even with proper venting. And, because you can't see, smell, or taste carbon monoxide, you may not be aware of its presence until it is too late.

To ensure that you become aware of deadly levels of carbon monoxide before they cause you undue harm, you should use a carbon monoxide meter. This type of meter will alert you to dangerous levels of carbon monoxide so that you can leave the area before you are overtaken by the gas.

Figure 10 shows a carbon monoxide meter. The meter provides an early warning of the buildup of carbon monoxide and other gases that cannot be smelled or otherwise detected.

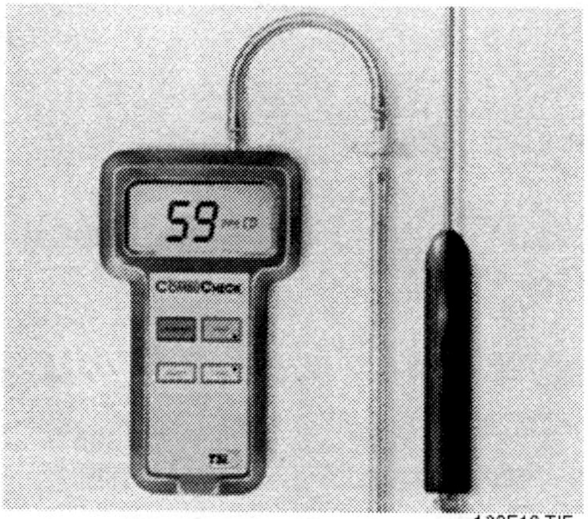

102F10.TIF

Figure 10. Carbon Monoxide Meter

The standard detector warns of combustible and hazardous gases and vapors with an audible alarm and visual meter indicator. It can detect low concentrations of carbon monoxide and alert you visually or audibly to its presence. The higher the gas concentration, the more frequently the monitor will buzz.

You should use a carbon monoxide meter whenever you are working in a heated area or in a confined space. Choose one that best fits your needs and learn how to operate it effectively. You should also calibrate it on a regular basis to make sure it is operating correctly. Remember, your life and the lives of others could depend on how well you know how to operate a carbon monoxide meter.

3.0.0 HANDLING CONCRETE SAFELY

Concrete construction requires the setting and stripping of forms, and the transporting, placing, consolidating, finishing, and curing of concrete. It may also involve sawing, grinding, and other methods of treating the hardened surface of the concrete. The following sections provide some guidelines for safe operation when performing these tasks.

3.1.0 MIXING CONCRETE

Generally, large amounts of concrete are mixed at batching or ready-mix plants, and only small amounts are mixed by hand at the job site. The active ingredient of concrete, portland cement, is **alkaline** in nature and therefore caustic. It is also **hygroscopic**, which means that it tends to absorb moisture. Prolonged contact between the fresh mix and skin can cause skin irritation and chemical burns to hands, feet, and any other exposed skin. It can also saturate a worker's clothes and carry alkaline or hygroscopic elements to the skin. In addition, the sand contained in fresh concrete is abrasive to the skin.

If a mobile mixer is used, it is important that you know how to operate the machine properly according to the manufacturer's instructions for that particular unit. Follow the usual precautions for working with powered equipment, such as wearing the proper protective clothing and keeping your hands away from pinch points and moving parts.

You should also follow precautions specific to the mixer, such as keeping all guards in place (including drum guards) and never sticking your hands or any solid object, such as a shovel, into the drum while the mixer is in operation.

Needless injury can be avoided by taking the following precautions:

- Keep cement products off your skin at all times by wearing appropriate personal protective equipment, including boots, gloves, and knee pads.
- Avoid rubbing your skin against cement products. Rubbing increases the chance of serious injury. If your skin does come in contact with any cement products, wash your skin promptly. An application of lanolin cream to the affected area can help relieve any pain. If a reaction persists, seek medical attention.
- Keep cement and cement products out of your eyes by wearing safety glasses. If any cement or cement mixture gets in your eye, flush it immediately and repeatedly with water, and consult a physician promptly.
- Rinse off any clothing that becomes saturated from contact with fresh concrete. Even indirect contact through clothing can cause severe burns. A prompt rinse with clean water will prevent continued contact with skin surfaces.
- End each workday with a bath or shower to remove any concrete from the skin.

A cement burn occurs with very little warning because little heat is sensed by the skin. If you experience any irritation or inflammation on your skin, or if you have persistent or severe discomfort, seek immediate medical attention.

3.2.0 TRANSPORTING CONCRETE

At most work sites, concrete is produced at central plants and brought to the point of placement by the truck; on other sites, mobile mixers are used. If trucks are used, it is important that the truck have clear access to the site and sufficient room to maneuver safely. A fully loaded truck mixer usually weighs over 30 tons. Therefore, if the truck route will bring it close to the side of an excavation, proper reinforcement may be needed to prevent the sides of the excavation from collapsing.

When the mixer truck must negotiate around barricades and through traffic, a spotter may be necessary to help the driver maneuver and back up. Make sure the driver has a clear understanding of where to maneuver before beginning the operation. If you are directing any backing maneuvers, discuss the hand signals to be used with the driver. Make sure you stand clear of the path of the truck and in view of the driver. If you can see his face in his side mirrors, he can also see you.

If the concrete cannot be discharged from the truck directly to the placement point, wheelbarrows, Georgia buggies, or power buggies may be used. Runways should be laid out prior to the beginning of the operation. They should be in a circular or loop path so that traffic can move continuously and safely. Plan ahead for adjustments as the placement progresses. Walkways or runways should always be supported from grade or from the deck form, not by reinforcing steel.

3.3.0 PLACING

Concrete can be placed satisfactorily using several methods, including chutes, conveyors, buggies, hoists, cranes, and pumps. The simplest way to place concrete is to have ready-mix trucks discharge directly into the forms. This helps to deliver the concrete uniformly to the formwork. It also prevents accidents caused by attempting to transport concrete across work sites in wheelbarrows, buckets, or conveyors. However, there are some precautions that should be taken when placing concrete.

- Protect yourself from splashed concrete by wearing the necessary clothing, boots, and gloves.
- Check to make sure the work site is clear of personnel and other obstructions before beginning placement.
- For all forms of placement that require moving the concrete distances of more than a few feet, be sure there is a method to quickly stop the flow of material in case of emergency. Make sure everyone understands the procedure and who is responsible for stopping the flow.
- Regulate the placement of the concrete so that it produces a balanced loading and prevents formwork failures.
- Take care not to damage or misalign the formwork.
- When pumping concrete, keep an eye on any flexible hose and ensure that it is free from kinks. A kink in the hose can cause pressure buildup, which can endanger nearby workers.

After the concrete is deposited in the desired area by the chute, it should be pushed, not lifted, into final position with a shovel. Remember, a cubic foot of concrete weighs approximately 150 pounds. Short-handled, square-end shovels or hoe-like tools called come-alongs should be used for spreading to reduce the likelihood of back or muscle injury.

3.4.0 FINISHING

Finishing usually involves some close contact with the fresh concrete. This could involve floating, troweling, edging, or applying some type of an architectural finish. When working with fresh concrete, care should be taken to avoid skin irritation or chemical burns. Prolonged contact between fresh concrete and skin surfaces, eyes, or clothing may result in burns that are quite severe, even third-degree burns. Portland cement is alkaline in nature, so wet concrete and other cement mixtures are strongly basic. Strong bases, like strong acids, are caustic and harmful to the skin. Portland cement is also hygroscopic, which means it will draw water away from any material it contacts, including the skin.

Waterproof pads should be used between fresh concrete surfaces and knees, elbows, and hands to protect the body during finishing operations. Eyes and skin that come in contact with fresh concrete should be flushed thoroughly with clean water. Clothing that becomes saturated from contact with fresh concrete should be rinsed out promptly with clear water to prevent continued contact with skin surfaces.

When working with fresh concrete, begin each day wearing clean clothing and conclude the day with a bath or shower.

Concrete can be finished in many ways depending on the effect desired. Some finishes, such as acid etching, involve the use of acid or other toxic materials. When applying these materials, it is important to follow both the manufacturer's guidelines for the specific agent and your company's guidelines for using toxic materials. Be sure to apply the acid evenly instead of sprinkling it over the surface, which could result in acid flying into the air and hitting either a person or a piece of equipment, causing injury or damage. When you are finished flushing the area, test the surface with litmus paper to ensure that the acid has been completely removed or neutralized.

3.5.0 CURING

The process of curing concrete involves keeping the newly-placed concrete moist and at a temperature of approximately 50°F or above. The problem with curing is that many curing components are toxic and most are highly flammable. Chemical curing compounds are among the most potentially dangerous substances a concrete craftworker will ever use. Therefore, they need to be treated with caution. Follow the manufacturer's recommendations or the directions found on the **Material Safety Data Sheets (MSDS)** maintained by your employer.

When working with curing compounds, it is important to follow all recommended safety procedures and wear appropriate personal protective equipment. The appropriate respirator will help to avoid lung damage, brain damage, or cancer—all of which can result from breathing chemical fumes. Using chemical curing compounds indoors in unventilated areas is very hazardous. You should never smoke when applying curing compounds or use them near an open flame.

3.6.0 FORMWORK

Forms are an essential part of concrete construction. They control the position, alignment, size, and shape of concrete, and serve as a temporary support structure for concrete while it is curing.

The formwork phase of a project can be the most costly as well as the most dangerous of the functions on a job site. The formwork must be capable of supporting all loads without collapsing and injuring workers. Concrete placed in or on formwork imposes loads on either the ground or an existing structure. These loads should be checked by the project structural engineer to ensure that the magnitude, location, and timing of the imposed loads do not exceed (with an appropriate safety factor) the structural capacities of the foundation and structure. Formwork designers must follow local code requirements as well as OSHA regulations for formwork.

3.6.1 Placing Concrete In Formwork

Formwork failures can be attributed to human error, substandard materials and equipment, omission, and basic inadequacy in design (under-design).

The most effective means of achieving safety in the use of forms is to have competent supervision and knowledgeable workers during erection and concreting. Formwork must be constructed exactly as designed, following a safe erection procedure, so that no members are temporarily overloaded. Construction procedures must be planned in advance to ensure the safety of personnel. The contractor should have a program of field safety inspection of formwork as the job progresses. Form watchers are needed to assure early recognition of possible form displacement or failure during concrete placement.

Before concrete placing begins, formwork must be inspected to see that it is in the correct location, at the correct elevation, and built so that it will produce the required finish and dimensions with adequate safety. Any foreign material must be removed from the forms. Make sure that all the wall ties and connection hardware are correctly installed. One missed form tie can cause a form to bulge or fail.

The concrete should be placed at or as near as possible to its final position in the forms. Concrete should not be dropped over 5 feet without using a drop chute (elephant trunk).

Unanticipated loads on multi-story floor slab formwork can be overlooked and lead to form sagging, or even to failure. Some examples of unanticipated loads are: placing concrete in a concentrated area; impact loads by dropping concrete from excessive height into the form; and stockpiling reinforcing bars in one spot.

One of the most dangerous periods in a construction project is when the concrete is being placed in the form. All personnel must be especially aware of unusual deflection or form movement during concreting that might warn of impending collapse.

Formwork should be continuously watched during and after concreting. Precautions must be taken to protect formwork watchers and maintain a safe area (and exit route) during concreting. If any serious weakness develops during the concreting that would endanger workers or cause undue distortion of the structure, work must be halted while the formwork is strengthened.

3.6.2 Form Removal

Before forms can be removed safely, ample time must be allowed for the concrete to obtain its required strength. It is advantageous to leave forms in place as long as possible to continue the curing period. The designer should specify the minimum strength requirements of various structural members before form removal.

In any case, forms should not be removed until the concrete is strong enough to satisfactorily carry the stress from both the dead load of the structure and any imposed construction loads. The concrete should be hard enough so that the surfaces will not be injured in any way when reasonable care is used in removing forms. In general, for concrete temperatures above 50°F, the side forms of reasonably thick, supported sections can be removed 12 to 24 hours after concreting. Beam and floor slab forms and supports (**shoring**) may be removed between 3 and 21 days depending on the size of the member and the strength gain of the concrete. For most conditions, it is better to rely on the strength of the concrete as determined by testing rather than to select arbitrarily the age at which forms may be removed.

4.0.0 WORK SITE SAFETY/ACCIDENT PREVENTION

Your safety and that of your co-workers should be the primary consideration in your work life. In order for a work site to be safe, each individual must act responsibly and follow established safety procedures. In addition, everyone should constantly be aware of conditions in the work area that may lead to injuries or illness, or that could endanger others.

Attitudes have much to do with work site safety and job performance. On the job it is your responsibility to exert every effort to avoid accidents and injuries. You must constantly be on the alert and exercise good judgement. You should contribute your share to overall good housekeeping in the work area, thus eliminating unsafe conditions. Appropriate personal protective equipment is made and provided to give you maximum protection. It should be worn whenever the work makes it necessary. Only through such a cooperative effort can a safety program in the best interest of all be established and preserved.

Here are attitudes you must develop if you are to make safety work for you, and if you are to do your work without accidents. Remember:

- Accidents can be prevented
- Preventing accidents is more important than blaming the people who cause them
- Safety rules and regulations are reasonable and important
- The people I work with will respect me for practicing safety
- Safety is a mark of intelligence and skill
- It is dumb to take chances
- I have a responsibility to do something about correcting any hazard that I see

Developing attitudes such as these will not only help you work safely and avoid accidents, they will make you a strong force for safety on any project.

4.1.0 WORK SITE SAFETY

Work site safety is the responsibility of everyone who is involved with the project as well as the workers at the job site. Your employer will provide you with appropriate personal protective equipment and ensure safe working conditions on the site, and will also keep you informed of any new or changing safety regulations. You need to act responsibly and safely and accept responsibility for your actions and how they may affect you, your fellow workers, and the public.

One of the first steps to becoming safety conscious is being able to recognize safe on-the-job practices. You need to familiarize yourself with these practices by attending safety meetings, studying company policies, and reading product and equipment manufacturers' literature.

One of the easiest ways to promote safety on the work site is to keep the work site clean and free of obstructions. This means planning your work to avoid unsafe conditions, maintaining a clean and litter-free area, and cleaning up after your work. The following actions will help to create a safe workplace:

- Maintain the job site in a clean and orderly condition. This includes bending all exposed nails and staples so they don't injure anyone; securing any and all air, water, and electric lines so they don't become tripping hazards; and keeping all aisles and passageways clean and clear of obstructions.
- Encourage and enforce personal hygiene. This includes washing your hands after handling materials to remove any residue or hazardous substances that could lead to accidents or illnesses. Change your clothes, gloves, or shoes if they become wet.
- Keep all work surfaces dry and slip-resistant. If you do spill something, clean it up immediately.
- Store combustible scrap and waste safely and dispose of it properly.
- Remove accumulations of dirt in order to prevent damage to equipment and reduce hazards around the site.
- Store flammable materials properly according to the manufacturer's instructions. Do not smoke around such materials.
- Be aware of—and work safely around—gas and electric lines.

Another way to promote safety is to maintain a good work attitude. This means conducting yourself in a professional manner and avoiding horseplay, pranks, or fighting. Practical jokes can be fatal and should be avoided. It is also important that you maintain your health and know your physical limitations. Do not attempt to operate any equipment or perform activities that are dangerous while under the influence of medication or drugs.

A good work attitude means that you will:

- Follow safety regulations and procedures at all times, including company and industry standards, and federal, state, and local laws.
- Participate regularly in safety meetings.
- Get your share of sleep and good food. A good night's sleep and a healthy meal can keep you alert and able to avoid accidents.
- Watch out for unsafe work methods or unsafe conditions and report them promptly to your supervisor.
- Report any injury or accident—serious or not—to your supervisor.
- Use appropriate safety warning devices to notify others of any danger zone.
- Use safety harnesses where necessary, and be sure the harness is tied off to an area that is free of the work platform.
- Assist other trades when necessary to maintain safe operations.
- Respect electricity. Never use electrical equipment in areas of excessive moisture unless you have taken all safeguards. Use only approved electric power tools and extensions that are grounded.
- Store materials and supplies safely in their proper places.
- Lift with your legs, not with your back.
- Know the location of fire-fighting and **first aid** equipment.
- Be conscious of others' safety and encourage others to do the same.
- Listen and adhere to the alarms or warning bells on mixers, pavers, and ready-mix trucks.

Just as there is a list of things to do to promote safety, there is also a list of things not to do:

- Do not attempt any work or repair for which you are not qualified.
- Do not perform hazardous work or work in a confined space alone.
- Never place yourself in a dangerous position or allow others to work in a dangerous position. For example, never use a crow bar by pulling it toward your body.
- Do not tamper with, remove, or damage any protective devices or safety devices that are installed on tools or equipment.
- Do not leave electrical wires or equipment in water. Do not use wet or metal extension handles around electrical equipment or power lines.
- Never enter any underground vault, manhole, silo, or enclosed area until you are certain that the air within contains no flammable, contaminated, or toxic vapor or gas.
- Never use compressed air to clean your clothes or yourself.
- Be aware of pinch points on forms and equipment and avoid being caught in them.

4.2.0 EQUIPMENT PROCEDURES

Aside from concrete equipment, work hazards include general site equipment used to perform other work activities. Special care needs to be taken around ladders and scaffolding, as noted in the next sections.

4.2.1 Ladders

Many construction sites are full of ladders and various types of scaffolding. In fact, the ladder is one of the most important tools on the job site. When selecting a ladder for a job, it is important to choose one that will extend at least 36 inches above the landing surface you are trying to reach. You should always place the base of the ladder so that the distance between the base and the wall is ¼ of the ladder length from the base elevation to the point where the ladder touches the wall, as illustrated in *Figure 11.*

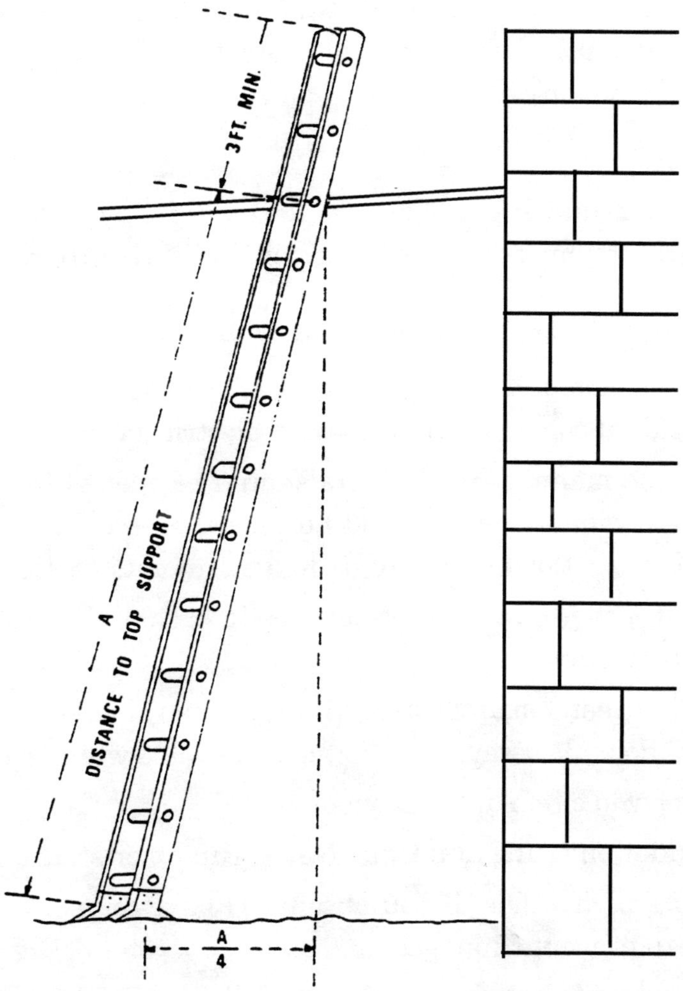

102F11.TIF

Figure 11. Positioning A Straight Ladder

If you set up a ladder in front of a door or entrance, make sure you lock the door or board it over. Otherwise, you may be knocked over if someone tries to open the door. You should also watch out for overhead electrical lines and stay as far away from them as possible. The minimum allowable distance to place a ladder from an electrical line is 10 feet.

The following four-point safety check should be performed before using a ladder:

- Check the specified workload of the ladder. Do not use a ladder if it has a smaller capacity than what you need.
- Check the ladder for damage such as slivers, cracks, or splits in the wood, as well as loose nails, screws, bolts, or rungs.
- Check the ladder to make sure it has a nonslip base and a safety top.
- Check various parts of the ladder, depending on what type you have. Stepladders, for example, should be checked for instability, which is often caused by loose or bent hinge spreaders, broken steps, or loose hinges. Extension ladders should be checked for defective or broken extension locks or rope deterioration. Wooden ladders should be checked to make sure they haven't been painted; paint could cover up a structural defect in a ladder.

If you find a defect in the ladder, take it out of service and place a tag-out tag on it. Send any ladder with a defect back to the repair department, as per OSHA requirements.

The following precautions should be kept in mind when setting up and using a ladder:

- Do not use a metal ladder when performing any type of electrical work or whenever there is a possibility that you might come into contact with electrical conductors.
- Place the ladder in a stable manner. Stepladders should be opened fully on a solid, even surface. Portable and extension ladders should be placed on a solid base in such a way that the base of the ladder is ¼ the distance of its length from its vertical support.
- Place the ladder so that it leaves 6 inches of clearance in back of the ladder and 30 inches of clearance in front of the ladder.
- Place the ladder so that it leans against a solid and immovable surface. Never place a ladder against a window, door, doorway, sash, loose or movable wall, or box.
- Face the ladder whenever you climb up or down on it.
- Climb or descend the ladder one rung at a time. Never run up or slide down a ladder.
- Do not use ladders during high winds. If you absolutely have to, make sure you lash the ladder securely in order to prevent slippage.
- Check to make sure the soles of your shoes are free of oil, mud, and grease.
- Keep both hands free so that you can hold the ladder securely while climbing. Use a rope to raise and lower any tools and materials that you might need.
- Never rest any tools or materials on the top of a ladder.

- Move the ladder in line with the work to be done. Never lean sideways away from the ladder in order to reach the work area.
- Never stand on the top two rungs of a ladder.
- Use ladders for only short periods of off-ground work. If you must work from a ladder for extended periods, use a lifeline fastened to a safety belt.
- Lay the ladder down on the ground when you have finished using it, unless it is anchored securely at the top and bottom where it is being used.
- Never use makeshift substitutes for ladders; never use stepladders for straight ladder work.

4.2.2 Scaffolds

Scaffolds are another piece of equipment often encountered on a job site. Scaffolds are elevated working platforms that support workers and materials. The two main types of scaffolding that you will encounter are tubular steel sectional scaffolding and adjustable tower scaffolding. Tubular steel scaffolding can be either fixed or mobile. One general safety rule for all types of scaffolds is that they have a minimum safety factor ratio of 4:1. This means that the scaffolding should be able to support at least four times the weight that will be placed on it.

A typical tubular steel scaffold is shown in *Figure 12*. The main part of the scaffold is the working platform. A working platform should have a guardrail system that includes a top rail, mid-rail, toeboard, and screening. To be safe and effective, the top rail should be approximately 42 inches high, the midrail should be located halfway between the toeboard and the top rail, and the toeboard should be a minimum of 4 inches high.

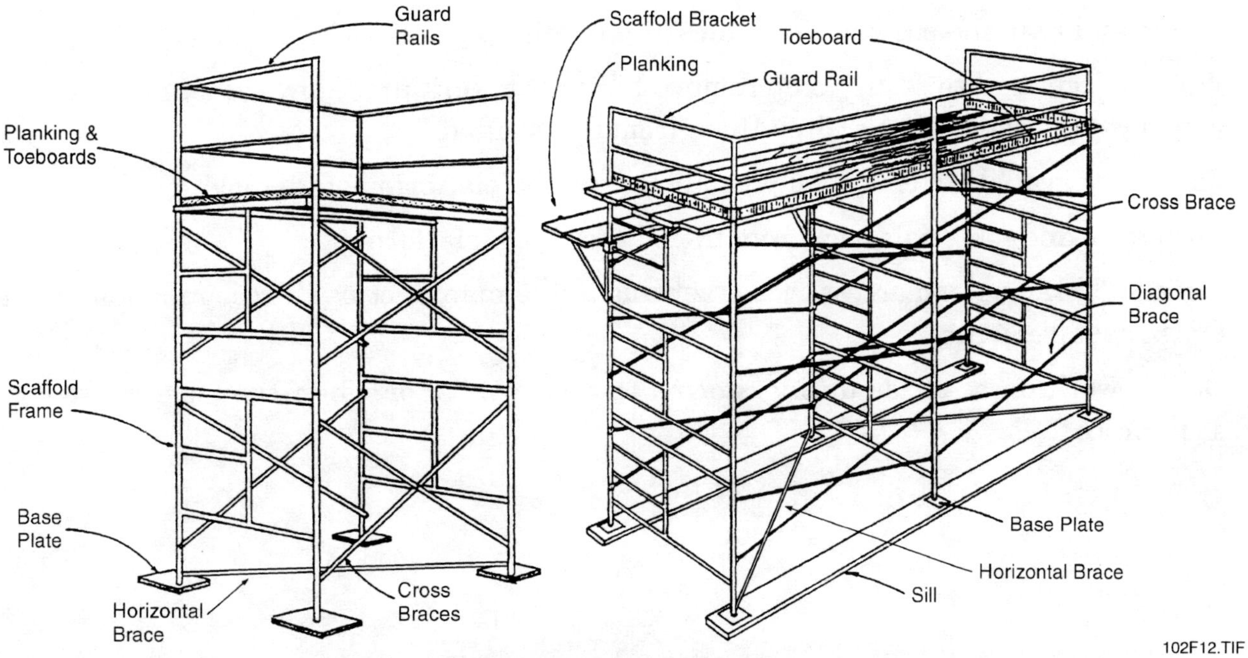

Figure 12. Tubular Scaffold

If workers will be passing or working under the scaffold, the area between the top rail and the toeboard must be screened. Finally, the platform planks must be laid closely together. For safety purposes, the ends of the planks should overlap at least 6 inches and no more than 12 inches.

Before using any scaffold, you must inspect it to make sure it is safe. Make sure the base is placed on firm ground or flooring. It should be level and plumb. Be sure there is safe access and exit to the work platform. Any defects, or dangerous or unsafe conditions, should be reported to your supervisor. If you are using a ladder to get on and off the scaffold, it is also important to follow all the ladder safety guidelines explained earlier in this module. Never climb on the cross braces of the scaffold. Braces are not made to support your weight.

When several people are working on a scaffold at the same time, you must make sure that everyone spaces themselves to distribute the weight evenly. When materials are delivered to the scaffold by forklift, keep everyone in the clear and be careful not to overload the scaffold.

Faulty design and inadequate construction are sometimes involved, but in most cases scaffold accidents are caused by careless maintenance and improper use. Keep scaffolds safe by observing the following basic procedures:

- Always raise the scaffold to the desired height. Do not try to increase height by using ladders, boxes, or sawhorses on top of a scaffold.
- Always keep scaffold planks clear of extra tools and materials.
- Always clean up any slippery substances that might get spilled on a scaffold.
- Use extreme caution when working near power lines.
- Anchor free-standing scaffolds by guy wires in order to prevent tipping or sliding.
- Locate and read the posted safety rules and regulations for scaffolding use.
- When objects and tools are to be removed from the work area, lower them to the ground with a rope. Never throw or drop them from the scaffold.
- For mobile scaffolds, remove all personnel from the platform before moving.
- Keep tools and materials back from the edge of work platforms.
- Install all bracing and accessories according to the manufacturer's recommendations and OSHA requirements.
- Do not work on scaffolds during storms or high winds, or when they are covered with snow or ice.

4.3.0 TYPICAL ACCIDENTS

OSHA sets rules and regulations regarding the physical environment of the workplace, yet physical hazards on the job site have resulted in more accidents and damage than any other aspect of working in construction. According to OSHA, falls are the leading cause of worker fatalities in the construction industry in the United States. Each year an average of 150 to 200 workers are killed and more than 100,000 are injured as a result of falls at construction sites.

Hazards and accidents at a job site are caused by changing environmental conditions and human behavior. According to OSHA, 80 percent of accidents are caused by **unsafe acts** rather than **unsafe conditions**. However, you should be aware of the types of accidents that are caused by both. The following lists include the top twelve reasons in each category that accidents happen. If you acquaint yourself with these situations, a majority of accidents can be avoided.

The top twelve unsafe actions are:

- Unauthorized use or operation of equipment
- Failure to secure or tie down against unexpected movement
- Operating or working at an unsafe speed
- Failure to warn or signal as required
- Removing or making safety devices inoperative
- Using defective tools or equipment
- Using tools or equipment unsafely
- Standing in an unsafe place or taking an unsafe posture
- Servicing moving or working equipment
- Riding hazardous moving equipment
- Horseplay, distracting, startling, and kidding
- Failure to wear appropriate personal protective equipment

The top twelve unsafe conditions are:

- Lack of adequate guards or safety devices
- Lack of adequate warning system
- Fire and explosion hazards
- Unexpected movement hazards
- Poor housekeeping
- Protruding object hazards
- Close clearance and congestion hazards
- Hazardous atmospheric conditions
- Hazardous arrangement, placement, and storage
- Hazardous defects of tools and equipment
- Inadequate illumination, intense noise
- Hazardous personal attire

Safety procedures must be learned for each task a worker performs. Sometimes, new material or equipment is brought on the job and new safety procedures must be learned by everyone on the job. The important thing to remember is that safety *can* be learned, and what you learn about safety can help prevent accidents.

4.4.0 ACCIDENT PREVENTION AND FIRST AID

In order to prevent accidents, all work must be conducted in compliance with:

- Federal, state, and local laws
- Company policies and procedures
- Specific project requirements
- Manufacturer's recommendations
- Plain old common sense

How well you conduct your work in a hazard-free manner according to directions provided is a key factor in your performance, for which you are accountable. By your actions, you can prevent accidents and injuries and take pride in your accomplishments on the job, in the industry, and at home.

One way to prevent accidents is to learn as much about safety on the job as possible. Know industry safety standards and OSHA requirements as well as your company policies and procedures. Attend company-sponsored safety training and regularly scheduled toolbox meetings. Toolbox meetings are held for a purpose. Attend willingly and offer ideas for developing accident-free job sites. You should also seek advice from other, more experienced workers. They often know several ways to deal with the hazards of the workplace.

It is important to understand the obligations that exist for everyone's safety. An obligation is like a promise or a contract. In exchange for the benefits of your employment, you agree to work safely. In other words, you are obligated to work safely, and to make sure your co-workers are working safely. Your employer has the obligation to maintain a safe workplace for all employees.

One of the best things you can do for yourself and your co-workers is to learn basic first aid. First aid can be of immense value in the event of injury. It prepares you for what to do and what not to do in a medical emergency. Knowing how to give first aid can mean the difference between life or death, temporary or permanent disabilities, and rapid recovery or long hospitalization.

Figure 13 presents an example of directions for giving first aid. When you enter the job site, find out who has taken approved training so that you know who to go to in case there is an emergency. Know where to find the first aid kit and the phone number to call in case of an emergency.

102F13.TIF

Figure 13. General Directions For Giving First Aid

Figure 14 is an example of a typical emergency contact information sheet. When you call for emergency help, tell the operator the telephone number and location you are calling from, where the emergency is, what the emergency is, how many people need help, and what first aid has been given to the victim(s) so far.

Emergency Telephone Numbers

Rescue Squad ______________________

Fire Department ______________________

Police ______________________

Physician ______________________

Hospital Emergency Room ______________________

Poison Control Center ______________________

Additional Numbers ______________________

When you call for emergency help, tell:

1. WHERE the emergency situation is, with cross streets if possible.
2. TELEPHONE NUMBER you are calling from.
3. WHAT HAPPENED—heart attack, auto accident, fall, etc.
4. HOW MANY persons need help.
5. WHAT is being done for the victims.

NOTE: YOU HANG UP LAST. Let the person you called hang up first.

Record and post emergency telephone numbers in advance so that they will be readily available when you need them.

The programs of the American Red Cross are made possible by the voluntary services and financial support of the American people.

Poster 1023-A
Rev. Mar. 1983

American Red Cross poster

102F14.TIF

Figure 14. Emergency Contact Information

Job-related concrete finishing injuries that may require first aid include:

- Bleeding caused by cuts, punctures, and scrapes
- Burns caused by coming in contact with hot motors, electrical equipment, or open flames
- Chemical burns caused by contact with the concrete
- Loss of consciousness caused by a blow to the head or being overcome by fumes
- Respiratory failure or cardiac arrest caused by the inhalation of toxic gases or electrocution
- Eye injury caused by burns, cuts, or irritation from toxic fumes or gases

Quick and effective use of first aid procedures can save your life and the lives of your fellow workers. Even in situations that are not life-threatening, first aid can reduce the seriousness of an injury or prevent it from becoming life-threatening. For this reason, you should learn as much first aid as possible.

5.0.0 ENVIRONMENTAL CONDITIONS

More and more construction work is being accomplished during extreme weather conditions. Large jobs generally cannot wait for a day of good weather. The work has to be done to meet completion dates no matter what the temperature might be.

5.1.0 WORKING IN EXTREME COLD

One of the problems brought on by working in extreme cold is just that—extreme cold. Typically, the body cannot withstand extremely cold temperatures for prolonged periods of time. In order to work in the cold, you should wear protective clothing and cover all exposed areas of skin from head to toe.

Wind chill is the cooling effect of moving air over the body by taking into account both the temperature and wind speed. Generally, the higher the wind speed, the colder the wind chill. This means that if the temperature is 32°F and the wind is gusting, it will feel colder than 32°F because the wind is causing a cooling effect to the skin. *Figure 15* is a chart showing the effect increased wind speed has on actual temperature readings.

For a properly clothed person, a wind chill down to a value of minus 21°F should not cause serious danger. However, anything colder than this causes increasing danger, with the possibility of frostbite and the freezing of exposed flesh.

WIND-CHILL CHART							
Estimated Wind Speed MPH	ACTUAL THERMOMETER READING (DEGREES F) 40	30	20	10	0	-10	-20
10	28	16	4	-9	-21	-33	-46
20	18	4	-10	-25	-39	-53	-67
30	13	-2	-18	-33	-48	-63	-79
40	10	-6	-21	-37	-53	-69	-85
Wind speeds greater than 40 MPH have little additional effect	Little danger for properly clothed person			Increasing danger			Great danger
				Danger from freezing of exposed flesh			

102F15.EPS

Figure 15. Wind Chill Chart

5.1.1 Cold Weather Hazards

Common hazards related to cold weather work include immersion foot, trenchfoot, frostbite, and hypothermia.

- *Immersion foot* is caused by prolonged exposure to cool or cold water. It results in numbness, swelling, and superficial gangrene.
- *Trenchfoot* is caused by exposure of more than one day to moisture at or near freezing. It results in swelling and tissue degeneration.
- *Frostbite* is the actual freezing of tissue. It is the most common cold weather hazard. With frostbite, there is a potential for massive tissue loss and gangrene after 25 hours of continued exposure. Over time, the affected parts get numb and the joints become stiff and do not function properly. It is very painful when the feeling starts returning to frostbite-affected parts.
- *Hypothermia* is the progressive lowering of the body's core temperature. It is caused by exposure to temperatures below 32°F for a prolonged period of time. As the body temperature drops, organs tend to shut down, resulting in loss of consciousness and eventually death.

Symptoms of decreasing body temperature include shivering, clouded consciousness, and pupils that are dilated but react to light. Symptoms of frostbite include white or grayish skin and a hard skin surface with underlying soft tissue.

The following precautions should be taken when working in extreme cold:

- Wear protective cold weather clothing such as gloves, insulated/vapor-barrier boots, face masks, and hard hat liners in order to protect the entire body, especially the hands, feet, and head. It is generally better to wear several layers of light clothing rather than one layer of heavy clothing. This way, you can discard a layer or two if the temperature rises.
- Maintain a work rate that is sufficient to prevent heavy sweating. Too much sweating will cause you to dehydrate. It will also dampen your clothes, which could make you feel colder.
- Take break periods in a warm place.
- Work with a buddy. Check each other often for early warning signs of cold disorders.
- Drink warm beverages.
- Do not get your clothes wet. If you do, change into a dry set of clothes as soon as possible.
- Work in heated shelters as much as possible.
- Attend a training program addressing working in cold weather.

5.1.2 Working Around Heating Equipment

In addition to the hazards of frostbite and hypothermia, cold weather also brings the additional hazards of working with heaters and heating equipment, often in confined spaces. Some companies use propane-fired heaters, as in *Figure 16*, to heat enclosures while working or curing concrete. Heated enclosures are commonly used when the air temperature is near or below freezing. Heat may be supplied to enclosures by live steam, forced hot air, or portable heaters like the one shown.

102F16.TIF

Figure 16. Heated Enclosure

If you construct a temporary housing to protect the concrete, be sure to avoid the risk of fire by placing heating units away from any other flammable material. You should also vent the heating units to the outside. Provide sufficient space above and around the units in order to allow the free flow of warmed air. If the enclosure holds live steam, stay out while the concrete is curing.

The necessity of supplying heat to the concrete can create fire hazards that are a threat to the work area and to workers. Combustion heaters generate carbon monoxide, which can be dangerous. Make sure you follow your company's guidelines for working with or near heaters, and know what to do in case of a fire. Follow these basic guidelines:

- Store any flammable material outside away from heaters
- If propane heaters are used, keep the tanks outside and run lines to the heaters
- Keep fire extinguishers in good working order and placed in strategic locations
- Know where the shutoff valves are for any heating equipment
- Periodically check the enclosures with a carbon monoxide meter to detect unsafe levels of this gas

5.2.0 WORKING IN EXTREME HEAT

Just as the rigid scheduling of construction work requires people to work in extreme cold, it also requires people to work in extreme heat. This causes its own particular set of hazards. In general, heat emergencies are caused by the body becoming overheated. Normally, the body cools itself by sweating. However, when the outside temperature is equal to or greater than the body's own temperature, there is too much heat for the body to cool itself.

Reactions to hot weather vary with a person's age, general physical condition, and activity level. Older people, small children, and very active young people are the most likely to suffer from heat illness.

The human body is capable of adapting itself to varying conditions, but the combined effect of heat and heavy physical exertion may be too much for the body's natural temperature regulation. Heat can make some people sick, and in some cases, it can be deadly.

The body's heat regulation system tries to maintain a fairly consistent temperature of about 98°F. To cool the body, the heart pumps blood out to the extremities (arms and legs) where the heat is dispersed. Another way the body eliminates heat is through the cooling evaporation of sweat. A cool breeze and low humidity help body cooling. High humidity interferes with sweating and adds to discomfort.

5.2.1 Heat Syncopy Or Heat Fainting

This is common among people who are not acclimated to hot weather. It normally occurs soon after someone goes out into the heat to work or exercise. The blood pressure falls suddenly as blood flows to the extremities to cool the body and not enough goes to the head. The person feels giddy, light-headed, and then passes out. The best first aid is to have the person lie down to rest and elevate the feet.

5.2.2 Heat Cramps

These muscular cramps result from exposure to high temperatures for a relatively long period of time, coupled with physical exertion. Heavy sweating and dehydration usually precede the cramps.

5.2.3 Heat Exhaustion

This occurs after some hours in the sun, and is related to loss of water and salt through sweating without sufficient fluid replacement. Common symptoms include a subnormal body temperature, pallor, dizziness, profuse sweating, cool, moist skin, nausea, and headaches. The best first aid is to have the person rest in a cool place and drink water or fruit juice. If the person does not feel better in an hour or two, or cannot hold liquids, take them to a hospital emergency room.

5.2.4 Heat Stroke

This condition can prove to be fatal, so immediate medical attention is necessary. It is caused by a complete breakdown of the body's heat regulating system, with heat production exceeding heat loss. Common symptoms include hot, dry skin, little or no sweating, severe headache, visual disturbance, rapid temperature rise, loss of consciousness, flushed face, weakness, and confusion.

First aid treatment should be aimed at cooling the body quickly and getting the victim to the hospital. Remove the person from the source of heat, loosen the clothing, and try to cool the body as rapidly as possible using a cool bath or cold compresses. Do not give water to a heat stroke victim.

When the temperature begins to soar upward, take a few precautions to minimize your body's level of heat stress.

- Try to take frequent rest periods.
- Space heavier workloads or workouts throughout the day, rather than concentrating them into a few consecutive hours.
- Avoid heavy activity during the hottest part of the day.

- Drink plenty of water or other fluids.
- Wear loose, light clothing that reflects the sun's rays and allows sweat to evaporate. A hat will add further protection.

5.3.0 OTHER ENVIRONMENTAL CONDITIONS

Just as schedules and deadlines often require concrete to be placed in severe heat or cold, they may also require it to be placed during bad weather or at night. During these times, it is necessary to take extra safety precautions. During storms, for example, it is important to keep all electrical tools and equipment dry. If there is lightning in the immediate area, take cover away from equipment and trees. When working at night, make sure there is sufficient light to enable you to see the work site clearly. Dark areas can be full of hazards.

Some general environmental hazards directly related to the concrete finishing process are given below:

- Remember that curing concrete generates heat. This may create a hazard by raising the temperature around the concrete on a hot day. Be sure to adjust your routine and workload with this in mind.
- Concrete often needs heat added during cold weather. Forms and reinforcing steel can be heated, and mix water itself may be heated. This is another potential hazard; be wary of burns from unexpected heat sources on cold days.
- Concrete is sometimes cured by live steam inside a temporary housing. Stay out of such a housing. Know what is happening inside any temporary housing before entering.
- Concrete may contain metal reinforcement that may not be covered or grounded. Long metal handles on bullfloats and straightedges can act as electrical conductors. Lightning may be an unexpected hazard. Stop work and take shelter during electrical storms.

6.0.0 OTHER SAFETY REQUIREMENTS

In addition to site hazards, concrete finishing can bring hazards from dangerous or toxic substances. Improper use of hand and power tools can also bring danger. The next sections discuss these hazards.

6.1.0 HAZARDOUS MATERIALS LABELING

Concrete construction requires the use of many different materials that could be harmful. Chemicals in these materials could enter the body through absorption, inhalation, or swallowing. This poses a serious problem for workers. Exposure to chemicals can cause or contribute to many serious health defects, such as heart ailments, kidney and lung damage, sterility, cancer, burns, and rashes. Some chemicals may also be safety hazards, having the potential to cause fires, explosions, and other serious accidents.

OSHA has a requirement called "Hazard Communication," with the basic goal of reducing the incidence of chemical-source illness and injuries by informing employers and employees about work hazards and how to protect themselves. As part of this requirement, chemical manufacturers, importers, and distributors must be sure that all containers of hazardous chemicals are labeled, tagged, or marked with the chemical's identity, appropriate hazard warnings, and the name and address of the manufacturer or other responsible party.

Your employer is responsible to ensure that each container in the workplace is properly labeled, tagged, or marked with the identity of hazardous chemicals contained therein, and with the hazard warnings appropriate for employee protection. Your responsibility is to adhere to the warnings on these labels, and to use the materials appropriately.

Before working with any hazardous material, you should:

- Put on the appropriate personal protective equipment that is recommended by the manufacturer.
- Verify that all containers are clearly labeled as to their contents, with the appropriate hazardous warning and the name and address of the manufacturer listed. No container of hazardous materials should be accepted without proper labeling in place.
- Know the hazardous properties of the chemicals you work with. Read the labels on all containers and follow the manufacturer's instructions for using them.
- Know emergency response procedures and practices. For example, you should know your company's emergency call procedures regarding whom to call and when.
- Do not transfer hazardous materials to temporary containers other than for immediate use. At the close of work, leftover amounts should be returned to the original container or turned over to your immediate supervisor for proper disposal.
- Store chemicals in a safe manner and according to the manufacturer's recommendations. Keep chemical containers closed when not in use.
- Know where your company keeps hazard communication program records for all hazardous materials or substances. Your site should have a Material Safety Data Sheet (MSDS) file for all hazardous materials or substances.
- Know the first aid treatment for any chemical you work with, and be prepared to carry out that treatment immediately if required.
- At regular intervals, inspect containers and pipelines that hold corrosive materials. Report leaks immediately.
- When using corrosive materials, know where the closest shower and eye fountain are located. Keep the shower and eye fountain unobstructed and in good working condition.

6.2.0 TOXIC SUBSTANCES

There are a number of toxic substances a concrete finisher might encounter on a job. Toxic substances can be in solid, liquid, or gas form. They can be inhaled, absorbed through the skin, ingested, or contracted through open cuts and sores on the skin.

Some chemicals may only affect the bodily areas they contact, while other chemicals may penetrate into one area and then affect other internal organs. That is why it is important to know the substances you work with and what their properties are.

Acids and bases can easily damage skin and eyes. They can be in the form of liquids, solid granules, powders, vapors, or gases. Commonly used acids include sulfuric acid, hydrochloric acid, muriatic acid, and nitric acid. Some common bases are lye (sodium hydroxide) and potash (potassium hydroxide).

Always follow these rules when working with acids and bases:

- Know what chemicals you are working with and how strong (concentrated) they are.
- Use appropriate personal protective equipment as required.
- In case of skin or eye contact, flush with cool water for at least 15 minutes, but do not rub the skin or eyes.
- Always add acid to water to prevent splattering which can occur if water is added to acid.
- Keep acids and bases apart, store separately, and clean up spills promptly.
- Check with your supervisor if you need more information.

Cement and concrete are alkali compounds in their dry and wet form. As dust and powder, they can cause damage to the skin and eyes when they react with moisture in your body. The more lime cement contains, the more irritating it is to the skin.

Polyurethanes are used for a variety of concrete treatments. They contain a chemical that is a low-grade form of cyanide. When polyurethane is sprayed, it releases this chemical into the air. This is why it is crucial that you use an approved respiratory device when spraying polyurethane at work. You also need to take care to not get polyurethane on your skin. If you do get it on your skin, do not use solvents to wash it off. Solvents will break down the material and carry it through your skin and into the bloodstream.

Epoxy grout is another toxic substance that is used often in concrete finishing. It contains silica, which is very dangerous if inhaled. It is also an irritant to skin and a hazard to eyes. The hardener in epoxy can also cause severe chemical burns. When working with epoxy, it is important to wear safety goggles, chemical-resistant gloves, full-coverage clothing, and a respiratory device. In case of skin contact or respiratory problems, follow recommended first aid procedures. For eye contact, prompt medical attention is essential.

6.3.0 CARE AND MAINTENANCE OF HAND TOOLS

Becoming familiar with the basic tools and materials of your trade and understanding how to handle them safely are important steps toward working safely. The need to keep tools clean and in good working order cannot be overstressed. Safe and proper use of tools will not only help preserve the tools in good working condition, but can also help to prevent accidents. All tools are capable of inflicting injury, even hand tools.

Concrete finishers must know how to use many different tools, and need to develop specific skills before these tools can be used efficiently. Faulty, damaged, or improperly used tools and equipment pose a substantial threat to the safety and well-being of all employees.

One of the most important factors regarding tool safety is the quality of your tools. Quality tools will last longer if they are used and maintained properly. It is also important to use each tool appropriately. If you use the proper tool for the job, the job will not require as much physical effort. Using an inappropriate tool, such as an undersized winch for lifting a heavy load, may result in equipment failure or injury to nearby personnel.

When using tools, you need to stay alert. This means positioning all cutting edges away from your body, keeping your fingers away from any and all cutting edges, and holding tools such as screwdrivers, wrenches, and chisels in such a way that you will not be hurt if they slip from your grasp. When working around electrical equipment, you should use tools with insulated or wooden handles. Check to make sure everyone is clear when using long-handled tools such as bullfloats and straightedges.

You also need to inspect tools frequently, to make sure they are in good condition. Do not use a file, for example, unless it has a handle. And never use broken tools, or tools with loose or broken handles or dull blades. If you find a broken or damaged tool, repair it promptly or tag it out. Dispose of broken tools that cannot be repaired.

The tools you use must be maintained in safe working order, and kept clean, dry, and sharp. The quality and quantity of work you do increases when you use clean, sharp tools in good condition. When tools are not cleaned before concrete has had time to harden on them, for example, the job of getting them back into readiness becomes more difficult and takes time away from the job at hand. Tools that have grease or oil on them may slip out of your hand, causing damage or harm.

When not in use, it is important to store tools properly. Do not carry tools in your pockets, and do not place tools where they can fall. Dropping a builder's level, laser level, or any small leveling instrument can affect its accuracy. After cleaning, tools that easily rust should be coated with a rust-preventing oil and stored in a dry place with their cutting edges protected from damage.

6.4.0 CARE AND MAINTENANCE OF POWER TOOLS

Power tools can be driven by electricity, air, or internal-combustion engines. Like hand tools, they should be operated in a safe manner at all times and stored properly when not in use. It is also very important to wear appropriate personal protective equipment and proper clothing when using power tools. Loose-fitting clothing that may become caught in the equipment should be avoided.

Power tools require additional safety precautions. If used improperly, power tools can cause fire and physical damage, as well as electric shock and other personal injuries very quickly. It is very important that you operate only those power tools that you have been trained to operate, and use power tools only for their intended purpose. You also need to be familiar with the correct operation and adjustment of power tools.

Guidelines for safe and proper operation of power tools include the following:

- Be sure you understand the tool's principles of operation.
- Read the manufacturer's operation manual before you begin.
- Check to make sure any electric power tool is properly grounded before use.
- Assume a safe and comfortable position before starting work.
- Do not distract others or let anyone distract you while operating a power tool.
- Disconnect power before performing maintenance or changing accessories.
- Do not use dull or broken accessories. When trowel machine blades become ragged, for example, they should be changed. Change all the blades on a tool at the same time.
- Do not leave power tools running unattended.
- Report any unusual noises, sounds, or vibrations to your supervisor.
- Ensure that all guards are in place, all parts are securely attached, and hoses and couplings are in good condition. Do not use tools that have missing guards or safety devices.
- Use a proper extension cord of sufficient size to service the particular electric tool you are using.
- Do not operate electrical power tools if your hands or feet are wet.
- Keep a firm grip on the power float or trowel machine handle while operating the machine.
- Be sure there is proper ventilation before operating gasoline-powered equipment indoors.
- Keep a fire extinguisher handy when filling and operating gasoline-powered equipment. OSHA requires that gasoline-powered equipment be turned off prior to refueling.
- Make sure all extension cords, lines, and hoses are placed out of the way so workers do not trip over them and they will not be damaged.

- Check the insulation on power cords and the condition of plugs and sockets every day. If cords, plugs, or sockets are frayed, worn, cut, or broken, repair them before using.
- Clean chips from grinders and other equipment with a brush. Do not clean equipment with your hands or with a rag. Never try to clean equipment while it is running.
- Limit the exposure of connectors and tools to excessive moisture by using watertight or sealable connectors.
- Inform other workers in the area of what you are doing so they can take appropriate measures for their own protection.
- Wear rubber boots and gloves when using electric-powered tools on a wet surface.

6.5.0 SPECIFIC TOOL SAFETY

The tools most commonly used for placing and finishing concrete include straightedges, floats, trowels, vibrators, and saws. In addition, there are tools such as grinders, sprayers, and pumps that are used for specific jobs when required. Guidelines for properly using and maintaining these tools in safe working order are summarized in the following sections. Always check the manufacturer's operation manual and maintenance requirements for specific information.

6.5.1 Straightedges

Straightedges are rigid pieces of wood or metal that are used to strike off excess concrete and level the surface to proper grade as soon as concrete has been placed. They should be approximately 2 feet longer than the distance between edge forms so that they can be handled at a slight diagonal. A straightedge is usually worked by two people who push it back and forth in a sawing motion as they move it across the surface of the concrete.

Straightedges are made of a variety of materials. The simplest form is a 2 × 4 board eight to ten feet long like the one shown in *Figure 17*. Other materials used for straightedges include magnesium and aluminum.

102F17.TIF

Figure 17. Straightedge

Power straightedges are also commonly used for striking of large areas. *Figure 18* shows a picture of a power straightedge or screed with a motor mounted in the center of a beam spanning the forms.

102F18.TIF

Figure 18. Surface Vibrating Screed

Straightedges are long, cumbersome pieces of equipment to handle. Safety precautions to be taken with straightedges include lifting and moving each side so as to not hit other workers or get caught in pinch points when handling the equipment. Never try to move or lift this equipment alone. Even for simple 2 × 4's ten feet long or longer, have another worker lift the other end out of the concrete and help to carry it. Other types of straightedges should be handled by two or more people depending on the size and weight of the equipment.

6.5.2 Floats

Floats are wide, flat tools pulled and pushed across the surface of the concrete with the blade of the float against the surface. There are both manual and power type floats. Manual type floats, like the bullfloat shown in *Figure 19,* have long handles that protrude some distance behind the worker using it. It is important to stay alert and not interfere with other workers or hinder traffic. When working close to other machinery or electrical lines, use floats with insulated handles. When carrying the float, always watch for other workers so they are not hit with the end of the handle.

102F19.TIF

Figure 19. Bullfloat

Power floats are driven by gasoline engines. These machines have numerous moving parts and are heavy and cumbersome to move. Do not place hands or fingers in the operating mechanism when it is running or try to lift the machine by yourself. Always refer to the operator's manual for repair or adjustment of the equipment.

6.5.3 Trowels

There are two main categories of trowels. Hand trowels come in various shapes and sizes for different uses. They are made from top quality steel and usually fitted with wooden handles. Power trowels are powered by a gasoline engine mounted on top of a circular frame. The blades of the trowel are fitted to a shaft extending from the engine. As the blades rotate on the shaft, they smooth the soft concrete.

Hand trowels, like the ones shown in *Figure 20*, are used for sealing concrete surfaces, cleaning other tools before washing them, and cleaning out and patching honeycombs or voids. It is important to use the correct trowel for the job. It is also important to keep trowels in good condition.

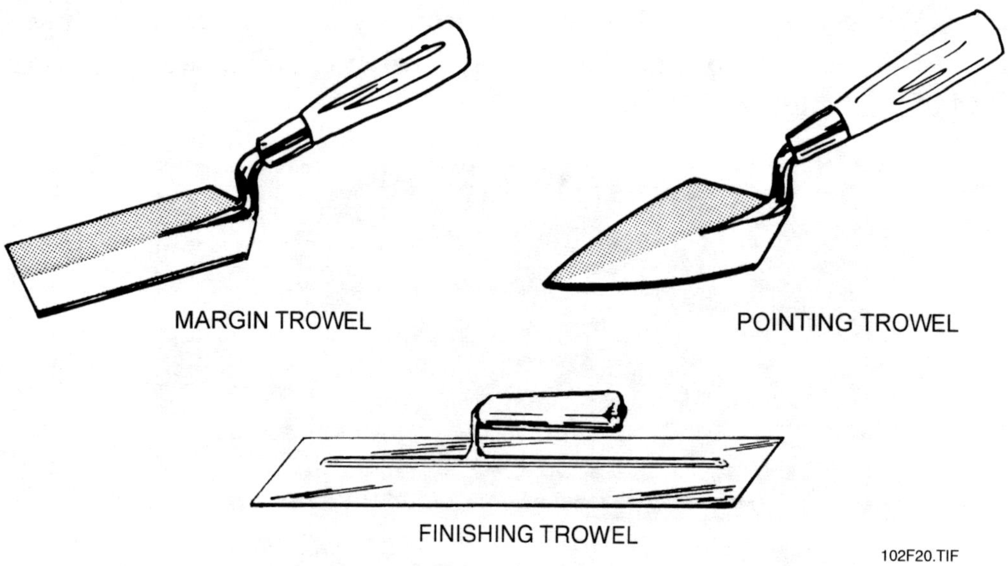

Figure 20. Trowels

Troweling concrete can wear down the edge of a trowel until it is razor-sharp. Such a trowel is hazardous. To prevent accidents, edges should be filed when they become worn.

When filing trowels, lay the file on a flat surface. Hold the trowel by the handle and stroke the trowel edge on the file. Do not hold the trowel with one hand and file it with the other, because one slip could cause the edge of the trowel to cut your hand.

It is also important to clean trowels after each use. If concrete is allowed to dry on the trowel, it will harden, making it difficult to remove.

Walk-behind power trowels have blades up to 48 inches in diameter. These larger machines are heavy and can create a large amount of force. If the blades penetrate a soft spot in the concrete, the trowel handle can throw the operator. Some machines have a safety feature which stops them automatically when this happens. The operator must always be in control of the power trowel so there is no danger to workers close by.

6.5.4 Vibrators

A concrete vibrator is a mechanical oscillating device of high frequency that is used to consolidate fresh concrete. Vibrators are available in three types: internal, external, and surface.

The internal vibrator is inserted into the concrete so the vibrating motion settles the concrete in the enclosed area. This machine can usually be handled by one person. Be careful when pulling the vibrator out of the fresh concrete so that strain on the upper body is minimized. If the vibrator gets stuck on reinforcing bars, applying extra force to the pull cord could damage the rebar, the vibrator, or your back.

External vibrators operate on the same principle as a hand or electric hammer. Some have a flat piece of 4" × 4" hard rubber attached to an iron plate on the end. This is held against the forms and the vibration of the operating tool causes the plastic concrete to move into the voids. The vibrator must not be held in one place too long or it will cause structural defects.

CAUTION: Do not hold a vibrating plate or other vibrating mechanism against your body. Injury to your skin or internal organs could result.

The surface vibrator employs a portable horizontal platform on which a vibrating mechanism is mounted. Some machines combine the surface vibrator with the power straightedge and accomplish both jobs at the same time. These machines usually require two people to operate, one at either end. Watch out for pinch points and moving parts where clothing and fingers can get caught. As this equipment is heavy, never attempt to lift or shove one of these machines into place. Use the proper lifting equipment to move or relocate the machine.

6.5.5 Concrete Saws

Concrete power saws are available in electric and gasoline models ranging from 3 to 60 horsepower. They are used to construct various types of joints as well as to define limits of breakout work in hardened concrete.

You should study the operater's manual to learn everything about the saw before using it. Your own safety and the safety of those working around you will be improved if you take the time to know the machine. Before using a saw, take the following precautions:

- Check the water hose and connections and make sure there is enough water pressure
- Make sure the guiding gauge will stay in alignment
- Ensure that the motor has been serviced properly
- Wear rubber boots to keep your feet dry

Wear appropriate personal protective equipment such as a face shield, goggles, and a dust mask to protect yourself from flying particles and dust caused by the sawing.

6.5.6 Concrete Pumps

Concrete pumps are used extensively in modern construction operations. They make it possible to convey and place concrete rapidly and economically even in hard-to-reach areas. Pumps come in many different sizes and models. Most systems use pistons to create the pressure to force the concrete through the line; some models use compressed air. The main point to remember is that pumped concrete is under pressure. This requires extra safety measures during set-up of the equipment and during operation.

The operation and maintenance of concrete pumping equipment is a specialized activity that requires additional training in pumping procedures, as well as in the operation of the specific type of equipment. When working around a pumping operation, you should be aware of the following general safety requirements:

- The pumping equipment must be in perfect technical condition, and control of the operation should be in the hands of a well-trained operator.
- The pipeline must be laid carefully and should be free of tension. It must meet the requirements of the site and be adequately dimensioned. All couplings must be tight.
- Bends in the pipe increase the pressure in the system. Make sure there are no unnecessary bends or dents in the pipe. End hoses should be in good shape and not have any rips or tears.
- Make sure everyone involved in the pumping operation knows their duties and can communicate with the operator, either through hand signals or by radio.
- Do not look into, or place your hand into, the open end of a hose or pipe connected to a concrete pump.
- Stay clear of the operation if your assistance is not needed.

6.5.7 Sprayers

Spraying equipment is used for several different work activities in the concrete construction industry, primarily for moistening forms, applying curing compound, and cleaning hardened surfaces. Sprayers come in many different sizes and models. They can be hand or power driven.

When using sprayers, it is particularly important to wear appropriate personal protective equipment, including a hard hat, gloves, safety shoes or boots, face and eye protection, and full coverage clothing. Because eye and lung protection is so important when spraying any chemical compound, you also need to use an approved respiratory device and safety goggles.

Always read or review the operating procedures supplied by the manufacturer of the equipment before beginning operation. Filling reservoirs with spraying compounds should follow recommended procedures and OSHA safety requirements. When using the equipment, never point the nozzle toward another person or yourself. After completing the work or stopping for any reason, always release the pressure in the reservoir of the applicator or turn off the pumping mechanism.

To keep the spraying equipment in serviceable condition, you need to clean it thoroughly after each use with water or recommended solvent. Make sure you disassemble all nozzles, screens, and filters, and never allow materials to dry or build up in hoses, wands, or nozzles, as this will severely restrict the flow of materials. Follow the manufacturer's recommendations for lubrication and maintenance schedules.

6.5.8 Surface Grinders

Grinders are used in concrete finishing to smooth out rough areas in hardened concrete and to repair defects that might occur. Most grinders operate on gas or electricity, although some are powered by compressed air. It is important to follow the safety procedures in the operator's manual for the type of grinder you are using.

Before operating a grinder, you must be familiar with the basic safety guidelines for its power source. Be sure to wear appropriate personal protective equipment. When working with dry grinders, for example, it is important to wear goggles to protect your eyes from flying particles. Also, a dust mask must be worn to protect your lungs from dangerous chemicals like silica and asbestos that may be in the dust released into the air. If inhaled, these chemicals can cause severe scarring of the lungs, loss of breathing ability, or cancer. When working with a floor grinder, rubber boots and gloves are necessary, because you will be standing in water.

Dust-causing materials are safer to work with when they are damp. Before grinding a concrete surface, clean up the work area with a damp broom or dampen the area with a fine water mist, if possible. If the area is wet rather than damp, you should use a pneumatic or gas-powered tool rather than an electric tool. Dust can also be kept to a minimum by vacuuming before and during the grinding operation; however, protective equipment is still necessary.

Before using an electric grinder, make sure you have a sufficient supply of heavy-duty safety cord. The cord should be long enough to give you freedom of movement while working. Position the cord so that neither you nor other workers will trip over it. Remember to check all cords and extensions for fraying, especially at connections. If you will be grinding above the normal shoulder-high level for extended periods of time, use scaffolding. This will prevent fatigue that may lead to unnecessary accidents. For gas-powered grinders, make the recommended safety checks. Have an adequate supply of fuel available and follow the manufacturer's operating instructions.

To keep your grinder running smoothly and efficiently, be sure to remove the wheel and wipe off the grinder with a clean rag. Keep the air vents and motor housing free of dust and dirt by blowing with compressed air. Excessive buildup of metal dust or oil can lead to shorting of the electrical current from internal parts to exposed metal parts. After cleaning, return the grinder, wheels, and cord to the proper storage area. Never set the grinder in oil or grease; the wheel can absorb these substances and be thrown out of balance. You should never raise or lower the grinder by its cord. This can cause electrical wires to pull loose from the motor. Be careful not to drop or bump the grinder. This can cause damage to the wheel, shaft, housing, and guard.

Avoid prolonged use of the equipment under load. Overloading causes a considerable reduction in speed and efficiency. If the grinder becomes hot, run it at no load for a minute or two until the fan cools it to normal working temperature. Do not switch the grinder off and on under load, as this will reduce the life of the switch and cause a possible short in the wiring.

To ensure maximum life and effectiveness, the grinder should be returned to the repair department for regular cleaning, inspection, and service. This is especially important when the tool is used in heavy industrial conditions. A defective grinding wheel could fly apart at full rpm and cause extensive injury.

Any tool showing signs of malfunction or damage must be removed from service immediately and tagged for repair.

SUMMARY

Safety is everyone's responsibility. You can't assume that someone else is taking care of it. For a safe workplace, you and every worker on the job must be safety conscious and practice safe work habits. In order to work safely, you need to learn the safety equipment and clothing required for each job, the safety rules and procedures, general work site safety, and environmental safety as it affects each job.

Safety equipment and clothing, when worn properly, can help protect you and your fellow workers from harm. Typical safety gear includes a hard hat, safety goggles, hearing protectors, boots, gloves, and knee pads. When working in areas of dust or toxic fumes, it will also be necessary to wear a respirator or dust mask. Other equipment, such as a safety harness, may be needed for fall protection.

Of course, it is not enough to have the proper safety equipment. You also need to act responsibly and think about what you are doing. To prevent accidents associated with long hair, for example, you should confine your hair in your hard hat. You should also remove all jewelry and be sure to wear close-fitting clothing that will not get caught in any machinery. When working around heaters, you should be sure to use a carbon monoxide meter to warn of dangerous levels of this deadly gas.

Each step in the concrete construction process carries its own hazards. In order for you to work safely, you need to be aware of these hazards so that you can reduce the possibility of an accident. When forming concrete, for example, you should make sure to use only those forms that are designed for the specific job. When mixing, it is doubly important to wear the proper safety clothing and protect yourself from the caustic effects of concrete. When performing finishing work, you need to follow specific guidelines for dealing with any toxic materials in the curing process.

Each work site also has its own hazards. That is why it is so important to follow general safety precautions like keeping the area clean and free of obstructions, and familiarizing yourself with the location of fire-fighting and first aid equipment. You need to do everything possible to make the workplace as safe as it can be for yourself and others.

In order to prevent accidents, all work must be conducted in compliance with federal, state, and local laws, company policies and procedures, manufacturer's recommendations, and plain old common sense. If you don't already know it, learn first aid and CPR. Take a course so that when an accident does happen, you know what to do.

You also need to follow safe work habits when working in extreme temperatures. Hot and cold weather cause hazards that need to be prevented if at all possible. When working in extreme cold, for example, you should be sure to wear layers of clothing so that you don't expose yourself to hypothermia or frostbite. When working in extreme heat, be sure to drink plenty of fluids and take frequent breaks in a cool area.

Hazardous materials and toxic substances require their own safety precautions and guidelines. Exposure to chemicals can cause heart ailments, kidney and lung damage, and cancer. Toxic substances can cause coma, shock, or allergic reactions. Familiarize yourself with the chemicals and substances that are used at your facility and learn what safety precautions need to be followed when working with them.

Finally, familiarize yourself with the tools you'll be using on the work site. Read the operator's manual for each tool and be sure to operate the tool only after you have received proper training. For safe and efficient operation, it is also important to maintain your tools in good working condition. This will ensure that they are always safe and ready to use when you need them.

References

For advanced study of the topics covered in this task module, the following reference materials are suggested:

Concrete Construction and Estimating, Craig Avery, Craftsman Book Company, Carlsbad, CA, 1991.

Concrete Manual, International Conference of Building Officials, Whittier, CA, 1992.

Concrete Parking Lots: Eight Steps to Quality Construction, Videotape, Portland Cement Association, Skokie, IL.

Concrete Technology, Third Edition, George R. White, Delmar Publishers, Inc., Albany, NY, 1991.

Design and Control of Concrete Mixtures, Thirteenth Edition, Portland Cement Association, Skokie, IL, 1988.

Finishing Concrete Flatwork, Videotape 32:00 minutes, Portland Cement Association, Skokie, IL, 1991.

Guide for Concrete Floor and Slab Construction, ACI 302.1R-96, American Concrete Institute, Farmington Hills, MI, 1997.

The Contractor's Guide to Quality Construction, American Concrete Institute, Farmington Hills, MI, 1997.

ACKNOWLEDGMENTS

Figures 1, 8, 11, and 15 courtesy of Roy Jorgensen Associates, Inc.
Figures 2-7, 9, 12, and 20 courtesy of Associated General Contractors
Figure 10 courtesy of TSI, Inc.
Figures 13 and 14 courtesy of the American Red Cross
Figures 16, 18, and 19 courtesy of Portland Cement Association
Figure 17 courtesy of Baker Concrete Construction

SI* (MODERN METRIC) CONVERSION FACTORS

APPROXIMATE CONVERSIONS TO SI UNITS

Symbol	When You Know	Multiply By	To Find	Symbol
		LENGTH		
in	inches	25.4	millimeters	mm
ft	feet	0.305	meters	m
yd	yards	0.914	meters	m
mi	miles	1.61	kilometers	km
		AREA		
in^2	square inches	645.2	square millimeters	mm^2
ft^2	square feet	0.093	square meters	m^2
yd^2	square yards	0.836	square meters	m^2
ac	acres	0.405	hectares	ha
mi^2	square miles	2.59	square kilometers	km^2
		VOLUME		
fl oz	fluid ounces	29.57	milliliters	mL
gal	gallons	3.785	liters	L
ft^3	cubic feet	0.028	cubic meters	m^3
yd^3	cubic yards	0.765	cubic meters	m^3
NOTE: Volumes greater than 1000 l should be shown in m^3.				
		MASS		
oz	ounces	28.35	grams	g
lb	pounds	0.454	kilograms	kg
T	short tons (2000 lb)	0.907	megagrams (or "metric ton")	Mg (or "t")
		TEMPERATURE (exact)		
°F	Fahrenheit temperature	5(F-32)/9 or (F-32)/1.8	Celsius temperature	°C
		FORCE AND PRESSURE OR STRESS		
lbf	poundforce	4.45	newtons	N
lbf/in^2	poundforce per square inch	6.89	kilopascals	kPa

APPROXIMATE CONVERSIONS FROM SI UNITS

Symbol	When You Know	Multiply By	To Find	Symbol
		LENGTH		
mm	millimeters	0.039	inches	in
m	meters	3.28	feet	ft
m	meters	1.09	yards	yd
km	kilometers	0.621	miles	mi
		AREA		
mm^2	square millimeters	0.0016	square inches	in^2
m^2	square meters	10.764	square feet	ft^2
m^2	square meters	1.195	square yards	yd^2
ha	hectares	2.47	acres	ac
km^2	square kilometers	0.386	square miles	mi^2
		VOLUME		
mL	milliliters	0.034	fluid ounces	fl oz
L	liters	0.264	gallons	gal
m^3	cubic meters	35.71	cubic feet	ft^3
m^3	cubic meters	1.307	cubic yards	yd^3
		MASS		
g	grams	0.035	ounces	oz
kg	kilograms	2.202	pounds	lb
Mg (or "t")	megagrams (or "metric ton")	1.103	short tons (2000 lb)	T
		TEMPERATURE (exact)		
°C	Celsius temperature	1.8C + 32	Fahrenheit temperature	°F
		FORCE AND PRESSURE OR STRESS		
N	newtons	0.225	poundforce	lbf
kPa	kilopascals	0.145	poundforce per square inch	lbf/in^2

* SI is the symbol for the International System of Units. Appropriate rounding should be made to comply with Section ASTM E380.

REVIEW/PRACTICE QUESTIONS

1. Safety is _____.
   a. rules and regulations
   b. defined by OSHA requirements
   c. an attitude
   d. someone else's job

2. It is important to wear a hard hat on the job site _____.
   a. when its very cold
   b. at all times
   c. to look professional
   d. even if its cracked

3. Which of the following is appropriate to wear while performing concrete finishing?
   a. Baggy jeans
   b. A sleeveless t-shirt
   c. High-top sneakers
   d. A close-fitting, long-sleeved shirt

4. When working with wet concrete, be sure to wear _____ on your hands.
   a. rubber gloves
   b. leather gloves
   c. hand lotion
   d. nothing

5. The active ingredient of concrete is _____.
   a. caustic
   b. toxic
   c. acidic
   d. neutral

6. When using a mobile mixer to mix concrete, you should _____.
   a. periodically stick a shovel in the mixer to test the mix
   b. keep your hands and clothing away from all moving parts
   c. keep it running at all times
   d. remove all safety guards to provide quick access

7. The most convenient way to place concrete is to _____.
   a. have ready-mix trucks discharge it directly into forms
   b. transport it on a conveyor belt
   c. use wheelbarrows
   d. carry it in small buckets

8. Hydrochloric acid for acid etching should be _____.

   a. applied evenly onto the surface
   b. dumped in a pile on the surface
   c. combed into the surface
   d. sprinkled haphazardly onto the surface

9. A good work attitude towards safety means that you will _____.

   a. never take responsibility for hazardous situations
   b. conduct yourself in a safe, professional manner
   c. always work by yourself
   d. never drop a tool

10. Which of the following is a hazardous procedure?

   a. Assisting other trades when necessary to maintain safe operations.
   b. Storing materials and supplies in their proper places.
   c. Lifting heavy objects with your legs.
   d. Using compressed air to clean your clothes.

11. What is the leading cause of fatalities in the construction industry?

   a. Falls
   b. Heat exhaustion
   c. Heart attacks
   d. Improper tool usage

12. A properly-clothed person should be able to work with a wind chill of _____ without adverse effects from the cold.

   a. -79°F
   b. -50°F
   c. -36°F
   d. -21°F

13. The most common cold weather hazard is _____.

   a. trenchfoot
   b. hypothermia
   c. frostbite
   d. immersion foot

14. Which of the following is *not* a symptom of heat stroke?

   a. Hot dry skin
   b. Body spasms
   c. Weakness
   d. Confusion

15. Normally, the body cools itself by _____.

    a. sweating
    b. shaking
    c. fainting
    d. sleeping

16. The basic goal of OSHA's Hazard Communication standard is to _____.

    a. increase general public awareness of safety hazards
    b. inform the general public of typical safety policies
    c. reduce the incidence of chemical-source illness and injuries
    d. eliminate all safety hazards

17. Hazardous materials should *not* be _____.

    a. transferred to temporary containers
    b. clearly labeled
    c. marked with hazard warnings
    d. identified

18. When using hand tools, it is important to _____.

    a. keep cutting edges away from your body
    b. always use tools with metal handles
    c. use whatever tool is handy
    d. make sure to wear rubber gloves

19. When using power tools, it is important to _____.

    a. remove all guards before operation
    b. keep the tool running if you have to walk away
    c. report unusual noises or vibrations to the supervisor
    d. clean the tool with a rag

20. What should you do to trowels when the edges become worn?

    a. Discard them
    b. File them
    c. Sandblast them
    d. Spray them

ANSWERS TO REVIEW/PRACTICE QUESTIONS

Answer	Section Reference
1. c	1.1.0
2. b	2.1.0
3. d	2.2.0
4. a	2.1.0
5. a	3.1.0
6. b	3.2.0
7. a	3.3.0
8. a	3.4.0
9. b	4.1.0
10. d	4.1.0
11. a	4.3.0
12. d	5.1.1
13. c	5.1.1
14. b	5.2.4
15. a	5.2.0
16. c	6.1.0
17. a	6.1.0
18. a	6.3.0
19. c	6.4.0
20. b	6.5.3

NCCER CRAFT TRAINING USER UPDATES

The NCCER makes every effort to keep these manuals up-to-date and free of technical errors. We appreciate your help in this process. If you have an idea for improving this manual, or if you find an error, a typographical mistake, or an inaccuracy in the NCCER's Craft Training Manuals, please write us, using this form or a photocopy. Be sure to include the exact module number, page number, a description of the problem, and the correction, if possible. Your input will be brought to the attention of the Technical Review Committee. Thank you for your assistance.

Instructors – If you found that additional materials were necessary in order to teach this module effectively, please let us know so that we may include them in the Equipment/Materials list in the Instructor's Guide.

Write: Curriculum and Revision Department
National Center for Construction Education and Research
P.O. Box 141104
Gainesville, FL 32614-1104

Fax: 352-334-0932

Craft ______________________ Module Name ______________________

Module Number ______________________ Page Number(s) ______________________

Description of Problem

____

____

____

____

(Optional) Correction of Problem

____

____

____

(Optional) Your Name and Address

____

____

____

notes

Properties of Concrete

Module 23103

Concrete Finishing Trainee Task Module 23103

NATIONAL
CENTER FOR
CONSTRUCTION
EDUCATION AND
RESEARCH

PROPERTIES OF CONCRETE

OBJECTIVES

Upon completion of this module, the trainee will be able to:

1. Describe the properties of concrete.
2. Explain how the properties of concrete are used in construction.
3. Determine how the ingedients of concrete influence mix, placement, finishing, durability, and performance.
4. Describe quality control tests on concrete ingredients, fresh concrete, and hardened concrete.
5. Mix a test batch of concrete.
6. Perform a slump test.

Prerequisites:

Successful completion of the following Task Modules is recommended before beginning this Task Module: Core Curricula; Concrete Finishing Level 1, Modules 23101 and 23102.

Required Trainee Material:

1. Trainee Task Module
2. Paper and pencil
3. Appropriate Personal Protective Equipment

COURSE MAP

This course map shows all of the task modules in the first level of the Concrete Finishing curricula. The suggested training order begins at the bottom and proceeds up. Skill levels increase as a trainee advances on the course map. The training order may be adjusted by the local Training Program Sponsor.

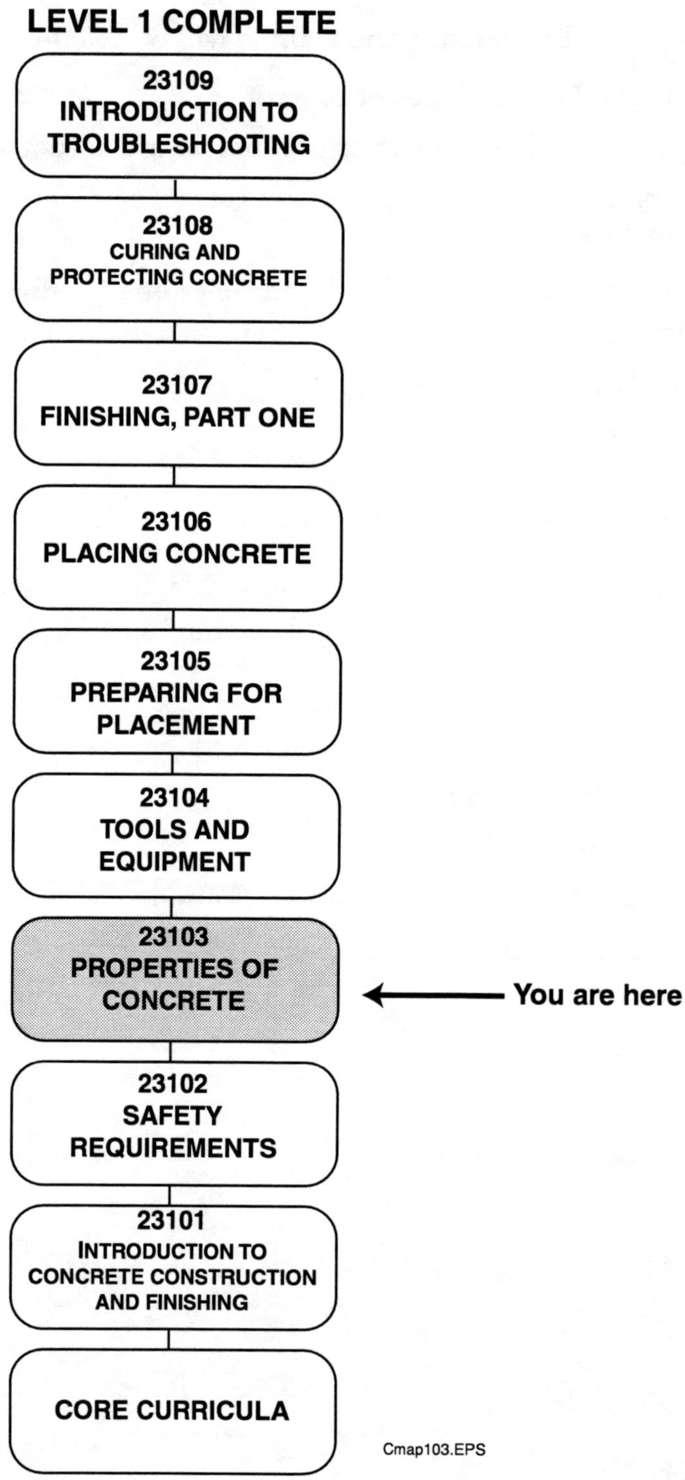

TABLE OF CONTENTS

Trade Terms Introduced In This Module

Absolute volume: A method of calculating the exact volume that each ingredient in a concrete mix will occupy per unit volume of concrete. This method uses specific gravity values to develop proportions or amounts of cement, aggregates, and water in a concrete mix.

Air content: The volume of air voids in concrete. Entrained air adds to the durability of hardened concrete in freeze-thaw exposures and to the workability of fresh concrete.

American Society for Testing and Materials (ASTM): The professional association that develops standards and specifications for materials of all types. It publishes material specifications, establishes test methods, and certifies equipment used for measurements.

Batch: The quantity of concrete or mortar that is mixed at one time.

Batching: Weighing or volumetrically measuring and introducing into the mixer the ingredients for a batch of concrete or mortar.

Bleeding: The movement of water within fresh concrete toward the top surface and its collection there; caused by settling of the heavier materials.

Cold joint: A joint (usually visible) in a concrete wall or floor where the fresh concrete has bonded imperfectly or not at all to the previously placed concrete, because too much time has elapsed between placements.

Deformed bar: A steel reinforcing bar with raised deformations on the surface to provide an adequate bond with the concrete.

Deicer: A chemical, such as sodium or calcium chloride, used to melt snow and ice on pavements.

Flexure: A combination of compressive and tensile stress in the cross-section of a structural member, induced by a vertical loading. The flexure shows as the resulting bending in the member.

Reinforcement: Bars or wire or strands embedded in concrete in such a manner that the reinforcement and concrete act together to resist forces on the concrete (see *deformed bars*).

Slump: A measure of consistency or fluidity of concrete. Low slump indicates stiff concrete; high slump indicates wet, fluid concrete. A range of 0-12 inches indicates the amount of slump.

Workability: The property of freshly mixed concrete, mortar, or paste that determines the ease with which it can be mixed, placed, compacted, and finished.

1.0.0 INTRODUCTION

As a concrete finisher you should know how concrete is made and how it behaves so that you can safely and efficiently place and finish concrete. It is also important to understand the properties of concrete and how they affect the strength, texture, set time, and **workability** of the mix.

This module provides information on the components of concrete, how concrete is produced, the critical properties it must have, and the quality control tests that measure the properties of fresh concrete.

2.0.0 COMPONENTS

The components of concrete are cement, water, aggregates, and admixtures. Aggregates are composed of both fine and coarse particles. Admixtures are chemicals that modify the plastic or the hardened properties of concrete. *Figure 1* shows the combination of ingredients by volume that make up a typical concrete mix.

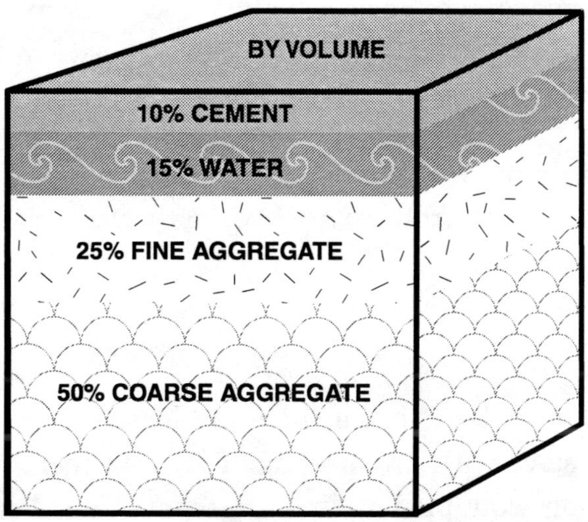

Figure 1. A Cubic Foot Of Typical Concrete

2.1.0 AGGREGATES

Aggregates make up approximately 60 to 75 percent of the concrete mix by volume. Aggregates must meet certain standards. They should be clean, hard, strong, and durable, and be of the proper shape and size. Aggregates should be free of absorbed chemicals, as well as particles of clay or other materials that might affect hydration or the bond between the cement paste and aggregate.

Sources of aggregate used in concrete include sand and gravel deposits, crushed rock, and recycled concrete. Sometimes, large-sized gravel is put through a rock crusher to reduce its particle size or improve particle shape.

Aggregate gradation measures the maximum particle size, the range of particle sizes, and the amount of particles in each size. Gradation of the aggregate is important because of its effect on the workability, economy, and strength of the concrete mix. This is done with the sieve test.

Concrete mixes that contain too much coarse material are difficult to finish. Mixes that contain too much fine material will be easier to finish but will require more cement to coat the many fine particles. In general, a good gradation for concrete includes a full range of coarse and fine particles. On an **absolute volume** basis, mixes typically contain 40 to 50 percent coarse aggregate and 20 to 30 percent fine aggregate.

2.1.1 Coarse Aggregate

Coarse aggregate is sized from ¼ of an inch to 1½ inches in diameter. It may be gravel, crushed rock, or crushed recycled concrete. Gravel usually has rounded edges. Crushed rock will have rough, angular surfaces. Aggregate used in concrete mixes for general construction will have a maximum size ranging from ½ of an inch to 1½ inches. In massive structures, such as dams, the maximum aggregate size may be 6 inches or more. For normal construction practices, aggregate must be of a size small enough to fit between the reinforcing steel or any formwork. Coarse aggregate is usually no larger than ⅓ of the smallest form dimension.

Natural gravel is found in current and former stream beds, glacial deposits, and lakes. Crushed aggregate can be made from gravel, quarry rock, recycled concrete, and blast furnace slag.

2.1.2 Fine Aggregate

Fine aggregate consists of particles ¼ of an inch and less in size. For testing purposes, they must pass through a wire sieve with an opening of ⅜ of an inch. Natural and manufactured sands are the most common examples of fine aggregates used in concrete manufacturing. They have particles ranging in size from ¼ of an inch to those small enough to pass through a sieve having 100 openings in an inch. A mixture made only with cement paste, fine aggregates, and water can be used as grout or flowable fill; however, these mixes are not used for concrete slabs or other formed structures.

Natural sand deposits are found in stream beds, lakes, rivers, glaciated areas, and naturally occurring wind deposits. Manufactured sand is produced from crushed materials such as gravel, quarry rock, recycled concrete, and blast furnace slag. Sand found at the seashore is not used for concrete because of its salt (sodium chloride) content. Salt causes reinforcing steel to corrode.

2.1.3 Aggregate Tests

A number of tests can be made to see if an aggregate is suitable for use in concrete. Aggregate characteristics that are most important include the following:

- *Hardness* – to resist grinding
- *Strength* – to stand up under loads
- *Soundness* – to remain whole during freezing and thawing
- *Maximum size* – to fit in the forms and go between the **reinforcement**
- *Uniform gradation* – to obtain workability, economy, and strength
- *No impurities* – to eliminate harmful effects on concrete

The tests to determine the characteristics of an aggregate are made in material laboratories by technicians. Material suppliers will provide the results of such tests upon request. Often, additional tests are made from the stockpiled material prior to mixing it for concrete. The tests are standard tests following procedures and methods prescribed by the **American Society for Testing Materials (ASTM).** Tests normally conducted on aggregates include those shown in *Figure 2,* which lists the characteristics for testing, their significance, and the ASTM reference. One of the most common tests is ASTM C 136, *Sieve Analysis*. This test defines the sizes of aggregate either as an individual size or in a blended configuration.

CHARACTERISTIC	SIGNIFICANCE IN CONCRETE	TEST METHOD
Soundness	Strength, durability, appearance	ASTM C 88
Chemical stability	Alkali-silica reaction, popouts, disintegration, appearance	ASTM C 227, C 289 C 295, or C 342
Abrasion resistance	Wear resistance of floors, hardness	ASTM C 131 or C 535
Grading, or sieve analysis	Workability, density, economy, shrinkage	ASTM C 136
Maximum size of aggregate	Economy, shrinkage, density, strength	Inspection
Percentage of crushed particles	Workability, economy, strength	Separate and count particles
Particle shape	Workability, economy, shrinkage, strength	Inspection
Surface texture	Bond, strength, durability	Inspection
Specific gravity	Durability, density, need for mix computations	ASTM C 127 or C 128
Absorption	Durability, need for mix computations and control	ASTM C 127 or C 128
Moisture content	Need for mix computations and control	ASTM C 70 or C 566 or other

103F02.EPS

Figure 2. Characteristics And Tests Of Aggregates

2.2.0 CEMENTS

Portland cement, by definition, is a type of hydraulic cement. That is, the cement sets and hardens by chemically reacting with water. The reaction process is called hydration. Different types of cement are made to meet different job requirements and for other specific applications. There are five basic types of portland cement and a variety of other cements for special uses.

- Type I Normal Cement is a general purpose cement that is used when conditions are normal. Normal conditions occur most of the time. Abnormal conditions include the presence of sulfate, the need for early strength, or a need to reduce heat of hydration.
- Type II Moderate Sulfate Resistant is used where there is a need to protect against moderate sulfate attack, as in drainage structures where sulfate in groundwater is present but not severe. Sulfate can attack and disintegrate concrete.
- Type III High Early Strength is used where concrete must develop high strength in a short time, usually within about one week. Uses include situations where forms must be removed early, or where a structure has to be put into service quickly in cold weather or in precast plants to speed production.
- Type IV Low Heat of Hydration releases less heat as it hydrates. This is used in massive structures such as dams where excessive heat could cause cracking.
- Type V High Sulfate Resistant is used only in concrete exposed to severe sulfate conditions or in soils or groundwater with a high sulfate content.
- Type IA, Type IIA, and Type IIIA Air Entraining correspond to Type I, II, and III portland cement, except that an air-entraining agent was added during the manufacturing process.

Blended cements are a result of the use of manufacturing by-product materials in portland cement concrete. Their ingredients include recovered materials such as slag and fly ash. The most common types of blended cement include the following:

- Type IS is a blended cement that contains blast furnace slag. From 25 to 75 percent slag may be substituted for the portland cement. It acts and performs similarly to Type I.
- Type IP is a blended cement that contains a pozzolan for 15 to 40 percent of the cement. The most common pozzolan is fly ash.
- Types M, N, and S Masonry Cement are designed for use in the mortar used in masonry construction. These cements are not used in concrete construction.
- Expansive Cements are used to produce shrinkage-compensating concrete.

2.3.0 MIXING WATER

Mixing water has two main purposes: it reacts with the cement during hydration, and it lubricates the concrete particles so that fresh concrete can be worked. In general, any water that is safe to drink can be used as mixing water.

Water that is not safe to drink may also be suitable for making concrete. Such sources must be evaluated and tested to determine suitability. Tests need to determine if there are impurities. Next, the water's effect on the concrete should be tested. Impurities can affect strength and time of set, cause staining, and corrode reinforcing steel.

2.4.0 ADMIXTURES

An admixture is defined by ASTM as "a material other than water, aggregates, and hydraulic cement that is used as an ingredient of concrete or mortar and is added to the **batch** immediately before or during mixing."

Admixtures can be used to modify both the plastic and hardened properties of concrete.

Admixtures in plastic concrete:

- Increase workability without adding water
- Change the rate of set
- Reduce **bleeding**
- Reduce segregation
- Improve pumpability
- Reduce the rate of **slump** loss
- Improve finishing ease

Admixtures in hardened concrete:

- Accelerate the rate of strength gain
- Increase ultimate strength
- Increase resistance to freeze-thaw damage
- Increase resistance to sulfate attack
- Reduce permeability
- Inhibit corrosion of reinforcing steel

Figure 3 provides a comprehensive list of different admixtures and their effects on concrete. While there are many different types of admixtures, they generally fall into these basic categories: chemical, mineral, and air entraining.

Type of admixture	Desired effect	Material
Accelerators (ASTM C 494, Type C)	Accelerate setting and early-strength development	Calcium chloride (ASTM D 98) Triethanolamine, sodium thiocyanate, calcium formate, calcium nitrite, calcium nitrate
Air detrainers	Decrease air content	Tributyl phosphate, dibutyl phthalate, octyl alcohol, water-insoluble esters of carbonic and boric acid, silicones
Air-entraining admixtures (ASTM C 260)	Improve durability in environments of freeze-thaw, deicers, sulfate, and alkali reactivity Improve workability	Salts of wood resins (Vinsol resin) Some synthetic detergents Salts of sulfonated lignin Salts of petroleum acids Salts of proteinaceous material Fatty and resinous acids and their salts Alkylbenzene sulfonates Salts of sulfonated hydrocarbons
Alkali-reactivity reducers	Reduce alkali-reactivity expansion	Pozzolans (fly ash, silica fume), blast-furnace slag, salts of lithium and barium, air-entraining agents
Bonding admixtures	Increase bond strength	Rubber, polyvinyl chloride, polyvinyl acetate, acrylics, butadiene-styrene copolymers.
Coloring agents	Colored concrete	Modified carbon black, iron oxide, phthalocyanine, umber, chromium oxide, titanium oxide, cobalt blue (ASTM C 979)
Corrosion inhibitors	Reduce steel corrosion activity in a chloride environment	Calcium nitrite, sodium nitrite, sodium benzoate, certain phosphates or fluosilicates, fluoaluminates
Dampproofing admixtures	Retard moisture penetration into dry concrete	Soaps of calcium or ammonium stearate or oleate Butyl stearate Petroleum products
Finely divided mineral admixtures		
Cementitious	Hydraulic properties Partial cement replacement	Ground granulated blast-furnace slag (ASTM C 989) Natural cement Hydraulic hydrated lime (ASTM C 141)
Pozzolans	Pozzolanic activity Improve workability, plasticity, sulfate resistance; reduce alkali reactivity, permeability, heat of hydration Partial cement replacement Filler	Diatomaceous earth, opaline cherts, clays, shales, volcanic tuffs, pumicites (ASTM C 618, Class N); fly ash (ASTM C 618, Classes F and C), silica fume
Pozzolanic and cementitious	Same as cementitious and pozzolan categories	High calcium fly ash (ASTM C 618, Class C) Ground granulated blast-furnace slag (ASTM C 989)
Nominally inert	Improve workability Filler	Marble, dolomite, quartz, granite
Fungicides, germicides, and insecticides	Inhibit or control bacterial and fungal growth	Polyhalogenated phenols Dieldrin emulsions Copper compounds
Gas formers	Cause expansion before setting	Aluminum powder Resin soap and vegetable or animal glue Saponin Hydrolized protein
Grouting agents	Adjust grout properties for specific applications	See Air-entraining admixtures, Accelerators, Retarders, Workability agents
Permeability reducers	Decrease permeability	Silica fume Fly ash (ASTM C 618) Ground slag (ASTM C989) Natural pozzolans Water reducers Latex

103F03.TIF

Figure 3. Concrete Admixtures By Classification (1 of 2)

Type of admixture	Desired effect	Material
Pumping aids	Improve pumpability	Organic and synthetic polymers Organic flocculents Organic emulsions of paraffin, coal tar, asphalt, acrylics Bentonite and pyrogenic silicas Natural pozzolans (ASTM C 618, Class N) Fly ash (ASTM C 618, Classes F and C) Hydrated lime (ASTM C 141)
Retarders (ASTM C 494, Type B)	Retard setting time	Lignin Borax Sugars Tartaric acid and salts
Superplasticizers* (ASTM C 1017, Type 1)	Flowing concrete Reduce water-cement ratio	Sulfonated melamine formaldehyde condensates Sulfonated naphthalene formaldehyde condensates Lignosulfonates
Superplasticizer* and retarder (ASTM C 1017, Type 2)	Flowing concrete with retarded set Reduce water	See Superplasticizers and also Water reducers
Water reducer (ASTM C 494, Type A)	Reduce water demand at least 5%	Lignosulfonates Hydroxylated carboxylic acids Carbohydrates (Also tend to retard set so accelerator is often added)
Water reducer and accelerator (ASTM C 494, Type E)	Reduce water (minimum 5%) and accelerate set	See Water reducer, Type A (Accelerator is added)
Water reducer and retarder (ASTM C 494, Type D)	Reduce water (minimum 5%) and retard set	See Water reducer, Type A
Water reducer—high range (ASTM C 494, Type F)	Reduce water demand (minimum 12%)	See Superplasticizers
Water reducer—high range—and retarder (ASTM C 494, Type G)	Reduce water demand (minimum 12%) and retard set	See Superplasticizers and also Water reducers
Workability agents	Improve workability	Air-entraining admixtures Finely divided admixtures, except silica fume Water reducers

*Superplasticizers are also referred to as high-range water reducers or plasticizers. These admixtures often meet both ASTM C 494 and C 1017 specifications simultaneously.

103F03A.TIF

Figure 3. Concrete Admixtures By Classification (2 of 2)

2.4.1 Chemical Admixtures

Chemical admixtures retard or accelerate the hardening of concrete or reduce the amount of water needed in the mix. The admixtures are designated by type (A through G) as follows:

- Type A – water-reducing admixtures
- Type B – set-retarding admixtures
- Type C – set-accelerating admixtures
- Type D – water-reducing and set-retarding admixtures
- Type E – water-reducing and set-accelerating admixtures
- Type F – high range water-reducing, set normal
- Type G – high range water-reducing, set retarding

Retarding admixtures are used primarily to offset the accelerating effect of high temperature and to keep concrete workable during the entire placing operation. This helps to avoid **cold joints** in the finished structure. It also helps to prevent damage from load deflection during placement by keeping the concrete plastic until placement is completed.

Water reducers are dispersion agents. They break up clumps of cement to allow the water to react more fully with the cement and increase the efficiency of the hydration process. Other positive effects include workability, improved finishing, reduced permeability, improved durability, and increased strength in the hardened concrete. Different water-reducing chemicals will affect the **air content**, slump, bleed water, rate of set, strength, shrinkage, and resistance to freeze-thaw. Trial batches to determine the effect that water reducers will have on the concrete should be made prior to placement, and the mix adjusted accordingly.

Another class of water-reducing admixture is the high range water reducer (HRWR). These admixtures are commonly referred to as superplasticizers. HRWRs can reduce the amount of water in a mix by 12 to 25 percent while maintaining desired workability. However, they are expensive compared to normal water reducers. Fresh concrete which contains HRWR may lose its slump after 30 to 60 minutes. Workability then becomes more difficult because the concrete is stiffer.

Accelerators are commonly used in cold weather to speed the delayed set resulting from cooler temperatures and slower chemical reactions. Accelerators allow earlier finishing and increase early strength development. Their disadvantage is that they could increase the potential for corrosion of the reinforcing steel. Calcium chloride is a commonly used accelerator, but its use should be limited, as it can cause corrosion.

2.4.2 Mineral Admixtures

There are three common mineral admixtures which can enhance concrete workability, durability, and strength. They are fly ash, ground blast-furnace slag, and silica fume. These are the materials used to make blended concretes.

Fly ash is a by-product of coal-burning electric generation plants. *Figure 4* shows the process which produces and recovers the fly ash. Fly ash is used to partially replace the cement in concrete. It improves the workability of the mix, reduces bleeding, and increases the ultimate strength of the concrete. Because the amount of cement is reduced, alkali reactivity is usually reduced and sulfate resistance is improved. One possible disadvantage is a reduced rate of strength gain, but the long-term strength of fly ash will generally equal or exceed non-fly ash concretes.

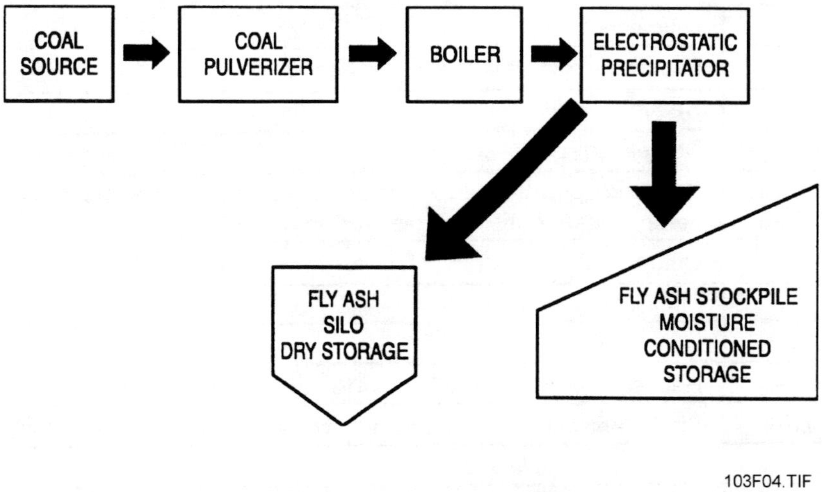

Figure 4. Fly Ash Production

Granulated blast furnace slag is a glassy by-product of manufacturing iron. The granular slag can be substituted for a portion of the portland cement concrete. It can reduce the permeability of the concrete and lower early strength gain.

Silica fume is a by-product of the production of ferrosilicon and certain specialty steels called ferrosilicon alloys. It is the almost pure silicon that reduces permeability of the concrete, provides early strength and higher ultimate strength, and improves durability. However, it requires the use of an HRWR for workability. It is frequently used to achieve very high strengths in concrete structures. A compressive strength of 20,000 psi is possible.

2.4.3 Air-Entraining Admixtures

Air-entraining agents (AEAs) are a special category of admixture. They affect durability and workability. They may or may not affect water content or set, depending on local conditions.

Air-entraining admixtures are chemicals used for producing air entrainment. They form microscopic air bubbles in the concrete, which create tiny voids when the concrete hardens. These voids allow space for water trapped in the concrete to expand as it freezes, increasing the durability of the concrete against freeze-thaw damage. In addition, the air entrainment can increase watertightness and improve scaling resistance to deicing chemicals and sulfates. In plastic concrete, the use of air-entraining admixtures can improve workability and reduce segregation and bleeding. However, increasing the air content of the mix reduces the compressive strength of the concrete. This loss must be considered in the mix design.

Figure 5 is a chart that summarizes the effect of entrained air on various concrete properties.

Properties	Effect
Abrasion	Little effect; increased strength increases abrasion resistance
Absorption	Little effect
Alkali-silica reactivity	Expansion decreases with increased air
Bleeding	Reduced significantly
Bond to steel	Decreased
Compressive strength	Reduced approximately 2% to 6% per percentage point increase in air; harsh or lean mixes may gain strength
Creep	Little effect
Deicer scaling	Significantly reduced
Fatigue	Little effect
Flexural strength	Reduced approximately 2% to 4% per percentage point increase in air
Freeze-thaw resistance	Significantly improved resistance to water-saturated freeze-thaw deterioration
Heat of hydration	No significant effect
Modulus or elasticity (static)	Decreases with increased air approximately 105,000 to 200,000 psi per percentage point of air
Permeability	Little effect; reduced water-cement ratio reduces permeability
Scaling	Significantly reduced
Shrinkage (drying)	Little effect
Slump	Increases with air approximately 1 inch per 1/2 to 1 percentage point of air
Specific heat	No effect
Sulfate resistance	Significantly improved
Temperature of wet concrete	No effect
Thermal conductivity	Decreases 1% to 3% per percentage point increase of air
Thermal diffusivity	Decreases 1.6% per percentage point increase of air
Unit weight	Decreases with increased air
Water demand of wet concrete for equal slump	Decreases with increased air; approximately 5 to 10 pounds per cubic yard per percentage point of air
Watertightness	Increases slightly; reduced water-cement ratio increases watertightness
Workability	Increases with increased air

Note: The table information may not apply to all situations.

103F05.EPS

Figure 5. Effects Of Entrained Air On Concrete Properties

3.0.0 CONCRETE PRODUCTION

The objective of concrete production is to produce a uniform mix that meets the specifications of the specific job. The production process involves **batching** and mixing the material whether it is done mechanically or by hand.

3.1.0 BATCHING

Batching is the first step in the physical production of concrete. Batching consists of the accurate measurement of the coarse aggregate, fine aggregate, cement, water, and admixture necessary to produce a particular amount of concrete. A batch is normally only a portion of the total concrete needed for a job. For example, if 70 cubic yards of concrete were needed for a floor slab and a ready-mix truck could deliver 7 cubic yards per trip, then 10 batches would be required. Each batch must be consistent with the other batches in order to produce concrete with the same properties. Errors in the measurement of any ingredient during batching will cause variation in the workability, strength, and durability of the concrete.

In large plants, batching operations begin with the transportation of the raw ingredients from their storage area, or stockpiles, to storage bins, then to the weighing station. In the storage areas there will be several piles of different sized aggregate. There should not be any mixing of aggregate in the stockpiles. The aggregates in each batch must be properly proportioned, and blending of the stockpiles will change the proportions of each type of aggregate. When handling aggregate, take care to prevent segregation of the graded particles. Larger particles tend to segregate from small particles.

Batching is done by one of three methods.

- Automatic batching systems control the flow of material from a single starting switch which initiates the total automatic batching operation.
- Semi-automatic batching systems require an operator to manually open the storage bin gates to start the batching process. The gates shut automatically when the correct weight of materials flows into the mixing drum.
- Manual batching is done by hand and is required only for small jobs.

In most instances, the batching of solids will be done by weight rather than volume. Water and admixtures are generally metered. *Figure 6* shows the control panel for weight-batching equipment in a typical ready-mix plant. Today, most controls for batching plants have been computerized and automated. In this type of operation, tolerances for accuracy of weight or volume measurement would be plus or minus 1 percent for water and cement, plus or minus 2 percent for aggregates, and plus or minus 3 percent for admixtures.

103F06.TIF

Figure 6. Control Panel For Weight-Batching Equipment

3.2.0 MIXING

Thorough mixing is required for the production of concrete, with all ingredients uniformly distributed. For most jobs, concrete is mixed either in a truck-mounted mixer or at a central mixer located at a plant or on the job site. Portable concrete mixing machines are available with a capacity for holding 1 cubic foot and going up to 1 cubic yard. Truck mixers can start as low as half a cubic yard and go to up to 10 cubic yards.

Regardless of the mixing method, there is a mixing cycle that is followed. First, the mixer is charged with ingredients in the proper sequence. Once mixed, the concrete is discharged from the mixer. Finally, the mixer is cleaned out and returned to a condition of readiness to accept another batch and start the cycle again.

Mixing starts with the first revolution of the mixing drum. In stationary mixers, this happens immediately after charging. In truck mixers, the mixing can be completed in transit or may not start until the truck arrives at the placement site. *Figure 7* shows a batch plant with truck mixers being changed. All mixing takes place in the truck. Truck mixing is the most common type of concrete mixing. Each truck is required to have a manufacturer's certified rating plate.

103F07.TIF

Figure 7. Truck Mixing

Capacity, mixing, and agitating specifications are indicated on the rating plate. The ASTM provides mixing specifications for truck mixers as follows:

- The volume of concrete in a truck cannot be greater than 80 percent of the drum capacity if it is mixed before entering the truck. If the batch is mixed in the truck, it cannot exceed 63 percent of the drum capacity.
- Only 70 to 100 revolutions at mixing speed are required for mixing.
- After mixing, the drum must rotate at agitating speed.
- The mix must be discharged before 300 drum revolutions occur after water is introduced into the mix.

All trucks must be equipped with a revolution counter. Over-mixing of the concrete is not desirable. Too much mixing can lead to strength loss, temperature rise, slump loss, and release of entrained air.

In a central mix plant, the mixing is done partially or completely in the plant. Central mixing plants are usually set up as a business to provide concrete to a wide range of customers at a geographic location. They are stationary. There are four general types: nontilting, tilting, vertical shaft, and horizontal shaft. The tilting type is the most common. Once mixed, the concrete is delivered to the placement site by truck. *Figure 8* shows a central mixing plant of the tilting type loading a batch of concrete into a dump truck for use on a highway project.

103F08.TIF

Figure 8. Central Mixing

Mobile batching mixers are trucks that carry uncombined ingredients and a mixer. They batch materials by volume, continuously mixing concrete as the dry ingredients and water are fed into the mixer. Mixing is rapid and occurs immediately before discharge. *Figure 9* shows a mobile batch mixer furnishing concrete to a pump driven placing system.

103F09.TIF

Figure 9. Mobile Batcher

Mobile batching mixers are used for small quantities of concrete when on-site mixing is advantageous. The trucks range in size of capacity from 4 to 12 cubic yards.

3.3.0 MIXING YOUR OWN CONCRETE

For jobs requiring less than 1 cubic yard of concrete, you may have to mix your own batch in a small, portable power mixer or by hand in a wheelbarrow or box. If you will be using air-entrained concrete, you must use a power mixer because you will not be able to hand mix it vigorously enough to entrain the air.

Regardless of the mixer used, the cement, sand, coarse aggregate, and water must be present in the correct proportions. *Figure 10* gives a mix formula for an average strength concrete. This formula will work using either cubic feet or buckets as units of measure.

	With or Without Air Entrainment			With Air Entrainment	Without Air Entrainment
Aggregate Size, Inches	Cement	Water *	Coarse Aggregate	Sand	Sand
3/8	1	1/2	1-1/2	2-1/4	2-1/2
1/2	1	1/2	2	2-1/4	2-1/2
3/4	1	1/2	2-1/2	2-1/4	2-1/2
1	1	1/2	2-3/4	2-1/4	2-1/2
1-1/2	1	1/2	3	2-1/4	2-1/2

Unit of measure is either cubic foot or bucket.

*** 7.48 gallons of water equal 1 cubic foot.**

103F10.EPS

Figure 10. Concrete Proportions By Volume

Note that the proportion of coarse aggregate changes with the size of the aggregate, and the amount of sand changes with the presence of air-entraining admixtures.

The formula is given in volume, not weight. Any size container—a 5-gallon bucket, a cubic foot measure, a gallon pail—can be used, as long as it is used for all the ingredients. For any batch size, the most important thing is to keep the proportions of the ingredients the same.

Note that the volume of the mixed concrete will be about ⅔ that of the ingredients.

3.3.1 Materials

Materials need to be kept free of contamination. Cement should be kept dry. Store bags on pallets, off the ground, and covered with plastic or tarpaulins to keep the rain off. Aggregate and sand must be kept free of contamination from dirt and protected from rain.

When batching your own concrete, obtain the fine and course aggregates separately from a reliable source. If there is a ready-mix producer in your area, it is preferable to purchase aggregates there, as they will make sure the aggregates are the correct sizes and are suitable for making concrete.

Store aggregates on a clean, hard surface, if possible, and not on the ground where they may become contaminated by dirt and mud. Cover aggregate piles with canvas or plastic tarpaulins to prevent them from becoming wet in case of rain.

Free moisture in aggregate becomes an important factor for mixing small batches of concrete. Free moisture in sand can vary considerably and is a large part of the reason for variation in slump in a particular batch of concrete. Therefore, adjustment needs to be made in the volume of water added, to account for too much or not enough moisture in the fine aggregate —particularly sand.

Sand used in concrete mix should be moist but not wet. With the proper free moisture, the sand will hold a ball shape without wetting your hand. Test the sand by picking up a handful and squeezing it into a ball, as shown in *Figure 11*.

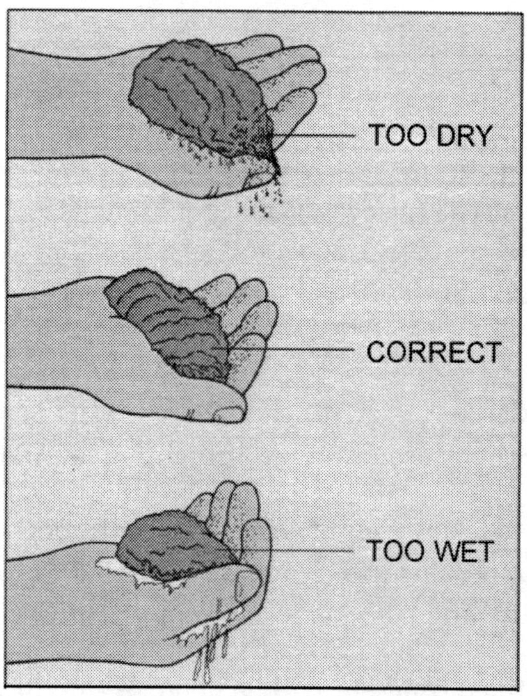

Figure 11. Testing Sand

If the sand is too dry, it will crumble and not hold a shape. If the sand is too wet, water will run out of the sand as it is squeezed. If the sand is moist enough, it will hold the ball shape without leaking or crumbling. Based on the moisture content of the aggregate, adjust the addition of water to the mix to account for water already present in the sand.

3.3.2 Mixing By Hand

Mixing concrete by hand is a tiring job. The mix becomes very heavy and hard to move as the water is added and hydration begins. If you must mix by hand, have all the ingredients close at hand, including the proper tools for the work.

These are the steps to follow when mixing concrete by hand using a plywood sheet or a mixing pan:

Step 1 Dampen the mixing pan or plywood sheet.

Step 2 Use a bucket or a cubic foot measure to measure out the ingredients. Fill the measure evenly and level off carefully.

Step 3 Measure the dry materials in this order: 1) gravel, 2) sand, and 3) cement.

Step 4 Place the measured dry materials in layers on top of each other in the middle of the mixing pan.

Step 5 Mix the dry ingredients thoroughly with a hoe or flat-end shovel.

Step 6 Measure the water.

Step 7 Make a shallow depression in the center of the mix, then pour in some of the water. Mix by pulling the dry materials into the water.

Step 8 Continue by adding water in small amounts, and keep mixing. Scrape any unmixed cement and aggregates from the bottom of the pile.

CAUTION: When mixing concrete, remember that it is caustic. Be sure to wear Appropriate Personal Protective Equipment.

Test the mix by using your shovel to make ridges in the mixed concrete, as in *Figure 12*. If the mix is too dry, the ridges will not be distinct. Correct by adding and mixing in a small amount of water. If the mix is too soupy, the ridges will not hold their shape, and water will seep out around the edges of the pile. Correct by adding and mixing in a small amount of sand.

3.3.3 Using A Power Mixer

Small portable mixers are powered by electric motors or small gasoline engines. Read the instructions and check out the operating controls before operating this equipment. Follow the manufacturer's instructions for proper and safe operation.

Figure 12. Test For Ridges

The size of a batch is usually about 60 percent of the capacity of the mixer to allow room for mixing without spilling. Do not overload the mixer. Charging the mixer poorly will result in unmixed material. Follow these procedures for mixing concrete in a power mixer:

Step 1 Measure the coarse aggregate and load it into the mixer.

Step 2 Measure and add half the water.

Step 3 Start the mixer, then add in order:
1) the sand, 2) the cement, and 3) the remaining water.

Step 4 Add any admixture with the second half of the water.

Step 5 Continue mixing for three minutes, or until the concrete is uniform in color.

After emptying the mixer, thoroughly clean it. Put a bucket of water and a bucket of coarse aggregate into the drum. Turn the drum on and let the stone scour the inside of the drum. Dump out the water and aggregate, and hose out the drum.

4.0.0 PROPERTIES OF CONCRETE

Concrete is the only building material that is manufactured at the building site. Concrete has certain properties that must be considered relative to its design, mixing, and placement. These properties include strength, workability, setting time, air content, and other performance characteristics needed for the application.

4.1.0 STRENGTH

Strength is the quality that allows concrete to withstand loads during service. The structural design is concerned with three types of strength: compression, tension, and **flexure**. Compressive strength is the ability to resist a crushing force. Tensile strength is the resistance to a pulling force. Flexural or bending strength is resistance to bending, a combination of compressive and tensile strength. *Figure 13* illustrates this general concept.

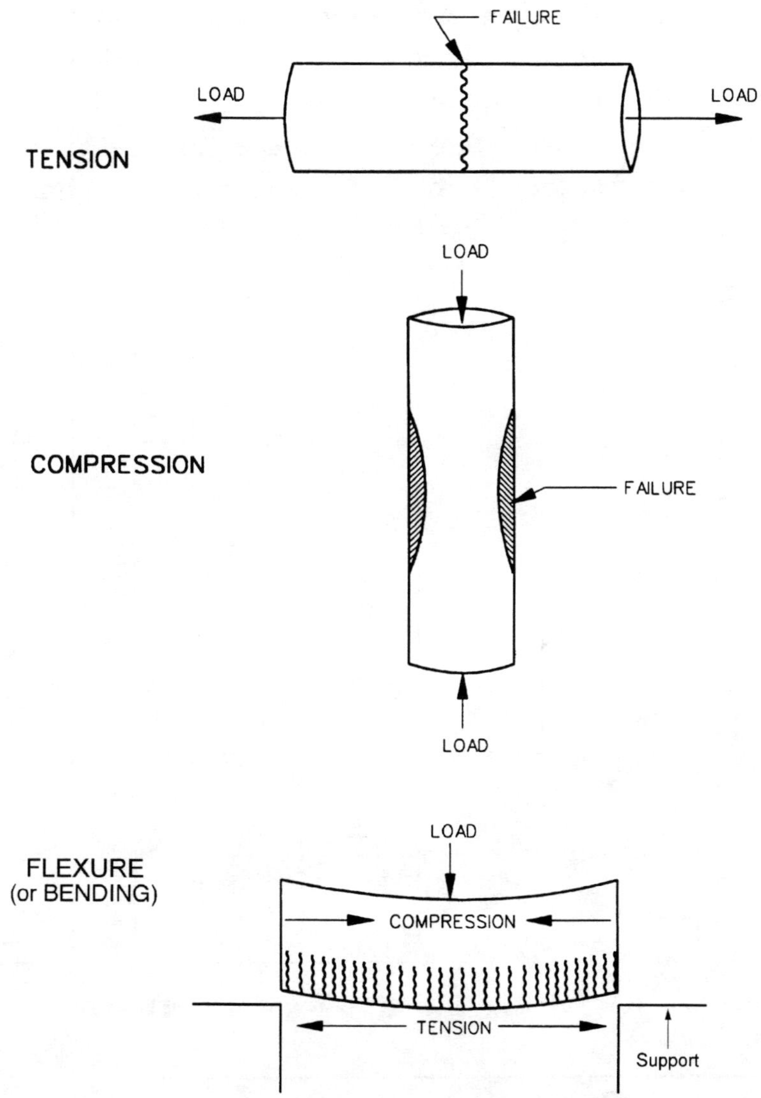

103F13.TIF

Figure 13. Types Of Strength

The compressive strength of concrete is specified in the design of bridges, buildings, and other weight-bearing structures. The flexural strength is usually specified in the design of pavements and slabs on grade. The correlation between compressive and flexural strength is usually established by laboratory testing.

To determine flexural or bending strength, tests are made on a small standard beam (6 inches × 6 inches × 20 inches) subjected to a load applied at the ⅓ points along the length. The total load divided by the cross-sectional area of the beam is the bending strength. It is expressed in pounds per square inch (psi) at an age of 28 days. Compressive strength is resistance to crushing. It is usually measured on a test cylinder of concrete, generally expressed in psi at an age of 28 days.

The main factors affecting concrete strength are its water-cement ratio and age. Increasing the proportion of water decreases strength. Age indicates the extent that hydration has progressed. *Figure 14* shows the compressive strength of concrete over a range of water-cement ratios at different ages for air-entrained and non-air-entrained concrete. As shown, the compressive strengths increase with age and decrease as the proportion of water increases. These two factors also affect tensile and flexural strength in a similar manner.

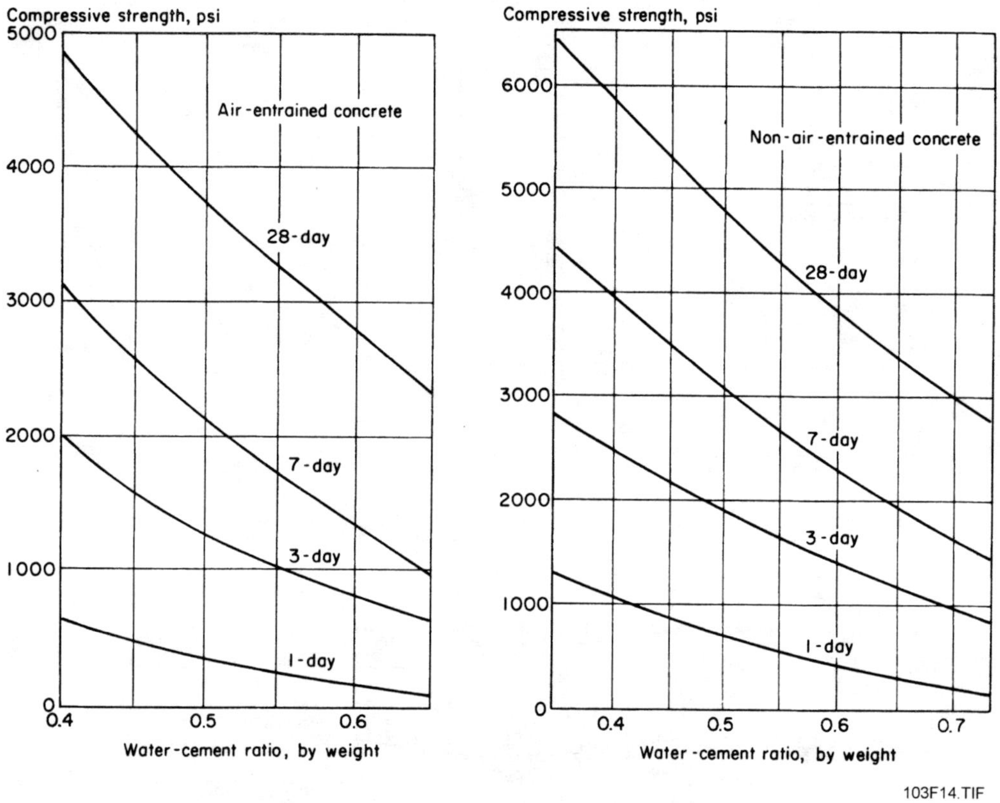

Figure 14. Concrete Age-Strength Relationships

Concrete is strong in compression and relatively weak in tension. Reinforcing steel is used to resist the tensile stress. Reinforcement must be the right kind, the right amount, and in the right place if it is to be of value. Designers are responsible for specifying the type of reinforcement, the size, the number of reinforcing bars, and the location of the reinforcing bars. The concrete craftworker is responsible for ensuring that the design and specifications are implemented. The craftworker must keep the reinforcing steel clean and ensure that it is correctly located in the forms.

Steel reinforcing bars are round and may be smooth or deformed. **Deformed bars** have ribs which help the concrete bond to the reinforcement. Almost all reinforcing steel bars used in the United States are the deformed type. Smooth reinforcement is used overseas. *Figure 15* shows some of the types of deformed bars available.

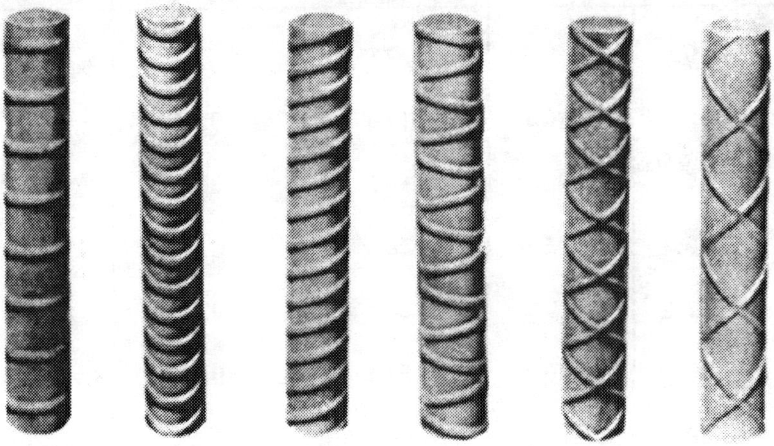

103F15.TIF

Figure 15. Various Types Of Deformed Bars

Bars are specified by a number representing either U.S. Customary or metric (SI) sizes. Each system has its own set of descriptions. *Appendix A* contains conversion tables for moving between the two sets of numbers. In the U.S. Customary system, the numbers correspond to the number of ⅛-inch increments in the diameter of the bar. For example, a number 3 bar has a diameter of ⅜ of an inch and a number 18 reinforcing bar is 2¼ inches in diameter. *Figure 16* is a table showing the available reinforcing bar number, weight, diameter, cross-sectional area and perimeter. Until the United States makes a complete conversion to the metric system, either designation may appear on reinforcing bars used in construction projects.

4.2.0 HYDRATION AND SETTING

The setting time of cement paste is defined to have an initial set and a final set. Initial set is the time after mixing that the paste begins to get stiff. Final set is the time that the paste can carry some load. Concretes generally set more slowly than cement paste tested in the laboratory because of their higher water content. The initial set of concrete normally occurs between 2 to 7 hours, and the final set from 4 to 10 hours. Setting times are affected by:

- Ingredients in the mix
- Fineness of the cement
- Water content
- Internal temperature of the concrete

BAR SIZE SI Metric	US Cust.	NOMINAL DIMENSIONS * DIAMETER mm	inches	CROSS-SECTIONAL AREA mm²	inches²	WEIGHT kg/m	lbs/ft
#10		9.5		71		0.560	
#13	#4	12.7	0.500	129	0.20	0.944	0.668
#16	#5	15.9	0.625	199	0.31	1.552	1.043
#19	#6	19.1	0.750	284	0.44	2.235	1.502
#22	#7	22.2	0.875	387	0.60	3.042	2.044
#25	#8	25.4	1.000	510	0.79	3.973	2.670
#29	#9	28.7	1.128	645	1.00	5.060	3.400
#32	#10	32.3	1.270	819	1.27	6.404	4.303
#36	#11	35.8	1.410	1006	1.56	7.907	5.313
#43	#14	43.0	1.693	1452	2.25	11.380	7.650
#57	#18	57.3	2.257	2591	4.00	20.240	13.600

103F16.EPS

Figure 16. Various Types Of Reinforcing Bars

Admixtures are often added to control the setting times of concrete. There are admixtures that will accelerate setting time and admixtures that will retard setting time, as noted.

Concrete setting times are important to know, as they determine the time of placing and finishing operations. As a result, they can affect the amount of work that can be completed in a given time. Under some weather conditions, it may be necessary to adjust the concrete mix design to improve the timing of finishing operations.

Hydration is the process that allows concrete made with portland cement to set and gain strength. Hydration of the cement compound produces heat and causes a rise in the temperature of the concrete. If the concrete gets too warm, it can cause cracking. Cracking may also occur when the surface cools faster than the inside and thermal stresses are set up. Temperature control measures are necessary when placing large masses of concrete such as mat foundations.

The fineness of cement also has a direct effect on the rate at which the cement hydrates. Finer cements hydrate faster, gain strength sooner, and generate more heat. The fineness of cement has increased over the years as a result of user demand for an increased rate of early strength gain. The extra fineness may require more water in the mix.

4.3.0 CONCRETE SPECIFICATIONS

Specifications describe how the work is to be done, what the end result should be, how the work is to be paid, and any testing that is required. Architects, engineers, and specifications writers produce contract documents that include specifications. *Figure 17* is an excerpt from a suggested concrete specification developed by the Portland Cement Association.

400. Concrete

401. Proportioning. Unless otherwise provided herein, each cubic yard of concrete shall contain 560 lb. of cement plus or minus 2% and the water-cement ratio by weight shall not exceed 0.50. In mild climates where freeze-thaw cycles and deicer use are infrequent, the cement content shall be not less than 520 lb. per cubic yard of concrete and the water-cement ratio by weight shall not exceed 0.53. The slump shall be adjusted to between 1 and 3 in.

The concrete shall contain the following percentages of entrained air:

Maximum size of coarse aggregate (in.)	Total air content (percent by volume)
1½	5 ± 1
¾–1	6 ± 1
⅜–½	7½ ± 1

After the materials have been accepted, the engineer shall determine the proportions, including the amount of air-entraining agent, if one is used, necessary to produce concrete of the required plasticity and workability. Any job site adjustments that may be necessary shall be ordered by the engineer.

If it becomes impossible to obtain concrete of the desired plasticity and workability with the proportions originally designated, the engineer shall change aggregate weights as required, maintaining the cement content originally designated. No change in the sources or character of the materials shall be made without due notice to the engineer.

Slumps shall be determined by AASHTO T119, ASTM C143, or CSA A23.2.20; and air content by AASHTO T152, ASTM C231, or CSA A23.2.19 for gravel and stone coarse aggregate and AASHTO T196, ASTM C173 or CSA A23.2.18 for slag and other highly porous coarse aggregate.

The contractor shall furnish the concrete necessary for casting test beams and cylinders. Test specimens shall be made, cured, and tested in accordance with AASHTO T22, ASTM C39, or CSA A23.2.13; and AASHTO T23, ASTM C31, or CSA A23.2.14; and AASHTO T97, ASTM C78, or CSA A23.2.15. Cement content shall be determined in accordance with AASHTO T121 and ASTM C138.

Figure 17. Section Of A Concrete Specification

This specification document describes in detail the materials and methods for constructing concrete streets. General provisions are given, together with requirements for earthwork, materials, concrete proportioning and mixing, forming, optional construction methods (slip form), placing and finishing, curing, jointing, measurement and payment, and opening to traffic. The excerpt shown in *Figure 17* is for proportioning the mix. Note that the specifications cover some of the material contained in this module.

5.0.0 TESTING AND QUALITY CONTROL

Quality control is performed by the people who create a product, while quality assurance is performed by the people who purchase a product. For concrete work, testing is important to check that the quality of the concrete meets or exceeds specifications. If the concrete does not meet specifications at the time of testing, corrections must be made or the concrete must be rejected. This testing effort is often referred to as quality control.

Quality control includes inspection to confirm that the process and methods are correct and the workmanship is acceptable. These inspections cover the preparation for placing, the placing work, and the finishing. Testing and inspecting for quality control is usually a contractor's responsibility.

The owner or architect will monitor the contractor's test results and observe the contractor's work in progress. The owner or architect may also test the fresh concrete, the preparations, the placing, and finishing. The owner's efforts are called quality assurance because they are assuring that both the work itself and the quality control effort were properly done.

5.1.0 TESTING

Prior to batching and mixing, the ingredients have been tested. The aggregates have been tested for gradation, hardness, abrasive resistance, and porosity. The water has been tested to make certain that it is suitable for concrete. The cement will be certified by the manufacturer. Trial mixes will have been made with the tested and certified ingredients. The strongest and most economical concrete mixture will have been chosen for the job. At the job site, more tests of the concrete will be made for the air content, unit weight and yield, temperature, and slump. Samples will be taken for testing compressive strength and possibly bending strength.

5.1.1 Air Content

Air entrainment increases the durability of concrete when it is exposed to freezing, thawing, and the application of **deicers**. Variation in air content will affect not only durability but also workability. High temperature, low slumps, and coarse sand tend to reduce air content. Frequent testing is needed to ensure that the proper air content of the concrete is maintained.

To test for air content, one of four devices is used. A pressure air meter device measures air content of normal-weight concrete. This is a practical method for field-testing all concretes except those made with highly porous and lightweight aggregates. A volumetric air meter is typically used for lightweight concrete. The Gravimetric method requires accurate knowledge of relative density and absolute volumes of concrete ingredients. It is impractical for a field test but can be used in the laboratory. The Chance air indicator is a very simple and inexpensive way to check the approximate air content of freshly mixed concrete. It is a pocket-size device that tests a mortar sample from the concrete. This test is not a substitute, however, for the more accurate tests mentioned earlier. *Figure 18* shows the pressure air meter and associated equipment. *Figure 19* shows the volumetric air meter and associated test equipment.

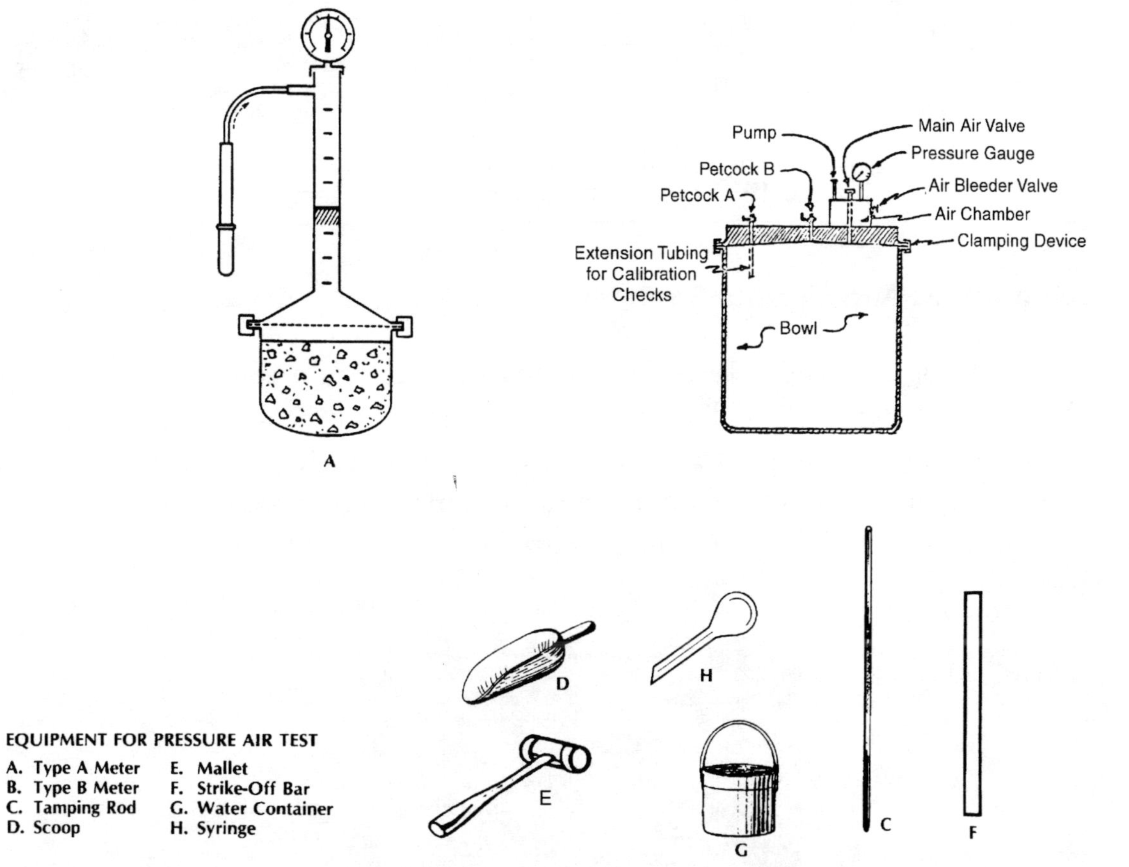

103F18.TIF

Figure 18. Equipment For Pressure Air Test

Using the pressure meter shown in *Figure 18*, the air content is determined by subjecting a sample of the freshly mixed concrete to a controlled air pressure and reading the air content on the dial.

The volumetric air meter, in *Figure 19*, is partially filled with a measured concrete sample, topped with water, then agitated to remove the air. The air content is read directly off a gauge.

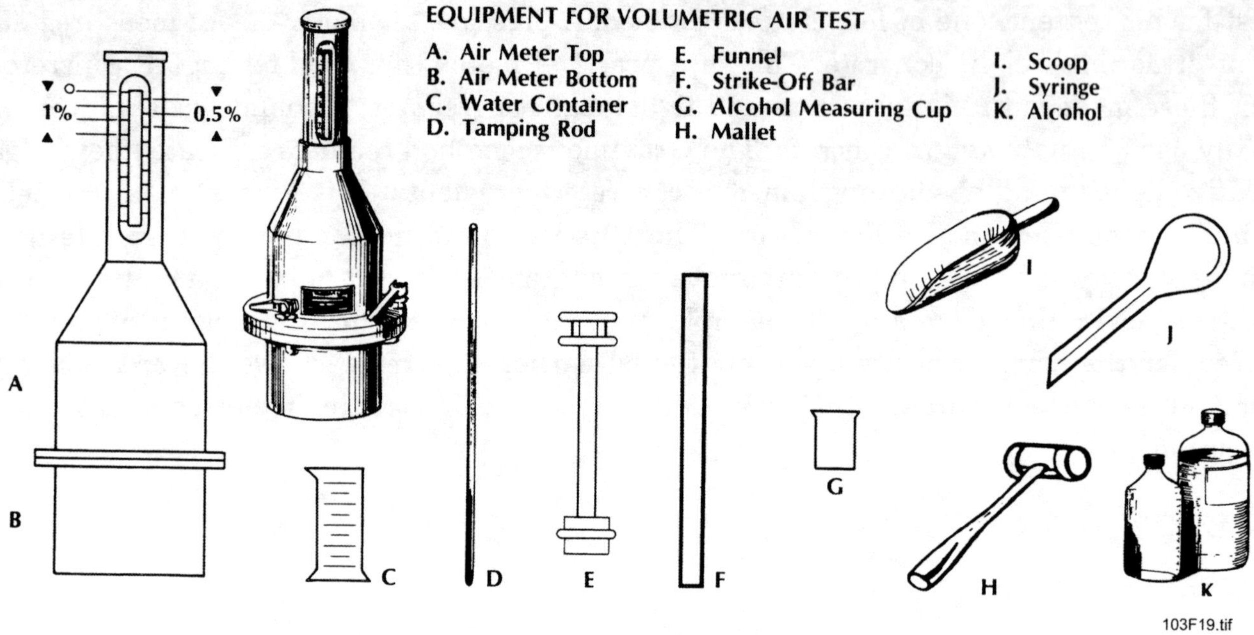

Figure 19. Equipment For Volumetric Air Test

5.1.2 Compressive And Flexural Strength

The compressive strength of concrete is determined by crushing a concrete cylinder made of mix taken from a batch on the job. The cylinders are normally 12 inches high × 6 inches in diameter. *Figure 20* shows a compressive test of a 6-inch × 12-inch concrete cylinder. The load on the cylinder is registered on the scale machine in a materials testing laboratory. Strength test results are used for quality control on delivered concrete and to estimate the strength of placed concrete.

103F20.TIF

Figure 20. Compressive Strength Test

For each placement, a minimum of four cylinders are made. One will be tested at 7 days to see how the strength gain is progressing. The remaining cylinders are tested at 28 days to determine the 28-day strength. Additional samples can be made for 3-day strength.

The cylinders must be stored under controlled temperature and moisture conditions while on the job site, during transportation to the materials laboratory, and while at the laboratory. Cylinders left at the job site and stored under the same conditions as the concrete placement from which they were sampled can be used to test the strength of placed concrete. Cylinders are generally left at the job site for one day before being transported to a testing laboratory. These cylinders are protected from evaporation and from temperature extremes.

5.1.3 Flexural Strength

Beams for testing flexural strength are normally 6 inches × 6 inches and 20 inches long. Bending strength is important for pavement design. The beam forms are filled in two equal layers, and each layer is rodded 25 times. The same requirements exist for the care, handling, and curing of the beams as exist for the compressive cylinders. At 28 days, the beams are tested in the materials laboratory. They are loaded at their ⅓ points until failure. The total load divided by the cross-sectional area provides the bending strength, measured in pounds per square inch.

5.1.4 Unit Weight

Unit weight is the weight of concrete for a given unit volume. For example, the unit weight of normal concrete is approximately 150 pounds per cubic foot. Yield is the volume of concrete produced from a known quantity of cement, water, aggregates, and admixtures. Yield is calculated from the measured unit weight. For example, the ingredients in a batch of concrete weigh 32,400 pounds, and the measured unit weight is 150 pounds per cubic foot. The yield for this batch would be 32,400 divided by 150, or 216 cubic feet. The cubic yard equivalent would be 216 divided by 27, or 8 cubic yards.

Yield computations are useful to check that the quantity of concrete delivered is equal to the quantity ordered. Yield can also be used to check air entrainment. To perform the unit weight test to determine yield, weigh a known volume of concrete and divide the weight by the volume to determine unit weight. Divide it into the batch weight and the yield is calculated.

5.1.5 Temperature

The temperature of the fresh concrete needs to be known. It is measured to check that internal concrete temperature is not significantly different from air temperature. High internal temperature may speed up set time, while low internal temperature may retard it. To take the temperature, insert a thermometer into a sample of concrete. Make sure that the thermometer has about 3 inches of concrete around it on all sides. Read the temperature and record it.

5.1.6 Slump

Slump is a measure of the concrete's consistency. The ingredient with the greatest effect on slump is the water content of the concrete. The higher the water content, the higher the slump. Because the strength of concrete depends on the water-cement ratio, a large slump means lower strength unless chemical admixtures (such as superplasticizers) were used to obtain the high slump.

Slump is an indication of workability, plasticity, or flowability of the concrete. It is a very useful indicator of batch-to-batch uniformity. In addition to changes in the water-cement ratio, changes in slump can also be attributed to air content, aggregate proportions and gradation, delivery time, and temperature.

5.2.0 SLUMP TESTING

The following steps are directions for performing a slump test. *Figure 21* shows the equipment needed to perform a slump test. The slump cone has a height of 12 inches and an open top and bottom. It sits on the base. The concrete is consolidated with the tamping rod. The strike-off bar is used to strike off any excessive concrete, and the riser is used to measure the slump.

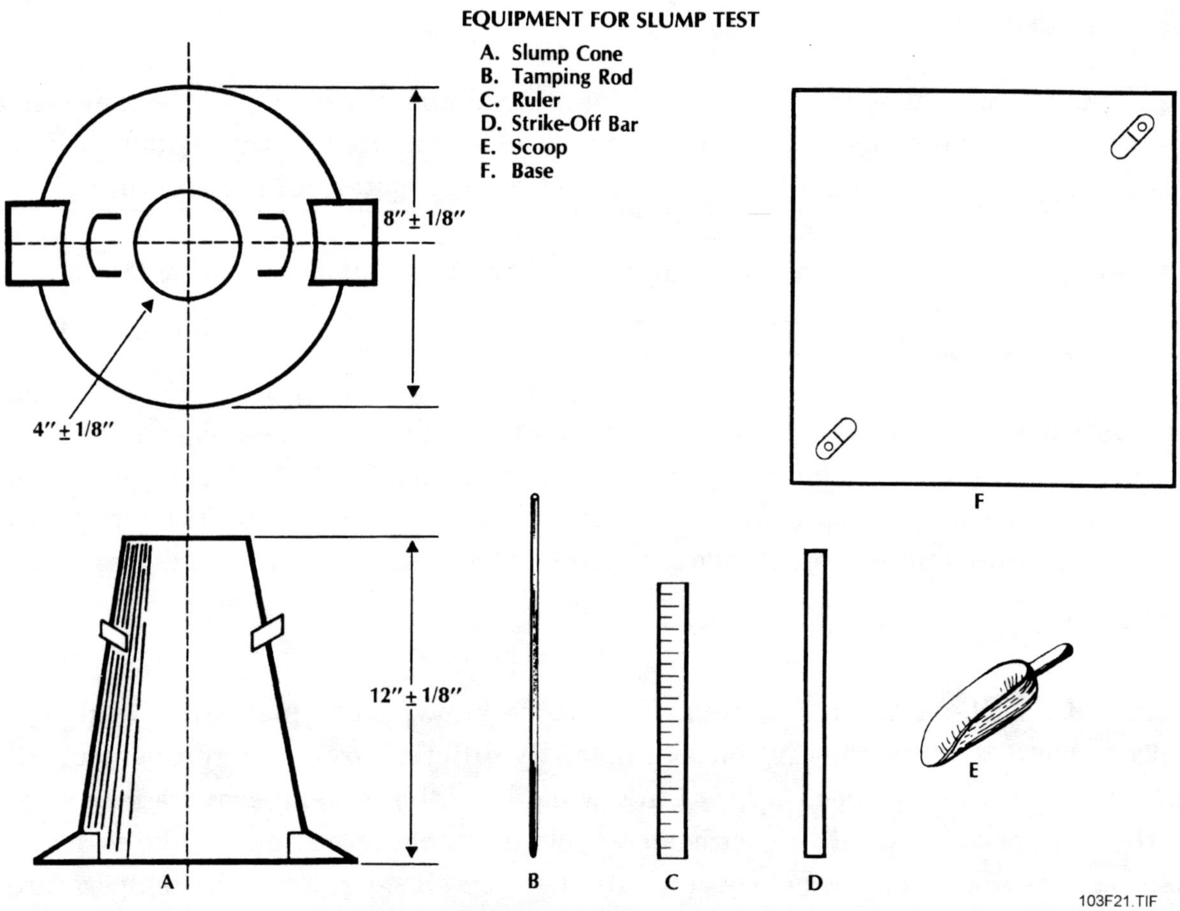

Figure 21. Equipment For Slump Test

To perform a slump test, start with the equipment and a sample of the freshly mixed concrete.

Step 1 Dampen the slump base if it is wood, so that it will not absorb water from the concrete.

Step 2 Put the slump cone, wide side down, on the base.

Step 3 Fill the slump cone with the sample concrete to about ⅓ of its volume.

Step 4 Rod the concrete 25 times.

Step 5 Fill the next ⅓ of the cone with concrete.

Step 6 Rod the concrete 25 times.

Step 7 Fill the final ⅓ of the cone with concrete.

Step 8 Rod the concrete 25 times.

Step 9 Strike off the top to level the concrete.

Step 10 Remove the cone by lifting straight upward, as in *Figure 22*. The concrete will slump, or slide down and spread out somewhat.

Step 11 Place the cone down on the base next to the concrete mound.

Step 12 Place the straightedge across the top edge of the slump cone.

Step 13 Put the riser on top of the concrete mound, and let the other end of it cross the straightedge, as in *Figure 23*. This will measure the distance the top of the sample has slumped or slid down below its original 12-inch height.

103F22.TIF

Figure 22. Lift Slump Cone

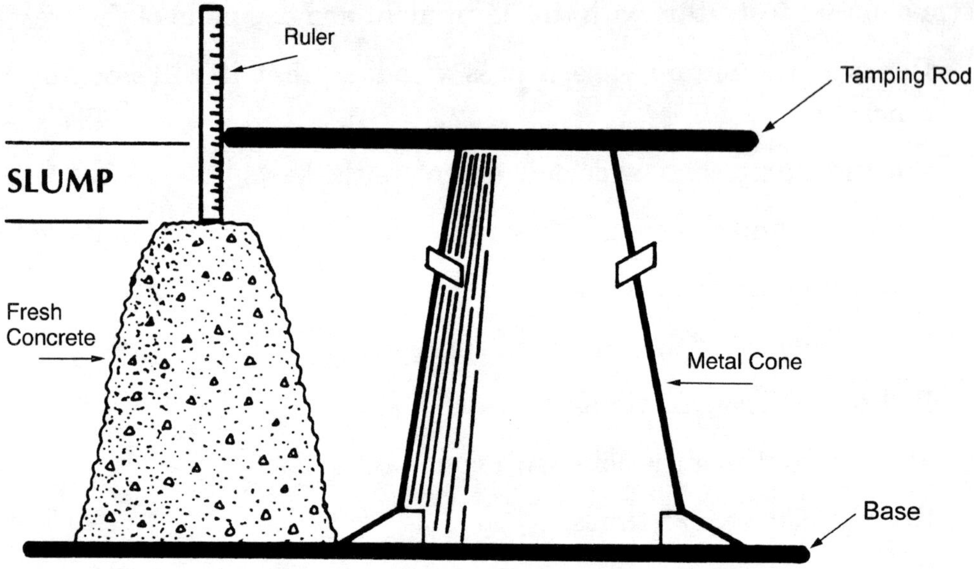

Figure 23. Slump Measurement

The amount the top of the sample slides below its original 12-inch height is called the slump. A small slump of 2 to 3 inches would indicate a stiff mix with a low water-cement ratio. A slump of several inches more would indicate a higher water-cement ratio. The measured slump should be consistent from one batch to the next. If it is not consistent, then something has changed in the batching or mixing process. The cause of these changes should be investigated.

5.3.0 INSPECTIONS

Usually, inspection of a concrete project is required by local building officials, as well as the owner or the owner's representative. The inspection is performed by a trained inspector who is thoroughly familiar with the entire job and its requirements. The inspector also needs a good knowledge of concrete and concrete construction practices. Concrete technicians are usually certified for this work.

5.3.1 Forms

It is clearly in the interest of the contractor and owner to have an inspection of the forms before concrete is placed. The inspection will provide a written record of the results with respect to accuracy, stability, and workmanship. Typically, the forms are inspected three times:

- Once when they are first erected to verify dimensions, location, and stability.
- Next, after the reinforcement is in place and embedded items installed. The inspection will verify completeness, accuracy, and satisfactory workmanship.

- The third inspection is made immediately before the concrete is placed. The purpose of this last form inspection is to make certain that the forms are stable; that reinforcement and other items have not been displaced; that the forms are clean; and that the ground is wetted.

Formwork should be continuously watched during and after concrete placement. A constant check of elevations, camber, and plumbness of the formwork should be made. If adjustments are necessary, they should be made before the concrete reaches its initial set. If any weakness of the forms should appear that would injure workers, or cause settlement or distortion, the work should be stopped until the problem is corrected.

5.3.2 Reinforcement

Reinforcement is inspected for the correct number of bars, bar size, bar shape, proper placement, and tie down. The steel must be clean and free of dirt, oil, or other contaminants. If the reinforcing bars are coated (epoxy or galvanized), the coatings must not be chipped or nicked. Coatings must be complete and continuous. Coatings can be patched on-site as necessary.

5.3.3 Batching And Mixing

The batch plant maintains records of the weight or volume of the ingredients mixed in each concrete batch. The trucks have counters to record the number of revolutions the concrete was mixed. The inspector checks each delivery ticket for time, date, and number of mixing revolutions. Next, samples are taken and tests are made for temperature, yield, slump, and air content. The last activity is to prepare sample test cylinders.

5.3.4 After Placement

Placement does not end the inspection process. The inspector observes the concrete placement to ensure that segregation of the concrete into aggregates and paste does not occur, that the concrete is consolidated correctly, and that the concrete is level.

The inspector observes the finishing process and checks that the concrete is finished correctly. The surface is checked with a straightedge to ensure that it is flat and level and has the finish specified in the contract documents.

The inspector also observes the curing process and monitors the concrete to be sure that it is kept continuously moist. In cold weather, the forms must be insulated and the temperature of the concrete kept well above freezing.

SUMMARY

Information presented in this module presents an in-depth treatment of concrete components, how the components are batched and mixed, the properties that must be achieved, and the quality control requirements on a project. The basic components of concrete are portland cement, aggregate, and water. Concrete contains about 40 to 50 percent coarse aggregate, 20 to 30 percent sand, 10 to 15 percent cement, and 15 to 20 percent water. Percentages are based on measurement by volume. Other ingredients can be added to the mix in order to obtain the characteristics and performance required for a particular job. These ingredients are referred to as admixtures. There are many types of admixtures; the most widely used are those that entrain air.

Depending on the size of the job, concrete is produced on-site, in central mix plants, and in truck mixers. Truck mixing is the most frequently used method of mixing concrete. Strength is the primary property that controls the concrete mix design. The strength of concrete depends mostly on the water-cement ratio and age. Compressive strength is important in most applications. Flexural strength is needed for pavements and slabs. Time of set determines the time available for placing and finishing the concrete. The rate of strength gain controls the time when the forms can be stripped or loads imposed on the concrete.

Quality control involves both testing and inspection activities. Tests are performed on the concrete ingredients, the freshly mixed concrete, and the hardened concrete. Tests performed at the placement site include slump tests and air content tests. Slump is an indicator of the consistency of the mix. It also indicates the relative amount of water in the mix. Air content tests are made to determine the amount of air in the mix. Inspection of the work as it progresses is part of quality control. Normal procedure is for the contractor to perform inspections of the work to assure compliance with plans and specifications. The owner will also make inspections to verify that the work is of the quality that the owner purchased.

References

For advanced study of topics covered in this task module, the following books are suggested:

Concrete Construction and Estimating, Craig Avery, Craftsman Book Company, Carlsbad, CA, 1991.

Concrete Manual, International Conference of Building Officials, Whittier, CA, 1992.

Concrete Parking Lots: Eight Steps to Quality Construction, Videotape, Portland Cement Association, Skokie, IL.

Concrete Technology, Third Edition, George R. White, Delmar Publishers, Inc., Albany, NY, 1991.

Design and Control of Concrete Mixtures, Thirteenth Edition, Portland Cement Association, Skokie, IL, 1988.

Finishing Concrete Flatwork, Videotape 32:00 minutes, Portland Cement Association, Skokie, IL, 1991.

Guide for Concrete Floor and Slab Construction, ACI 302.1R-96, American Concrete Institute, Farmington Hills, MI, 1997.

The Contractor's Guide to Quality Construction, American Concrete Institute, Farmington Hills, MI, 1997.

ACKNOWLEDGMENTS

Figures 1, 2, 10, 13, 16, 18, 19, 21, and 23 courtesy of Roy Jorgensen Associates, Inc.

Figures 3, 5-7, 9, 14, 17, 20, and 22 courtesy of Portland Cement Association

Figure 4 courtesy of Federal Highway Administration

Figures 6 and 8 courtesy of Scranton Gillette Communications, Inc.

Figures 11 and 12 courtesy of Creative Homeowner Press

Figure 15 courtesy of American Concrete Institute

SI* (MODERN METRIC) CONVERSION FACTORS

APPROXIMATE CONVERSIONS TO SI UNITS

Symbol	When You Know	Multiply By	To Find	Symbol
		LENGTH		
in	inches	25.4	millimeters	mm
ft	feet	0.305	meters	m
yd	yards	0.914	meters	m
mi	miles	1.61	kilometers	km
		AREA		
in^2	square inches	645.2	square millimeters	mm^2
ft^2	square feet	0.093	square meters	m^2
yd^2	square yards	0.836	square meters	m^2
ac	acres	0.405	hectares	ha
mi^2	square miles	2.59	square kilometers	km^2
		VOLUME		
fl oz	fluid ounces	29.57	milliliters	mL
gal	gallons	3.785	liters	L
ft^3	cubic feet	0.028	cubic meters	m^3
yd^3	cubic yards	0.765	cubic meters	m^3
NOTE: Volumes greater than 1000 l should be shown in m^3.				
		MASS		
oz	ounces	28.35	grams	g
lb	pounds	0.454	kilograms	kg
T	short tons (2000 lb)	0.907	megagrams (or "metric ton")	Mg (or "t")
		TEMPERATURE (exact)		
°F	Fahrenheit temperature	5(F-32)/9 or (F-32)/1.8	Celsius temperature	°C
		FORCE AND PRESSURE OR STRESS		
lbf	poundforce	4.45	newtons	N
lbf/in^2	poundforce per square inch	6.89	kilopascals	kPa

APPROXIMATE CONVERSIONS FROM SI UNITS

Symbol	When You Know	Multiply By	To Find	Symbol
		LENGTH		
mm	millimeters	0.039	inches	in
m	meters	3.28	feet	ft
m	meters	1.09	yards	yd
km	kilometers	0.621	miles	mi
		AREA		
mm^2	square millimeters	0.0016	square inches	in^2
m^2	square meters	10.764	square feet	ft^2
m^2	square meters	1.195	square yards	yd^2
ha	hectares	2.47	acres	ac
km^2	square kilometers	0.386	square miles	mi^2
		VOLUME		
mL	milliliters	0.034	fluid ounces	fl oz
L	liters	0.264	gallons	gal
m^3	cubic meters	35.71	cubic feet	ft^3
m^3	cubic meters	1.307	cubic yards	yd^3
		MASS		
g	grams	0.035	ounces	oz
kg	kilograms	2.202	pounds	lb
Mg (or "t")	megagrams (or "metric ton")	1.103	short tons (2000 lb)	T
		TEMPERATURE (exact)		
°C	Celsius temperature	1.8C + 32	Fahrenheit temperature	°F
		FORCE AND PRESSURE OR STRESS		
N	newtons	0.225	poundforce	lbf
kPa	kilopascals	0.145	poundforce per square inch	lbf/in^2

* SI is the symbol for the International System of Units. Appropriate rounding should be made to comply with Section ASTM E380.

REVIEW/PRACTICE QUESTIONS

1. Aggregates make up approximately _____ percent of concrete.

   a. 30 to 40
   b. 60 to 75
   c. 85 to 95
   d. 40 to 60

2. One source of aggregate is _____.

   a. oyster shells
   b. soil nodules
   c. coal
   d. crushed rock

3. Gradation of aggregate is determined by a _____.

   a. sieve test
   b. bending test
   c. compression test
   d. slump test

4. Coarse aggregate particles are larger than _____ of an inch.

   a. ¾
   b. ⅟16
   c. ½
   d. ¼

5. Crushed rock has _____ surfaces.

   a. rounded
   b. polished and angular
   c. rough and angular
   d. rough and rounded

6. Sources of coarse aggregate include _____.

   a. recycled asphalt
   b. recycled concrete
   c. recycled silica fume
   d. recycled glass

7. Aggregates in a concrete mix must be small enough to _____.

   a. float in the mix
   b. create the maximum bond with the paste
   c. fit between the reinforcing steel
   d. fit between the reinforcement ties

8. Type III cement is designated as _____.
   a. high early strength
   b. moderate sulfate resistant
   c. high sulfate resistant
   d. low heat of hydration

9. The difference between Type I and Type IA cement is the addition of an admixture for _____.
   a. retarding the time of set
   b. accelerating the time of set
   c. reducing water
   d. air entrainment

10. Mixing water has two main purposes: _____.
    a. hydration and strength
    b. hydration and lubrication of particles
    c. hydration and cleaning of forms
    d. lubrication and workability

11. The most commonly used admixture provides _____.
    a. fly ash
    b. calcium chloride
    c. air entrainment
    d. blast furnace slag

12. Fly ash is a by-product of _____ electric generation plants.
    a. nuclear
    b. coal-burning
    c. natural gas
    d. petroleum

13. A batch of concrete consists of the amount of concrete that can be _____ at one time.
    a. consolidated
    b. mixed
    c. discharged
    d. placed

14. Two factors that affect concrete strength are _____.
    a. its water-cement ratio and age
    b. final set and age
    c. water and air temperature
    d. slump and workability

15. Deformed type reinforcing steel is generally used in _____.
    a. the United States
    b. Europe
    c. Japan
    d. small concrete pipe

16. The largest US common measure reinforcing bar is a No. 18 which has a diameter of _____ inches.
    a. 2
    b. 2⅛
    c. 2¼
    d. 3¾

17. The setting time of concrete is affected by _____.
    a. aggregate size
    b. temperature
    c. compressive strength
    d. deicers and freeze-thaw

18. Fresh concrete is sampled and tested for _____.
    a. slump and air content
    b. aggregate shape and chemical composition
    c. compressive strength
    d. water-cement ratio and yield

19. Slump provides an indication of ______ between batches.
    a. workability
    b. flowability
    c. consistency
    d. set times

20. Inspectors should be knowledgeable of concrete properties and _____.
    a. concrete batching
    b. concrete transport
    c. concrete mixing
    d. concrete construction practices

ANSWERS TO REVIEW/PRACTICE QUESTIONS

Answer	Section Reference
1. b	2.1.0
2. d	2.1.0
3. a	2.1.3
4. d	2.1.1
5. c	2.1.1
6. b	2.1.0
7. c	2.1.1
8. a	2.2.0
9. d	2.2.0
10. b	2.3.0
11. c	2.4.0
12. b	2.4.2
13. b	3.1.0/Terms
14. a	4.1.0
15. a	4.1.0
16. c	4.1.0
17. b	5.1.5
18. a	5.1.0
19. c	5.1.6
20. d	5.3.0

NCCER CRAFT TRAINING USER UPDATES

The NCCER makes every effort to keep these manuals up-to-date and free of technical errors. We appreciate your help in this process. If you have an idea for improving this manual, or if you find an error, a typographical mistake, or an inaccuracy in the NCCER's Craft Training Manuals, please write us, using this form or a photocopy. Be sure to include the exact module number, page number, a description of the problem, and the correction, if possible. Your input will be brought to the attention of the Technical Review Committee. Thank you for your assistance.

Instructors – If you found that additional materials were necessary in order to teach this module effectively, please let us know so that we may include them in the Equipment/Materials list in the Instructor's Guide.

Write: Curriculum and Revision Department
National Center for Construction Education and Research
P.O. Box 141104
Gainesville, FL 32614-1104

Fax: 352-334-0932

Craft ____________________ Module Name ____________________

Module Number ____________________ Page Number(s) ____________________

Description of Problem ____________________

____

____

____

(Optional) Correction of Problem ____________________

____

____

(Optional) Your Name and Address ____________________

____

____

notes

Tools and Equipment

Module 23104

Concrete Finishing Trainee Task Module 23104

TOOLS AND EQUIPMENT

OBJECTIVES

Upon completion of this module, the trainee will be able to:

1. Name the tools used in placing and finishing concrete.
2. Name the power equipment used in placing and finishing concrete.
3. Describe how each tool is used.
4. Describe how the power equipment is used.
5. Associate trade terms with the appropriate tools and equipment.

Prerequisites

Successful completion of the following Task Modules is recommended before beginning study of this Task Module: Core Curricula; Concrete Finishing Level 1, Modules 23101 through 23103.

Required Trainee Materials

1. Trainee Task Module
2. Pencil and paper
3. Appropriate Personal Protective Equipment

AP

urse map shows all of the task modules in the first level of the Concrete Finishing icula. The suggested training order begins at the bottom and proceeds up. Skill levels icrease as a trainee advances on the course map. The training order may be adjusted by the local Training Program Sponsor.

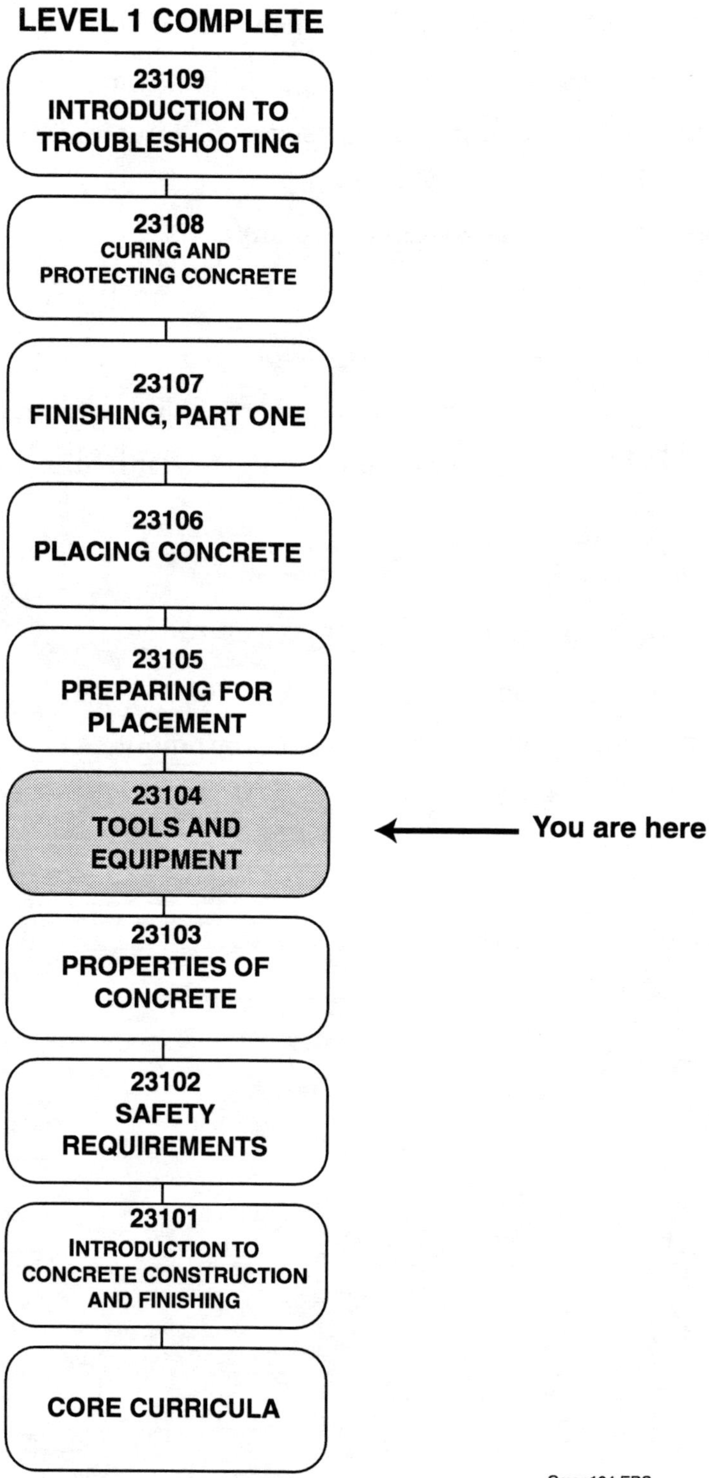

Cmap104.EPS

TABLE OF CONTENTS

TABLE OF CONTENTS (Continued)

Trade Terms Introduced In This Module

Admixture: A chemical or mineral other than water, cement, or aggregates added to concrete immediately before or during mixing to change its setting or curing time, or to reduce water, or to change overall properties of concrete.

Camber: To curve toward the middle; a convex curve on a road or bridge surface.

Compacting: The act of closely pressing, making dense, or firming by compressing the top surface of placed concrete.

Consolidating: Removing the entrapped air voids from concrete by pressing, rolling, rodding, vibrating, spading, or tamping.

Darbying: The act of moving a hand-held float across the surface of placed concrete in a sawing motion to smooth the surface.

Floating: The act of moving a flat-faced tool over the surface of concrete to consolidate and smooth. Performed prior to steel troweling.

Rodding: The act of packing tightly or consolidating by up-and-down prodding with rods in order to let air escape from placed concrete.

Screed: A manual or motorized straightedge tool used to strike off the surface of placed concrete to a predetermined grade.

Segregation: The unwanted separation (in mixed concrete) of the aggregates from the paste.

Spall: A chip, flake, or piece of concrete that has been broken or separated from the concrete surface, usually caused by expansion within the concrete.

Subgrade: A surface of earth or fill material compacted and leveled off to receive a deposit of concrete or the foundation of a structure.

1.0.0 INTRODUCTION

Tools are important for concrete work because concrete cannot be worked with bare hands. Its elastic physical qualities also make it difficult to put it where it needs to go. Concrete needs to be processed many times before it is finished. Concrete is also caustic, and can irritate or even burn the skin, as well as damage tools.

Concrete tools are the interface between the worker and the concrete. Because of the many stages involved in working concrete before it is finished, there are many special-purpose tools used. Most of these tools are used only for concrete construction. Some of the hand tools have not changed since Roman times, while some of the power tools and equipment use the most up-to-date technology. This module introduces the tools and equipment used to place and finish concrete.

This module introduces the hand and power tools and equipment that you will need to:

- Prepare the site and **subgrade**
- Place the concrete
- **Consolidate** the concrete
- Level the concrete
- Finish the concrete

By the end of this module, you will be able to identify each item, what it does, and how it is used in processing concrete.

1.1.0 TYPES OF TOOLS AND EQUIPMENT

In order to prepare the underlying earth and the subgrade, mechanical equipment such as a bulldozer, backhoe, or large earth-mover may be used. The surface must be **compacted** and leveled at a given elevation prior to placing forms and reinforcement steel. The concrete is mixed on-site or mixed off-site, then delivered to its placing location.

Placing tools are used to guide and move the concrete into the form or between the slab forms. **Screeding** tools are used to level and to smooth concrete to a given elevation. Finishing tools are used to further level the concrete and smooth the surface, or to finish it as specified. Hand tools and power equipment are used for these jobs. This module focuses first on placing and finishing hand tools, then follows up with power tools and equipment. Some of the tools are especially designed for use only with concrete construction and finishing. Remember that not every tool will be used on every job. Tool choices will depend on the type of concrete, the type of job, and/or other conditions.

1.2.0 SAFETY

Tool and equipment safety can be broken down into four activities:

- Following safe work practices and procedures
- Inspecting tools and equipment before use
- Using equipment and tools properly
- Keeping tools and equipment clean and properly maintained

Tool safety requires using tools properly and keeping them clean so they will not be damaged or ruined. A defective tool or piece of equipment can cause accidents, as well as slow down or stop work. All tools need to be kept clean. Accidents can also happen when tools and equipment are left in the way of other workers.

CAUTION: Concrete can be very caustic. It is important to protect your skin and eyes when mixing, placing, and working wet concrete. Also, wear safety glasses and a face shield when using grinding and cutting tools.

1.2.1 Cleaning Tools

All tools used in placing and finishing concrete must be cleaned immediately after use. Concrete will dry and harden on the tools if it is not washed off, making them unusable.

Power wash or hand wash placing and finishing tools thoroughly with water. If concrete has already started to harden, a concrete brush can be used. Be sure that the concrete is completely removed. Tools not used regularly can be coated lightly with a release agent or other approved coating to prevent concrete from adhering to the tool, and to prevent rusting. After cleaning tools, check to see that handles are secure and free from cracks or splinters. Oil pivot joints and sharpen blades and cutting edges as needed.

Store tools where people cannot trip over them and where they will be safe from damage. Clean, well-kept tools make for a safe work environment, and will help you to do a good job.

1.2.2 Equipment Maintenance

All motorized tools and equipment must also be cleaned after use.

CAUTION: Always turn off and disconnect power before cleaning or doing maintenance on power tools and equipment.

Bins must be hosed out. Conveyors, hoppers, belts, and scrapers need to be washed at the end of the day, and checked for smooth operation. Power floats and troweling tools need the blades checked frequently and replaced when worn. Rotors must be replaced even if only one blade is bent or broken; otherwise, the rotor will become unbalanced. Power screeds and tampers need their contact planes and edges cleaned and checked. Power screeds must be checked daily to ensure proper operation. Power screeds should be properly aligned and set to proper elevation prior to daily use.

Check with your supervisor when in doubt about specific care and maintenance for equipment. Also, it is a good safety practice to check that equipment will work properly before it is used. Check oil levels and safety switches to ensure proper working order.

Because power equipment is harder to control than hand tools, it is important to pay attention to what is happening around you. Hard hats, safety shoes, safety glasses, and other protective gear are required when working around equipment on all sites. They are critical to your well being.

2.0.0 PLACING AND FINISHING TOOLS

Placing and finishing tools affect the quality of slabs on grade. They are used to guide the concrete into place. They are used to consolidate, settle, and finish the surface, and to create grooves and joints in it. Both hand and power tools are used for these jobs. This section will cover hand tools.

2.1.0 RULES AND TAPES

Rules and tapes were introduced in the Core Curricula module, *Introduction to Hand Tools*. Rules include steel rules, steel measuring tapes, and folding rules. The best rules are accurate, durable, and easy to read. Rules and tapes are made of different materials and come in many different sizes. *Figure 1* shows the three most common types of rules you will find on the job site.

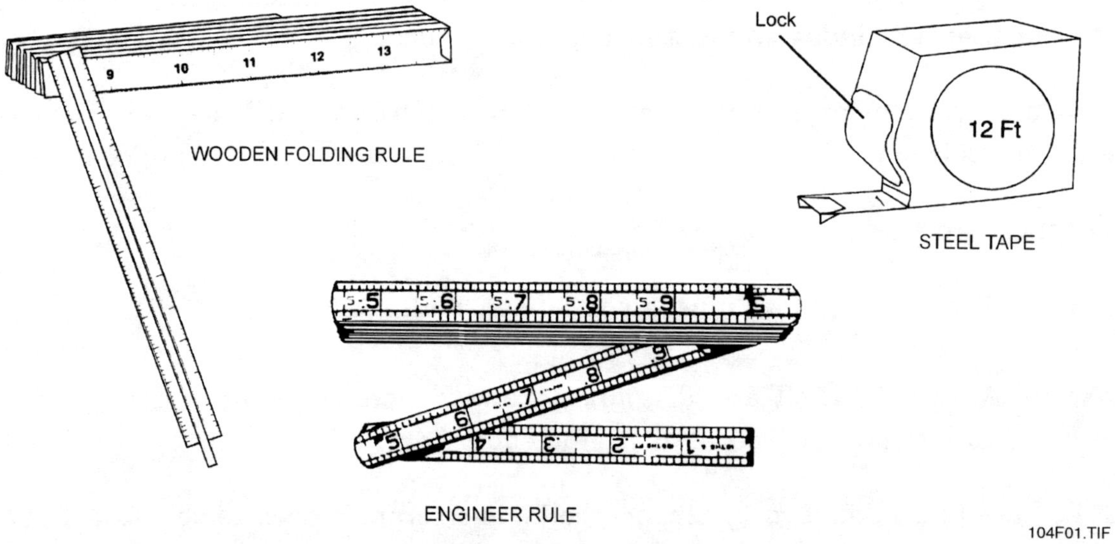

Figure 1. Three Common Rules

The wooden folding rule is used for measuring and marking lumber and other materials. Even though the total length of the rule can be up to 8 feet, it folds easily into a small space, making it convenient to carry and use. The pocket-sized steel tape is the most commonly used type of rule and comes in lengths up to 35 feet. When not in use, the flexible metal tape rolls up into the case, making it easy to store. Larger steel tapes measure up to 100 feet. The engineer rule is a wood or plastic folding rule with markings in feet and inches per foot. It is easy to read measurements such as "5 feet, 3 inches" directly off this rule without converting the figures.

Generally, rules have four sets of marks, one on each edge of both sides. Steel tapes have marks on only one side. The marks are of different lengths, with the longest lines showing inches. On one edge the inch is usually divided into eight equal spaces of ⅛-inch each. The other edge usually shows $\frac{1}{16}$-inch spaces. The ¼-inch and the ½-inch marks are slightly longer, to make reading easier. Some measures also show 32 and 64 spaces to the inch. As the construction industry shifts to the international metric measurement system, metric rules will be used on job sites. U.S. common measures of feet and inches convert to meters and centimeters. Appendix A gives conversions to and from metric measurements.

Rules are used for measuring lumber and other materials. Layout of any forms or other molding requires measuring and sizing using rules. Carefully read measure marks as you use rules. Clean rules when they get soiled by wiping them with a damp, oiled cloth.

2.2.0 LEVELS

Levels were introduced in the Core Curricula module, *Introduction to Hand Tools*. Spirit levels are commonly used at job sites. *Figure 2* shows the working parts of a spirit level. The two plumb tubes measure plumb, or vertical straightness. The center vial measures level, or horizontal straightness. These levels come in sizes from 24 inches to 78 inches. The longer the level, the greater the accuracy.

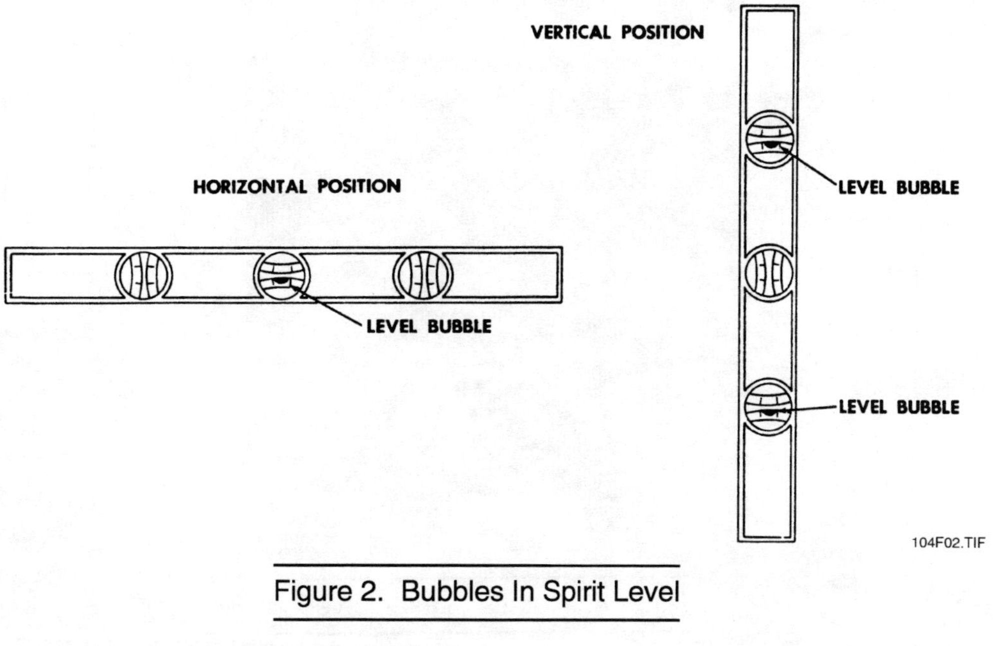

Figure 2. Bubbles In Spirit Level

Use a spirit level to check that forms are plumb or horizontal as needed. Rinse off your level if it comes in contact with concrete mix. Clean your level with a damp cloth so dust and grit do not build up on the windows, and so you will be able to easily see the marks.

A builder's level may be used to set elevations, establish lines, and turn angles. It cannot measure vertical angles as a survey transit or theodolite can. Elevations are read to $\frac{1}{100}$ of a foot so mathematical conversions may be needed. *Figure 3* shows a builder's level. Usually, a senior worker or the crew supervisor will use the level to locate a slab and check the grade. The builder's level requires a team of two workers, one to sight through the level and the other to hold the sighting rod.

104F03.TIF

Figure 3. Builder's Level

Laser levels are designed to establish grades and level over long distances. *Figure 4* shows a rotating laser level set up on a job site. Usually, a senior worker will use the laser level to establish and check grades. Laser levels are incorporated in many of the newer paving machines, and check the grade of the concrete as it is placed and screeded.

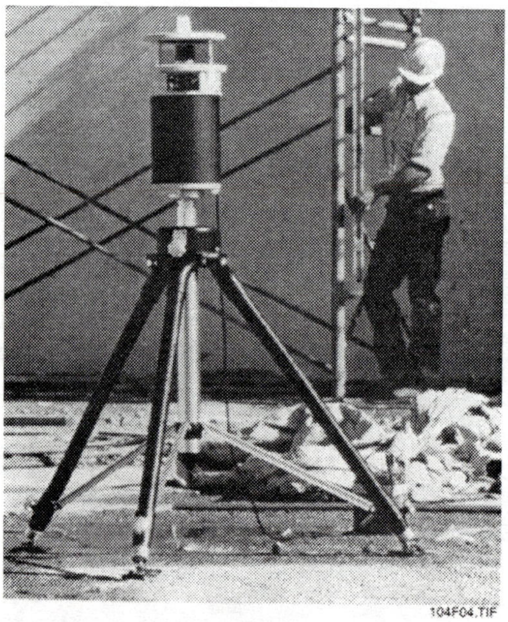

104F04.TIF

Figure 4. Rotating Laser Level

The advantage to using a laser level over the builder's level is that the laser level requires only one person to establish levels over an entire building site, rather than the two people required by the builder's level. The level points are established by placing a rod with a moveable sensor on the point to be leveled, as shown in *Figure 5*. The worker moves the sensor up or down on the rod until it aligns with the laser. The sensor will light up or make a sound when it aligns with the level laser beam, and a reading can be taken from the rod.

104F05.TIF

Figure 5. Level Rod With Laser Detector

2.3.0 STRAIGHTEDGES

The hand-held straightedge is a nonmotorized screed made of wood, magnesium, or steel. As shown in *Figure 6*, it has a rigid working surface. The straightedge may be from 2 to 20 feet long. It is usually 2 feet wider than the form it will be pulled across. The hand-held straightedge uses the top edges of the forms for rails, or guides. Note the builder's level set up in the background of the picture.

104F06.TIF

Figure 6. Hand-Held Straightedge Being Pulled Across A Form

Screeding with a straightedge strikes off the surface of the concrete to a given or preset grade. After the concrete is leveled, the straightedge is moved in a back and forth motion flat across the surface, as it is slowly being moved forward. This combined motion shapes and smooths the surface of the concrete. The contact edge of the straightedge may be tilted 15 to 25 degrees. This tilt prevents the straightedge from tearing out chunks of aggregate, and will help smooth as well as consolidate the top surface. Straightedges can be set for a **camber**, which is needed for sidewalks.

2.4.0 FLOATS

Floats are wide, flat tools pulled or pushed across the surface of placed concrete with the blade of the float flat against the surface. Keeping the blade flat prevents the concrete from sealing. **Floating** is usually done immediately after the concrete is **rodded** and screeded off, prior to the appearance of bleed water. Moving a float across the surface of fresh concrete accomplishes the following:

- Fills voids
- Brings mortar to the surface
- Embeds coarse aggregates into the surface
- Removes surface imperfections
- Produces textured surfaces

Hand floats are typically made out of wood, magnesium, rubber, or a combination of materials. Different types of floats are used for different operations. Wood floats are usually heavier and rougher than metal and are best for stiff, low-slump and large-aggregate concrete. Magnesium floats are typically used for air-entrained and lightweight concrete. All types of floats must have clean working surfaces, otherwise they will tear up the surface. The four main types of hand floats are described below.

2.4.1 Bullfloats

Bullfloats are made of wood or magnesium. They have wide, flat blades attached to long pole handles, usually with some type of swivel between the blade and the handle. A typical bullfloat is about 1 foot wide by 4 to 5 feet long, with a 10- or 12-foot handle. A magnesium bullfloat is shown in *Figure* 7. A magnesium bullfloat may have a reinforcing bar down its length, as magnesium is a less sturdy material than wood.

104F07.TIF

Figure 7. Magnesium Bullfloat With Pole Handle

Bullfloating is done immediately after rodding or screeding and before the bleed water comes to the surface. The bullfloat is pushed flat across the surface back and forth at right angles to the direction of the screeding.

CAUTION: Metal handles on bullfloats conduct electricity, so be especially careful when working near electrical equipment with this tool.

2.4.2 Darbies

Darbies are smaller than bullfloats and are used in tight spaces or for small jobs instead of bullfloats. Darbies may also be used to finish the edges of a deposit that has been bullfloated. *Figure 8* shows a wood darby with double hand-grips.

Figure 8. Wood Darby

A darby has a rectangular plane body of wood or magnesium, usually about 28 inches to 45 inches long, with a hand-grip handle. Some darbies may be as long as a bullfloat. A darby may taper from about 3½ inches at the heel to about 2¼ inches at the nose.

The darby is moved across the surface in a sawing arc motion instead of back and forth. Keep the darby as flat as possible to avoid sealing the surface. Do not bullfloat and **darby** the same surface.

2.4.3 Highway Straightedges

The highway straightedge is from 8 to 12 feet long, with a rectangular cross section. As shown in *Figure 9*, it has a 10- to 12-foot pole handle. A highway straightedge is used instead of a bullfloat in places where slab flatness is critical. This is for surfaces where specifications allow less than ¼ of an inch of deviation over a 10-foot span. These immense floats are used on highways and industrial slabs requiring precise leveling.

The highway straightedge is pushed and pulled across the surface by a concrete worker standing outside the form. The movement is at right angles to the direction of screeding. Using the highway straightedge takes practice and skill. Improperly used, the highway straightedge may dig into the concrete surface. Properly used, the highway straightedge can reflatten or restraighten a slab surface after other operations, such as power troweling, that may leave very slight irregularities.

104F09.TIF

Figure 9. Highway Straightedge

2.4.4 Hand Floats

A hand float is smaller than a darby and is used as part of final finishing. Hand floats may be made of wood, magnesium, or rubber. *Figure 10* shows a rubber float: a flat surface with a hand-grip handle.

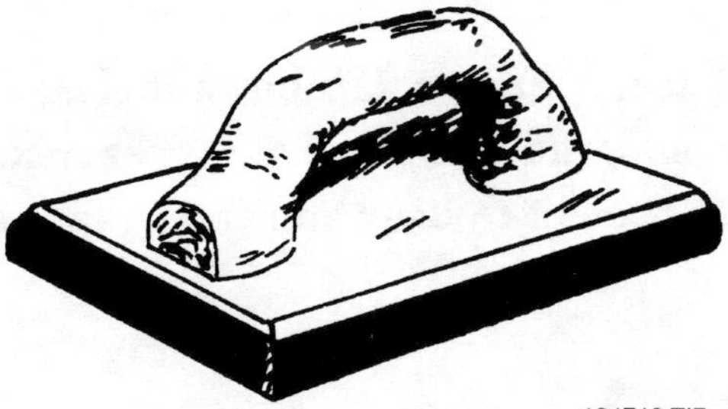

104F10.TIF

Figure 10. Rubber Float

The rubber or composition float is used for patching and repairing concrete, or for special finishing. The more dense the rubber foam, the smoother the concrete finish. Soft foam floats give roughly textured surfaces. Hard rubber floats make a hard, smooth finish on vertical surfaces. As with other types of floats, a wood hand float leaves a rougher surface than a magnesium float, and is used to float slop.

Hand floats are moved back and forth with the float surface flat against the concrete. Hand floating removes small blemishes and prepares the surface for troweling. It is generally done when the surface has no bleed water, while concrete may still be manipulated by hand.

2.5.0 SQUARE-END SHOVELS

Square-end shovels are larger than spades and have larger, square-ended spoons, as shown in *Figure 11.* These shovels are used to carry concrete and spread it inside a small form or marked area. They are also useful for mixing small batches of concrete in a wheelbarrow.

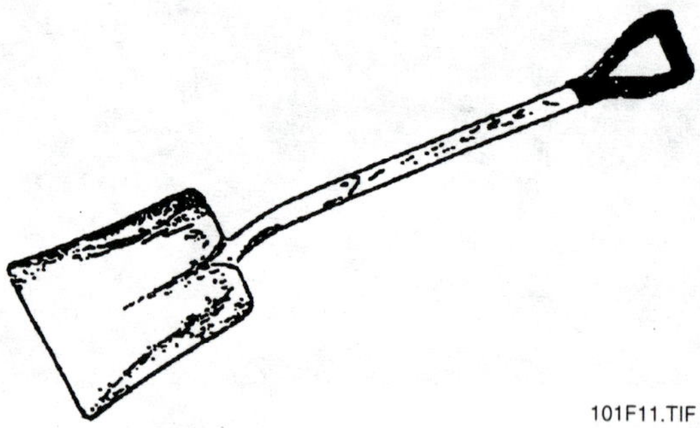

Figure 11. Square-End Shovel

When spreading mix, hold the end of the handle with one hand and grasp the handle near the shovel spoon with the other hand. Keep the bowl of the shovel facing you and drag the shovel along the surface until the concrete is spread. Remember to bend your legs to protect your back.

2.6.0 CONCRETE RAKES

Concrete rakes have very short teeth. *Figure 12* shows a close-up of these blunt teeth. This rake is used to spread and level subgrade, to remove excess mix, and to even out the surface for screeding. The concrete rake is easier to use than the square-end shovel for spreading mix over large areas.

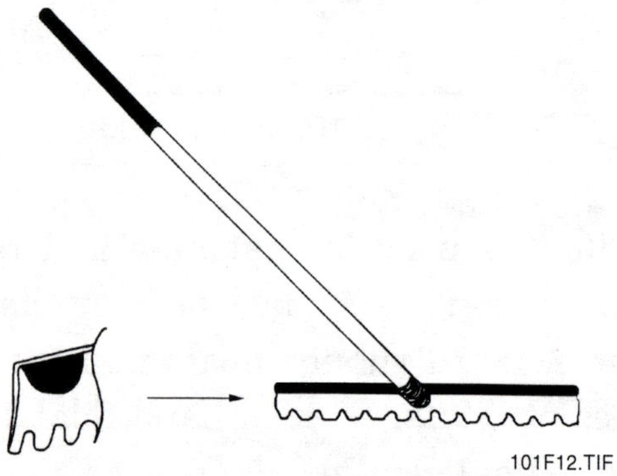

Figure 12. Concrete Rake

For removing excess mix, drag the rake across the surface with the teeth pointed down. Be careful not to rake so much that you **segregate** the mix. To remove excess concrete and even out the surface for the screeding operation, reverse the rake so that the flat edge is down. When used this way, the rake becomes a come-along or hoe.

2.7.0 COME-ALONGS

Come-alongs are shaped like very wide garden hoes, as shown in *Figure 13*. The come-along, with its 20-inch blade, spreads concrete when it is pushed or dragged across the form. Sometimes, the come-along is combined with the concrete rake.

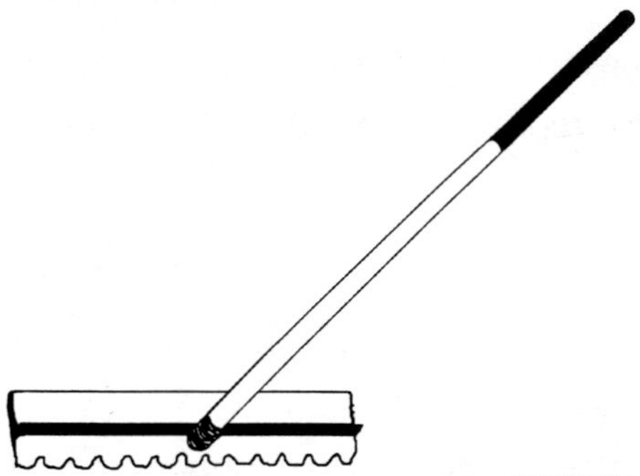

101F13.TIF

Figure 13. Come-Along Wiith Rake

2.8.0 FRESNO (ROLLERBUG) TAMPERS

The fresno or rollerbug tamper has one or two wire-mesh rollers about 5 inches in diameter. The joined rollers are 3 to 6 feet long, and are pushed by a long, light handle, as shown in *Figure 14*.

101F14.TIF

Figure 14. Fresno (Rollerbug) Tamper

After screeding, this tamper is pushed and pulled flat across the surface without being lifted. The direction of the force should not be downward. The rolling forces coarse aggregates below the surface. Roller tamping is easier and faster than hand tamping, but the concrete must be spread accurately and flatly graded beforehand.

2.9.0 TROWELS

Trowels are used only for final finishing, that is, after floating is complete and the bleed water has evaporated. Troweling seals the surface of the concrete and makes the surface stronger, denser, and more compact. The trowel is passed across the surface several times. For the first pass, the trowel is held almost flat and run in large sweeping strokes. The angle is increased slightly for each following pass, and shorter, narrower trowels can be used. Several kinds of trowels can be used, as shown in the following sections.

2.9.1 Finishing Trowels

The steel finishing trowel, shown in *Figure 15*, looks like a small float, but the blade is very thin and the handle is open-ended. This trowel may be 4 to 5 inches wide and 6 to 24 inches long. Since the first pass is the roughest, start with your longest and most worn trowel first and use shorter, narrower trowels on the following passes.

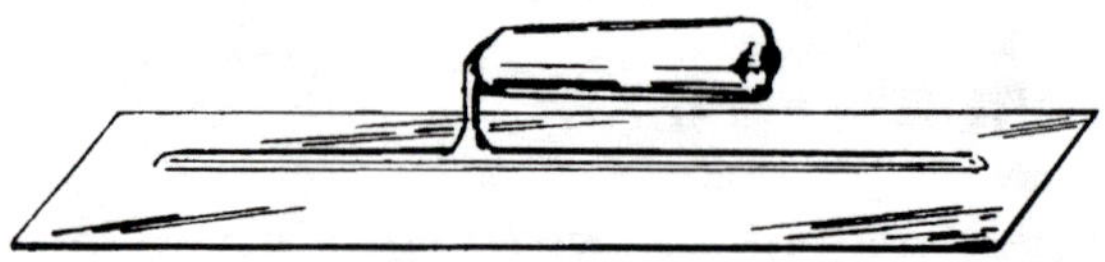

101F15.TIF

Figure 15. Finishing Trowel

2.9.2 Walking Or Fresno Trowels

The walking or fresno trowel is a large finishing trowel that looks like a bullfloat. Its blade may be about 5 inches wide and 2 to 5 feet long. Some blades are curved for finishing curved surfaces. Made of aluminum or steel, this trowel has a long pole handle with a locking adapter. The adapter is set for the troweling angle, and the pole is used to push the trowel while walking. The walking trowel is not meant to be used as a bullfloat. *Figure 16* shows the adapter of this trowel.

Figure 16. Walking Or Fresno Trowel

2.9.3 Pointing And Margin Trowels

Pointing and margin trowels, as shown in *Figure 17*, are small, stiff trowels used to finish and patch small areas and dislodge loose material from areas to be patched.

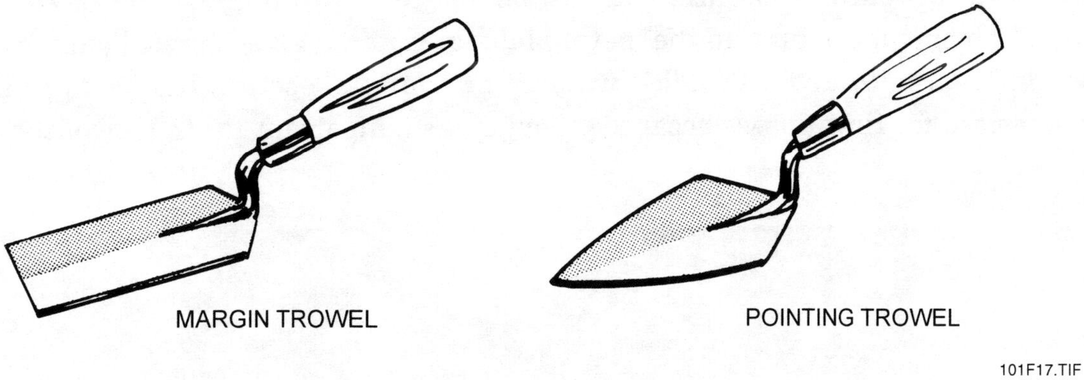

Figure 17. Pointing And Margin Trowels

The pointing trowel has a pointed blade that is 4 to 7 inches long. This can be used to exert a large amount of pressure in a small area, which is useful to pry out badly bonded or loose material. The point of this trowel is sometimes used to make the first scoring of grooves in slabs. The first pass with a trowel is followed with the scoring of an edger.

The margin trowel has a narrow, rectangular blade that is also useful for patching small areas and finishing hard-to-reach surfaces. Its blunt end is more useful for mixing and lifting small amounts of mix out of a bucket.

2.10.0 GROOVERS

Groovers are finishing tools. They score and shave the surface of the slab. They are the first tools used after floating is completed. Grooving or jointing is done in several passes, and at different times, to make sure these details are not lost as the slab dries.

Groovers or jointers come in two types: hand-held (*Figure 18*), and pole-handle (*Figure 19*).

101F18.TIF

Figure 18. Hand-Held Jointer Or Groover

The groover is a flat plate of bronze, magnesium, or steel with a groove bit or cutting edge underneath. The bit tapers up into the metal plate for rigidity. Pole-handled jointers, like the one shown in *Figure 19*, are also called walking groovers. These are harder to control, and should not be used for first passes because grooving is so important for the life of the slab.

104F19.TIF

Figure 19. Pole-Handle Groover Or Jointer

The jointer or groover makes a deep, narrow groove in the slab. This contraction joint must be at least as deep as ¼ the depth of the slab. This depth creates a weak section in the slab that will crack as the concrete cures or shrinks. Contraction joints control where the slab cracks and are easier to seal and maintain than random cracks. Therefore, grooving should be done before cracks are allowed to appear.

Groovers come in different sizes for working on different depths of slab. Usually, groovers will cut 1 inch deep for a 4-inch slab, 1½ inches deep for a 6-inch slab, and 2 inches deep for an 8-inch slab.

To cut a contraction joint, make two passes over the line with the groover keeping the blade flat. You can use a rule or board to guide it. Sometimes, a preliminary score is made with a pointing trowel. Grooving is done after bullfloating and after each step in the finishing process is completed. Steel groovers seal the surface, so they are used in the final finishing.

Some grooves are not contraction joints but only decorations because they are cut shallow. Decorative grooves channel water and help make the slab skid-resistant. These decorations can be made with a shallow groover, also called a cheater, or with a safety edger.

2.11.0 EDGERS

An edger is a tool with a flat blade and a curled-down lip on one side. Attached to a wooden handle, the blade can be made of bronze, magnesium, or steel in various sizes and shapes. Edgers are used to make neat, rounded edges next to the edge of forms. These rounded edges are less likely to chip or **spall**. Edging is not required on most floor slabs but is important on sidewalks, driveways, patios, and steps, wherever the edge of the slab is exposed to wear.

Edging is done before floating and with several passes. It is important to use a wide-bladed edger on the first pass so you do not depress the edge lower than the grade of the slab. Use narrow edgers for later passes, after the concrete has stiffened.

2.11.1 Basic Edgers

The basic edger, shown in *Figure 20,* is about 3 inches wide by 6 inches long with a ¼-inch curve radius and is commonly made of steel. Some edgers have turned up ends to prevent gouging the soft concrete. Edger plates are usually made of steel; bronze edgers are heavier and more durable.

101F20.TIF

Figure 20. Basic Edger

2.11.2 Sidewalk, Highway, And Windowsill Edgers

Sidewalk, highway, and windowsill edgers work like basic edgers. They are sized for the common dimensions of specific jobs. Their blades may run from 1½ inches to 8 feet long. Sometimes, these edgers have extended handles, like the windowsill edger shown in *Figure 21*. Windowsill edgers usually have a ⅛-inch curve radius; sidewalk edgers have a ¼-inch curve radius; pavement edgers have a ½-inch curve radius; and curb edgers have a ¾-inch curve radius.

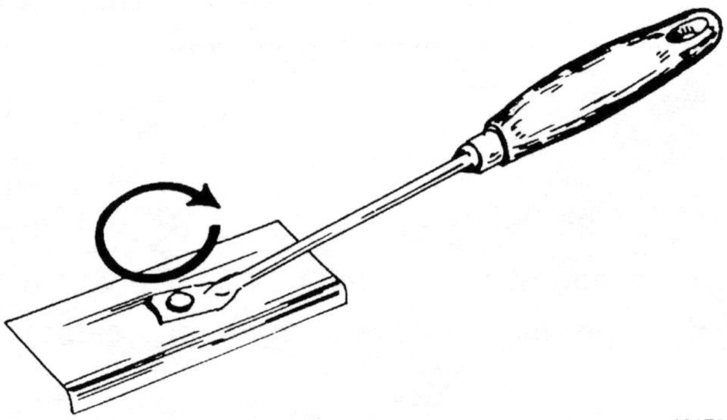

101F21.TIF

Figure 21. Windowsill Edger

2.11.3 Pair Edgers

Pair edgers are matched sets of inside and outside edging tools. They are used for stairs, curbs, corners, or other places that call for evenly matched outside and inside curves. Paired tools have the same curve radius at the lip. *Figure 22* shows a matched pair of edgers used for stairs.

101F22.TIF

Figure 22. Pair Edgers

2.11.4 Radius Edgers

The radius edger is very narrow and is used to shape curved or irregular edges. The lip is on the front of the blade rather than on the side. A typical radius edger is about 2 inches long and 2 inches wide, with a ⅜-inch curve radius. A radius edger is shown in *Figure 23*.

101F23.TIF

Figure 23. Radius Edger

2.11.5 Safety Step Edgers

The safety step edger creates a nonskid edge that is useful on outdoor stairs. This edger has a number of short bits set into the plate, parallel to the curved edge. The bits are similar to those on a groover. The bits score shallow grooves into the surface at the edge of the form as the slab is edged. *Figure 24* shows the bits under the edger plate.

Figure 24. Safety Step Edger

2.12.0 BRUSHES

Brushes are used for cleaning and for keeping surfaces damp. They are also used for finishing small areas by texturing the surfaces. Two general types of brushes used for concrete work are water brushes and texture brushes (*Figure 25*). Brushes can be used in place of brooms for finishing small, hard-to-reach surfaces.

Figure 25. Water And Texture Brushes

2.12.1 Water Brushes

Water brushes have long flexible bristles attached to a wooden handle. Sizes vary, but they are usually about 6 inches by 2 inches with bristles about 4 inches long. This type of brush is used to apply water and to dampen a small surface. This is also the brush used to clean tools and equipment. Do not use the same brush for dampening and for cleaning tools. It is important to clean the brush after it is used, so other surfaces will not become soiled.

2.12.2 Texture Brushes

Texture brushes have long stiff bristles attached to a wooden handle. They are usually larger than water brushes. Texture brushes can be used to score and texture fresh, small concrete surfaces. These brushes can also be used to make scratch coats for stucco. Since it is possible to tire your wrist when texture brushing, try brushing with your thumb pointed toward the top of the handle, not toward the bristles.

2.13.0 BROOMS

Brooms are important concrete tools because they can:

- Give finish texture to surfaces
- Expose aggregate
- Keep surfaces damp
- Clean up the site

Three types of brooms are used for concrete work. The following sections discuss the types of brooms and the work they are used for. As with other tools, brooms need to be rinsed clean after each use.

2.13.1 Finishing Brooms

The finishing broom, shown in *Figure 26*, has a thin row of stiff plastic or nylon bristles set into a holder with a long, pole-type handle. Finishing broom heads vary from 18 to 48 inches wide. For best results, rinse the broom clean in water after each pass. After rinsing, tap the broom to shake out excess water.

Broom-textured surfaces are attractive and nonslip, with a gritty or abrasive finish. The coarseness of a broom finish depends on when the sweeping is done and the type of broom. Coarse textures are created with a stiff-bristled broom on a freshly floated surface. Finer textures are created with a softer-bristled broom on a steel-troweled surface. Specially designed concrete finishing brooms give a sharp texture. The sweeping can be done in a pattern by sweeping in straight, curved, or wavy lines.

101F26.TIF

Figure 26. Finishing Broom

2.13.2 Industrial Push Brooms

The industrial push broom has many uses. The push broom can be used to clean dry surfaces, give a general cleanup, and make an exposed aggregate finish. The head of this broom is thicker than the finishing broom, and the bristles are softer. *Figure 27* shows a standard industrial push broom.

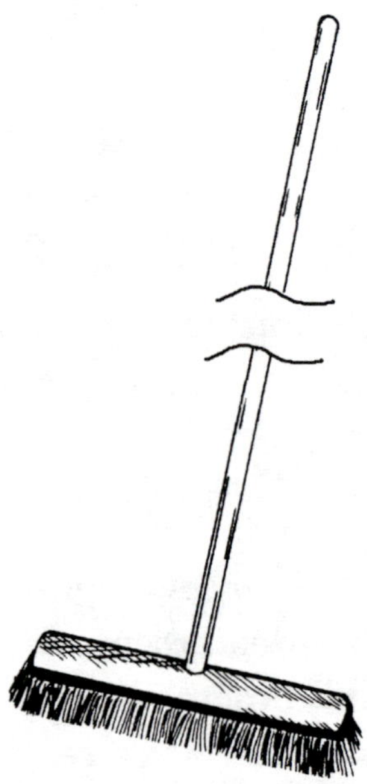

101F27.TIF

Figure 27. Industrial Push Broom

Any cleanup should be done with a broom reserved for only that purpose. Never clean up with finishing brooms or brooms used for exposing aggregate, or you will contaminate the finish work. Keep cleanup equipment separate from finishing equipment.

2.13.3 Water Brooms

A water broom is a push broom connected to a water hose, as shown in *Figure 28*. The hose sprays water as the surface is swept. The water broom is used for keeping surfaces damp to minimize dust. This broom is useful for general cleanup around the site, as it will not raise a dust cloud as you sweep. Be sure that the water broom used for cleaning is kept separate from the water broom used for finishing.

104F28.TIF

Figure 28. Water Broom

The water broom is also used to help create an exposed aggregate surface. Exposed aggregate finishes are made by embedding coarse aggregates into the concrete after it is placed. Then, excess mortar is removed from the surface. The first step is to lightly brush the slab with a stiff, nylon-bristled push broom. As the concrete hardens, it is brushed again, while spraying with a water broom. Make several passes to properly expose the aggregate. Work with the water broom until there is no visible cement film on the aggregate, and the water runs clean. If the water broom is used too early, it will displace the aggregate.

2.13.4 Rubbing Stones

Rubbing involves going over the surface of the hardened concrete to fill small surface voids, clean the surface, or smooth irregularities in the concrete. It can be used to produce a surface of uniform color or sandy texture, or to remove stains from hardened concrete. Rubbing does not improve or change the concrete structurally, only its appearance.

Several different techniques for rubbing are used, depending on the method specified or the results desired. Rubbing tools include different types of abrasive stones with hand grips, as shown in *Figure 29*. If the concrete surface is very fresh, rubbing can be done with burlap wrapped around a rubber float.

Figure 29. Rubbing With A Hand Stone

3.0.0 EQUIPMENT

Powered concrete tools and equipment are used for preparing and placing the concrete, as well as finishing it. They do everything hand tools do, but on a larger scale. Remember that as power equipment grows in size, it also grows as a potential hazard. Safety calls for alertness around equipment.

3.1.0 BATCH PLANTS

Batch plants measure and dry-mix the ingredients for concrete. *Figure 30* is a view of a combination batch plant and mixer. This batch mixer, used on site, combines batching and mixing operations. Note the large stationary mixer in front of the hopper. The mixer is situated so it can easily off-load ready-mix into trucks.

104F30.TIF

Figure 30. Batch Mixer

3.2.0 READY-MIX TRUCKS

The components of ready-mix concrete are accurately measured to job specifications, combined, then hauled to the job site in mixing trucks. Ready-mix trucks use front discharge or rear discharge chutes to unload the concrete at the site. These trucks are shown in *Figures 31* and *32*.

104F31.TIF

Figure 31. Front Discharge Truck

Front discharge trucks, like the one shown in *Figure 31*, are useful on sites where access and turnaround space is limited. It is easier for the driver to position the delivery chute on a front discharge truck.

As shown in *Figure 32*, rear discharge trucks must back up to the deposit site. The mix is off-loaded through a rear chute, as close to the deposit site as possible. This direct discharge is the simplest way to place concrete. *Figure 32* also shows finishers spreading the concrete as it comes out.

104F32.TIF

Figure 32. Rear Discharge Truck

The truck-mounted mixers are rotating drums that keep the concrete moving so it does not harden. There are time limits for mixing a batch of concrete: mixing occurs from 6 to 18 rotations per minute (rpms), and agitating 2 to 6 rpms. The agitation keeps the concrete from setting on the way to the job site.

When the truck discharges the ready-mix at the site, the mix may be discharged directly into the form, or onto a conveyor of some kind. There are several types of equipment used to convey concrete on-site.

3.3.0 PORTABLE MIXERS

Portable mixers include any mixer outside a concrete manufacturing plant. They may be either stationary or wheeled, tilting or nontilting. Not all portable mixers are small; their capacity may range from ⅓ of a cubic yard to 15 cubic yards. The smaller size typically is hand-held and used for mixing test batches, while the larger size may need a crane to move it.

ASTM C 94 requires that concrete be delivered and discharged within 90 minutes after mixing starts. For this reason, a portable mixer may be set up at a job site. Some large portable mixers are combined with batchers (see *Figure 30*), and operate in a fixed location on-site. Some portable mixers are placed on truck beds, while some have their own wheels.

The following four configurations are the most popular for portable mixers:

- A tilting mixer is a rotating cone-shaped or bowl-shaped drum that discharges by tilting the drum down from its horizontal axis.
- A nontilting mixer is a rotating cone-shaped or bowl-shaped drum that discharges by reversing the direction of the mix. The drum stays fixed on its horizontal axis.
- A turbine or vertical shaft mixer mixes by rotating blades or paddles on a vertical shaft. The mixing compartment may or may not rotate around its vertical axis.
- A pug mill or horizontal shaft mixer mixes by rotating plates attached to one or more horizontal shafts. This mixer is frequently used for mixing very stiff, nonplastic concrete.

Figure 33 shows a portable mixer. This gasoline-powered model discharges by tilting the drum downward.

104F33.TIF

Figure 33. Gasoline-Powered Portable Mixer

3.4.0 PAVING MACHINES

A paving machine is typically a large-scale, self-propelled combination of mixer with discharge chutes or conveyors, and screeds, plus some type of compactor. Used on highways and similar large-scale projects, paving machines combine several stages of concrete processing into one slow pass. The largest paving machines, as shown in *Figure 34*, do not include mixers. As shown in *Figure 34*, the ready-mix is dumped in lanes on the subgrade. The paving machine shifts, compacts, and screeds the concrete as it slowly passes.

104F34.TIF

Figure 34. Paving Machine

The paving machine may have attached, gang-mounted tube vibrators to consolidate the concrete as it is placed, or it may include roller compactors and motorized vibratory screed bars. The newest machines are combined with laser levels. Some paving machines include slip forms. These slowly advance as the concrete is placed, consolidated, and screeded behind the paving machine.

3.5.0 CHUTES

Chutes carry the concrete from the discharge point of the mixer to the deposit site. *Figure 35* shows lengths of chute joined to move the discharging concrete to a specific part of a form. Chutes operate by gravity feed, so they discharge downhill. They are metal tubes, usually open, with a semicircular cross-section. The end of the chute can be guided to control concrete deposit and to eliminate segregation.

104F35.TIF

Figure 35. Joined Discharge Chutes

Two special kinds of chutes are also used for concrete work.

3.5.1 Drop Chutes And Elephant Trunks

The slope of a chute needs to be limited in order to control the rate of deposit and to keep the concrete from segregating. Dropping the concrete more than 4 feet vertically from the end of the chute can cause segregation. This problem is avoided by adding a drop chute, as shown in *Figure 36*.

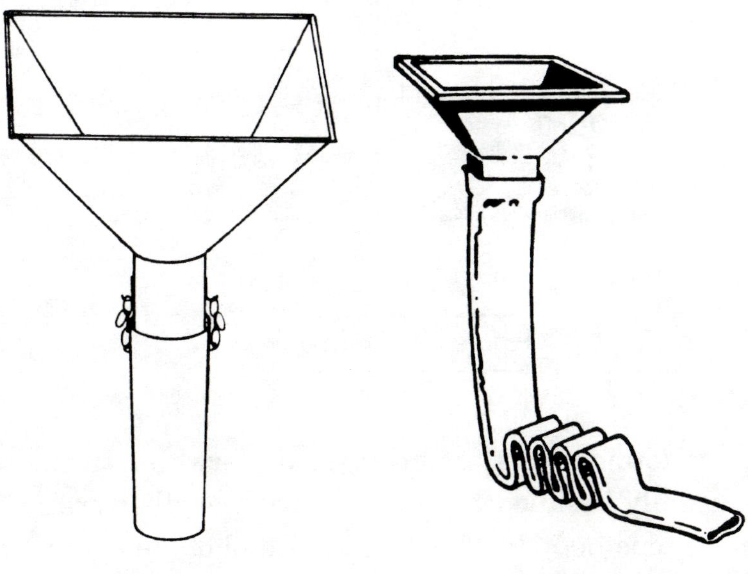

Drop Chute

Elephant Trunk Chute

104F36.TIF

Figure 36. Drop Chute And Elephant Trunk

A drop chute is a metal tube with a full circle cross-section, and with a hopper or funnel box at the top. An elephant trunk is a flexible drop chute made of canvas or plastic. Either one can be joined to the end of the conveying chute to control segregation when depositing concrete.

The end of the conveying chute goes into the hopper at the top of the drop chute. The discharge end of the drop chute can be guided to control where the concrete is placed.

3.5.2 Tremies

A tremie is a chute specially designed to place concrete under water. *Figure 37* shows a tremie, a closed chute with a full circle cross-section at least 10 inches in diameter. Its length is made of segments joined together by flanges and gaskets. It has a hopper or funnel box at the top.

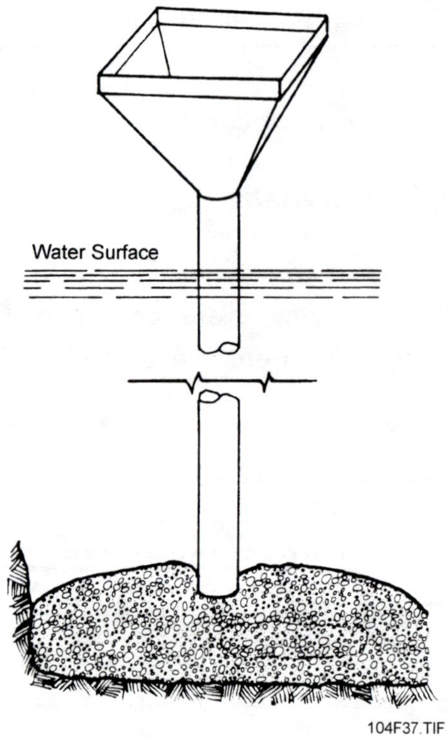

Figure 37. Tremie

Tremies are used to place foundations for bridges, underwater tunnels, and other deep projects. The tremie is usually handled by a crane, with another crane to handle loading concrete buckets into the tremie hopper. The discharge end of the tremie must be embedded in the fresh concrete, keeping a seal between the water and the concrete mass during placement.

3.6.0 BUGGIES AND BARROWS

Small wheeled containers are used to carry concrete over short distances. They are labor-intensive, but their advantage is speed: no delays waiting to rig chutes, conveyors, or other equipment to reach an awkward spot. Three commonly used carriers are wheelbarrows, hand buggies, and motorized buggies. The first two are hand-powered and require runways or paths. The motorized buggy can usually travel without a runway, but it does need a smooth path. These conveyances are described in the following sections.

3.6.1 Wheelbarrows

Wheelbarrows are valuable on small jobs or where placing conditions are constantly changing. *Figure 38* shows a small wheelbarrow in use. Note that the wheelbarrow has been run up a walkway from the ground to the edge of the form.

Figure 38. Wheelbarrow Discharging

The concrete may be unloaded directly or unloaded into a hopper or bucket, then off-loaded into the wheelbarrow. The wheelbarrow is pushed to the deposit site then tilted to discharge the concrete. However, wheelbarrows are slow, so they are mostly used to carry small amounts over short distances, less than 200 feet.

Some mixes may segregate when jarred in a wheelbarrow traveling over rough ground. Wheelbarrows may need walkways or runways to move smoothly and prevent segregation.

3.6.2 Hand Buggies

Hand buggies are large wheelbarrows that have two wheels in front, rather than one. They are superior to wheelbarrows because their two front wheels provide a better balance for the load. They are also easier to push because there is no need to lift. *Figure 39* shows a hand buggy, sometimes called a Georgia buggy. They are available in sizes to carry 6 to 11 cubic feet of concrete.

104F39.TIF

Figure 39. Georgia Buggy

3.6.3 Motorized Buggies

Buggies are a form of large, motorized wheelbarrow. Motorized buggies are self-propelling and most of them have a stand for the operator to ride upon. *Figure 40* shows a small motorized buggy traveling, while *Figure 41* shows it discharging. Buggies have motorized discharge, so they are easier to use and more efficient than wheelbarrows.

104F40.TIF

Figure 40. Motorized Buggy

104F41.TIF

Figure 41. Buggy Discharging

CAUTION: When driving a loaded buggy, remember that it has momentum and leave room to stop slowly.

3.7.0 HOISTS

Hoists are freestanding elevator lifts for small to medium-sized loads of materials and workers. *Figure 42* shows the parts of a portable materials hoist. Hoists vary in size and power. All hoists include a platform, lift cabling, and a gasoline, diesel, or electric motor. The platform may have a cage around it. The load is placed on the platform for lifting.

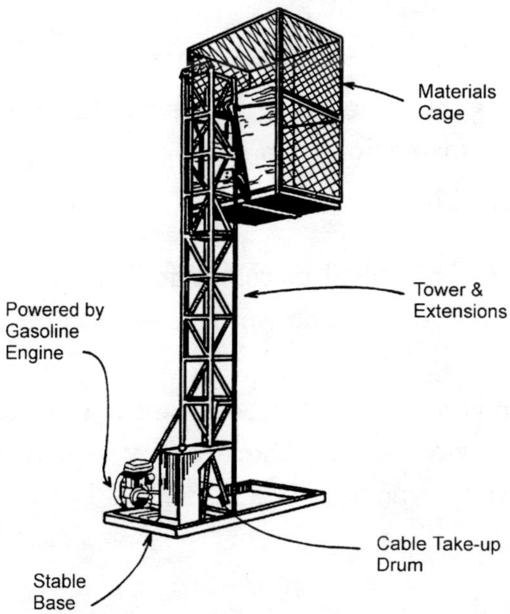

104F42.TIF

Figure 42. Portable Materials Hoist

Personnel hoists have guardrails, doors, and hand controls in addition to the features of the materials hoist. Materials hoists do not have the safety features found on personnel hoists, such as safety brakes.

CAUTION: Do not ride on materials hoists.

The lift framework for large hoists may be attached to the side of a building. Smaller, portable hoists may be mounted on a crawler or metal base and wheels to be moved about as needed.

3.8.0 CRANES

Cranes and derricks are motorized lifts for raising and lowering heavy weights and moving them over a limited horizontal distance.

Cranes can work in several ways. The three main configurations are as follows:

- Tower cranes stand alongside or in the middle of the building itself.
- Mobile cranes are mounted on crawlers or truck beds.
- Conventional cranes and derricks stand away from the building, with the distance determined by the length of the boom arm.

For concrete work, cranes are typically used to carry buckets of concrete from the mixer to the discharge site. Cranes are a necessary part of large-scale concrete projects, and may generate hazards. In high-rise construction, materials movement by crane poses the greatest safety hazard to workers on the job. Work safety calls for:

- Establishing clear pathways for materials movement
- Using a consistent set of signals to alert workers to movement (See the Core Curricula module, *Basic Rigging*, for information on hand signals.)
- Assigning individual responsibilities to each worker during materials movement

Several types of cranes are commonly used at construction sites. Concrete work usually does not include operating cranes, but does include working with concrete moved by cranes.

CAUTION: Workers around cranes need to be aware of materials movement at all times. Always watch the area overhead when working with cranes or any type of hoist.

3.8.1 Tower Cranes

Tower cranes or derricks, as shown in *Figure 43*, have a vertical tower and a horizontal swinging boom or jib. The boom can move in a full circular swing, unobstructed by the tower. The crane has a counterweight on the end of the boom to balance the load and a cab on the tower for the operator.

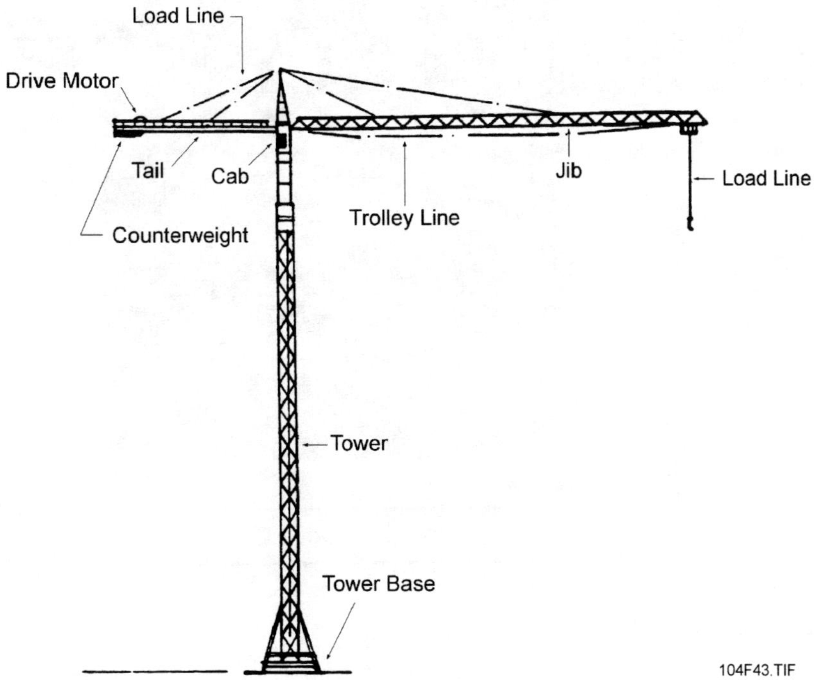

Figure 43. Tower Crane

Tower cranes are the most popular stationary cranes for construction because of their versatility and freedom of swing. Because they are self-balancing, they do not need to be tied to the structure. Tower cranes can be mounted in the middle of a high-rise building under construction. The tower can be moved from one level to the next as the structure rises.

3.8.2 Mobile Cranes

Mobile cranes lift with swinging booms raised and lowered by cable, as shown in *Figure 44*. The boom and the operator's cab can be mounted on a truck bed or on crawler tracks.

The lift capacity of a mobile crane depends on the stability of its footing, the strength of the boom, and the size of the counterweight. The counterweight is usually at the ground end of the boom. The main advantage of this crane is that it can move around the job site.

104F44.TIF

Figure 44. Mobile Crane

3.9.0 BUCKETS

Buckets are used to carry concrete from the mixer to the discharge point. *Figure 45* shows the most widely used bucket, the 1-cubic yard, bottom-dump bucket. The bucket is filled, then lifted by a crane or derrick to the discharge site, and guided as it discharges. The direction of deposit should always be toward the concrete already in place.

104F45.TIF

Figure 45. Bottom-Dump Bucket

Typical bottom-dump buckets range in size from ⅜ of a cubic yard up to 15 cubic yards. Their dump gates are usually manually operated by a lever. The speed of discharge is controlled by the gate opening.

A second type of common bucket is the larger lay-down bucket, shown in *Figure 46.* It is rectangular, with the dump gate on the side, not the bottom. The dump gate may be remotely controlled.

104F46.TIF

Figure 46. Lay-Down Bucket

The lay-down bucket gets its name because it lays flat to be filled. Its low loading height makes it easy to fill from truck-mounted mixers. As the bucket is lifted to the discharge site, it shifts from a horizontal orientation toward a vertical orientation. The side gate becomes the bottom gate as the bucket tilts for discharge.

A good size of bucket for a job is one that is big enough to hold one batch of concrete as mixed. Buckets can be moved by trucks, rail cars, helicopters, and overhead cableways, as well as cranes. It is important to avoid shaking and jarring the filled bucket as this causes segregation of the mix.

A third type of bucket is the hopper, or very large, short-term storage bucket. Hoppers are usually mounted on scaffolding legs. They are filled from the top by a pump, conveyor, or smaller bucket. They have gates at the bottom for discharging fresh concrete into buggies or conveyors.

3.10.0 CONVEYORS

Rigidly supported belt conveyors make it possible to deposit large volumes of concrete quickly, in obstructed areas. *Figure 47* shows a conveyor mounted on a truck, carrying concrete above site obstructions.

104F47.TIF

Figure 47. Conveyor

The pictured conveyor has two spans, called *flights*. The first flight is carrying concrete from the mixer truck to a second flight, with a telescoping support. A drop chute at the discharge end of the second flight helps prevent segregation.

Mechanically, a conveyor flight has a continuous belt powered by a chain or hydraulic drive run by a gasoline or electric engine. *Figure 48* shows these parts. The belt is usually about 16 inches wide and shaped into a trough by rollers or pans underneath it. Individual flights can be joined together by a swivel which supports the flight ends and guides the concrete from one belt to the next.

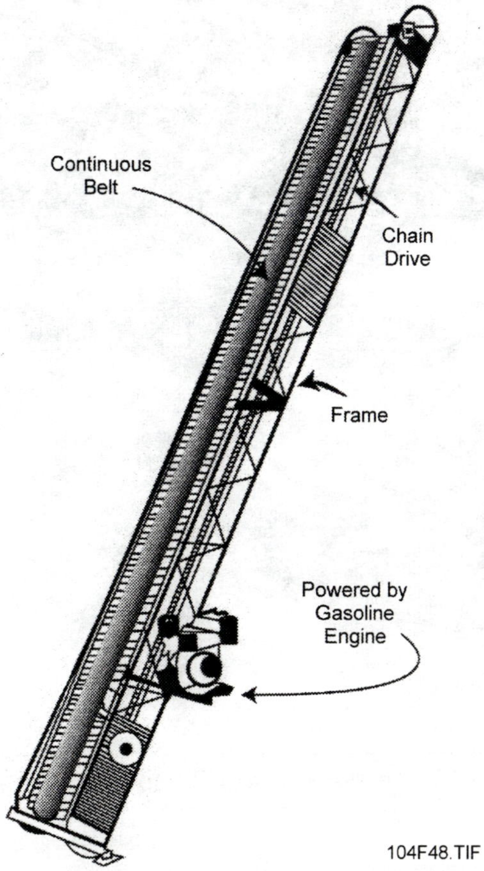

Figure 48. Conveyor Parts

Concrete can be deposited onto a conveyor belt from a hopper in a continuous ribbon. A conveyor can have several flights to move concrete horizontally and up or down. Conveyors can be mounted on truck mixers or their own truck beds; some can be towed to the deposit site.

3.11.0 PUMPS

Pumps convey concrete from a mixer through steel pipe or hose, directly into forms. The two major types of pumping systems move concrete by piston or by pneumatics. Large pumps can move concrete as much as 100 feet vertically and 300 feet horizontally. Pressure on the concrete may reach 2,500 psi. The largest and most powerful of these pumps can be used for placing concrete in tall buildings.

A more typical pump may be truck-mounted, with its pipeline supported by hydraulically-operated booms to extend delivery range. The truck-mounted pumps can be fed concrete by separate ready-mix trucks. The smaller truck-mounted pumps, as in *Figure 49*, combine the mixer and pump on one truck bed.

104F49.TIF

Figure 49. Truck-Mounted Pump And Mixer

For small jobs, a trailer pump, as in *Figure 50*, may be used. These pumps may be towed behind a truck and easily moved about the site.

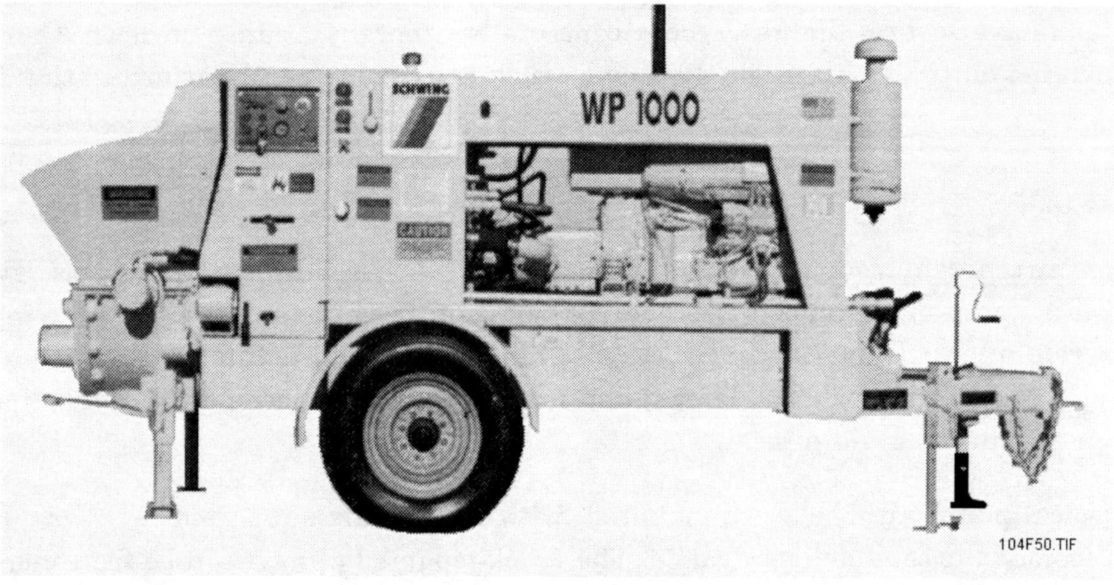

Figure 50. Trailer Pump

Unlike conveyors, pumps do not expose concrete to the weather. Pumps quickly move concrete to hard-to-reach deposit sites, but they require a continuous feed of concrete. A long delay in the supply may require breaking down the pipeline and cleaning it before pumping can start again.

Pumping requires a plastic mix, and some **admixtures** and aggregates are used specifically to improve the pumping flow. Elbow bends in pipeline slow the flow, so hose is laid in a wide-radius curve to lower resistance.

3.12.0 VIBRATORS

Mechanical vibration is used to consolidate concrete. When vibrated, newly placed concrete behaves temporarily like a liquid and settles in the form. Air voids rise to the surface during vibration so the form will be solidly filled. Before mechanical vibrators, concrete was consolidated by tamping with metal rods. Today, there are three types of mechanical vibrators used for concrete:

- Internal vibrators
- Form vibrators
- Surface vibrators

3.12.1 Internal Vibrators

Internal vibrators, called spud or poker vibrators, are constantly vibrating heads connected to a vibratory drive motor by a flexible shaft. The diameter of the vibrating shaft can range from ¾ of an inch to 6½ inches. The larger ones are gang-mounted on screeds or paving machines. The smaller poker vibrators can be carried on backpacks, as shown in *Figure 51.* The drive motor can be powered by electricity, gasoline, hydraulics, or pneumatics, depending on the size of the vibrator.

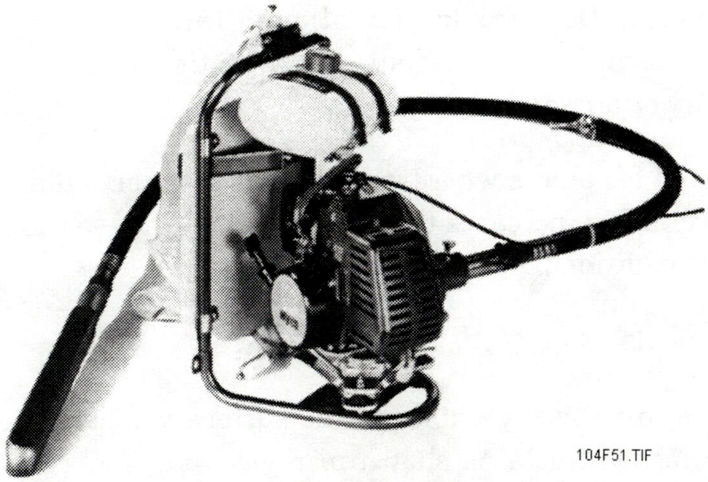

104F51.TIF

Figure 51. Internal Vibrator

The vibrator head is lowered into the wet concrete to the full depth of the layer being consolidated, and at least 6 inches into the underlying layer. The vibrator rod is inserted as close to vertical as possible. It is manipulated up and down for 5 to 15 seconds to knit the two layers together and allow air to escape. The head is drawn out slowly so that no voids are left in the concrete, and reinserted a short distance away.

The vibrator used must be the proper size for the form. It must be inserted so that its areas of action overlap, as shown at the far right of *Figure 52*.

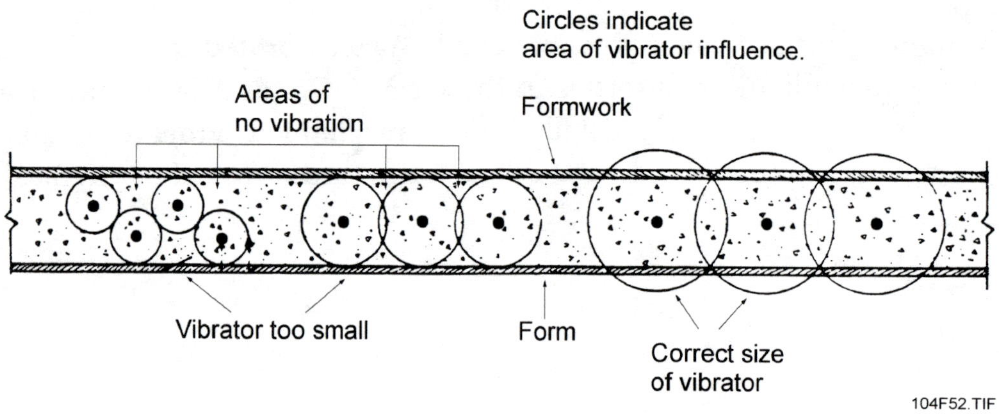

Figure 52. Vibrator Size And Placing

Figure 52 shows vibrator insertions as black dots, and the circles show the radius of the vibrator's action. Note at the left what happens when the vibrator chosen is too small for the form. When the vibrations are not properly overlapping, air pockets will be left, as shown on the left side of *Figure 52*.

3.12.2 Form Vibrators

Form vibrators are engines that provide vibration, attached to the outside of the form or mold. They vibrate the form, which transmits the vibration to the concrete. Form vibrators may have either a rotary or a reciprocating drive.

Form vibrators are useful for places where it is difficult to insert an internal vibrator, such as tunnel linings. Form vibration increases the stress on the form, so the form must be designed and built to handle this additional stress.

3.12.3 Vibrating Screeds

Motorized screeds are a third type of vibrator, the surface vibrator. Surface vibration can be used on slabs up to 8 inches thick. As shown in *Figure 53*, a vibrating screed consists of a single or double beam spanning the slab, much as a straightedge does.

104F53.TIF

Figure 53. Surface-Vibrating Screed

The black box perched in the middle of the beam in *Figure 53* is an eccentric-drive engine that gives the vibration to the beam. The screed beam itself is also powered to move slowly forward, flat against the surface. The vibratory screed also screeds the surface as it vibrates and moves forward. The strike-off operation becomes more efficient when the concrete is consolidated first.

The newest screeds combine the work of compacting and screeding with a laser, for a flatter screeded surface. *Figure 54* shows a new laser-guided screed with a telescoping boom arm. The screed beam moves forward as the laser levels on its edges measure the surface of the fresh concrete.

104F54.TIF

Figure 54. Laser Screed

3.13.0 CONCRETE SAWS

Concrete saws are used to cut contraction joints in slabs. They are radial-type saws with diamond-coated circular blades and blade irrigation systems. The walking saw, *Figure 55*, is mounted on a push frame with a cutting guide extending out before the blade. This saw is pushed over a premarked cutting line.

104F55.TIF

Figure 55. Walking Saw

Diamond-bladed saws require a stream of water to irrigate the blade, to prevent burning up the concrete. Diamond-bladed saws are typically used for cutting fresh concrete or "softcutting." Carborundum blades are softer and do not need irrigation. Carborundum blades are used in smaller saws without irrigating systems. They are also most commonly used in hand-held saws.

For small cuts, a hand-held saw (*Figure 56*) can be used. This saw also needs a premarked cutting line.

CAUTION: Be sure the guard is operating properly on the hand-held saw.

Hand-held sawing is done in two passes. A ⅛-inch wide blade makes the first cut to reach a quarter of the depth of the slab. A second cut is made with a ¼-inch to a ⅜-inch wide blade to reach an inch deep. This second cut makes a wider reservoir for the joint sealant that will be applied later.

Contraction joint sawing should be done within 6 to 8 hours after placing, so the slab will not be under tension as it is sawed.

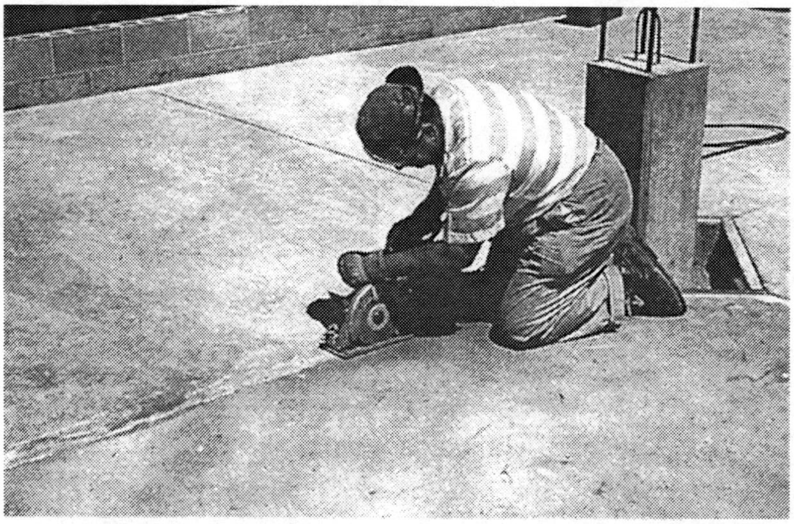

104F56.TIF

Figure 56. Hand-Held Saw

3.14.0 COMPACTORS (SUBGRADE)

A small subgrade compactor is a motorized, hand-held vibratory ramming plate, as shown in *Figure 57*. The compactor has an engine, a vibratory transmission shaft, and a compacting plate.

This compactor is related to a jackhammer. It is used for compacting the soil before forms are built or concrete is placed onto the earth. Soil compaction is required for foundations, slab work, driveways, and sidewalks.

104F57.TIF

Figure 57. Small Vibratory Compactor

CAUTION: Be aware of your toes in relation to the compactor. Wear steel-toed safety shoes.

Another type of compactor is shown in *Figure 58*. This larger compactor is steered from behind and is safer to use. It also uses a motor to drive a vibratory plate against the subgrade.

104F58.TIF

Figure 58. Vibratory Push Compactor

3.15.0 ROLLER COMPACTORS

Roller compactors, as shown in *Figure 59*, are upsized versions of lawn rollers. These riding rollers use the weight of the roller to compact concrete.

104F59.TIF

Figure 59. Riding Roller Compactor

The size of the roller ranges from those smaller than the model shown in *Figure 59* to some even larger. Some self-powered models can be attached to paving machines. These roller compactors can also be used in combination with a vibratory screed on a paving machine.

3.16.0 POWER FLOATS AND TROWELS

Power floats and trowels range in size from a small rotary trowel to large double- and triple-rotor riding machines. Some machines with combination blades can be used for floating and troweling. On other machines, blades can be set at different angles for floating or troweling. On other machines, rotors, or blade sets, can be changed for different tasks.

Figure 60 shows a self-propelled, rotary-blade power floating machine. The engine is mounted on the hub of the blade rotor, and the machine is steered from behind. The blades on this machine can also be set at different angles for troweling.

104F60.TIF

Figure 60. Small Power Floating Machine

Larger power floats and trowels may have two or three sets of rotor blades, and a riding seat, as shown in *Figure 61*. As with the smaller equipment, the rotors on the larger power finishing tools can also be adjusted to work as floats or trowels.

CAUTION: It is important to check all rotor blades frequently. All blades must be in the same condition, or the rotor will be thrown off balance. If one blade is bent or broken, all the blades on the rotor must be changed.

104F61.TIF

Figure 61. Riding Power Trowel

Still another kind of power float is the disk or Kelly float. This uses a 24- to 36-inch rotating solid disk to compact and float at the same time. The Kelly float has steel lift hammers that lift and drop as the disk rotates. The Kelly float is used on low-slump or zero-slump concrete, or on normal-slump concrete after the surface has stiffened.

3.17.0 GRINDING AND MILLING MACHINES

Some defects such as low spots and surface cracks can be repaired by grinding or milling. This smooths the damaged area so that it can be resurfaced, if necessary. Most grinding is done with hand-held power tools, as shown in *Figure 62*.

The two types of grinders used on floor slabs are diamond-disk grinders and stone grinders. Diamond-disk grinders work much faster than stone grinders. They are used with water to cool and lubricate the disk.

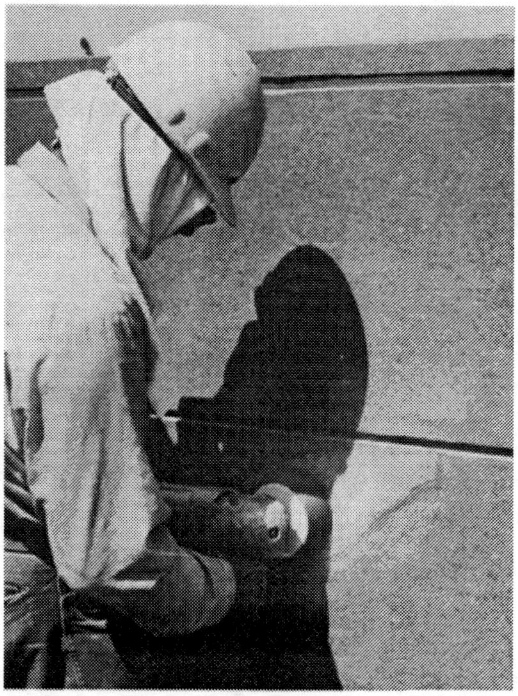

104F62.TIF

Figure 62. Hand-Held Grinder

If the repair must cover a large area, a milling machine can be used, as shown in *Figure 63*. To prevent possible damage, milling machines must be equipped with a device for stopping at a preset depth.

104F63.TIF

Figure 63. Milling Machine

SUMMARY

This module has introduced most of the hand tools, power tools, and equipment used to mix, place, consolidate, float, and finish concrete. The tools included can be used for jobs ranging from patching a walkway to building a high-rise industrial building or a highway. The experienced concrete finisher should know which tool to use for each job and how to use it properly. This would include its proper care and maintenance.

Safety is an important part of using any piece of equipment. Know the limits of the tools you use. Do not use any tool above its operating limits. This may cause damage to the tool or equipment as well as injury to the operator. If there is a problem with a piece of equipment it should be tagged out and returned to the shop for inspection and repairs. Unless you have been trained to repair the equipment, do not attempt to remove any guards or safety plates.

References

For advanced study of topics covered in this task module, the following references are suggested:

Concrete Construction and Estimating, Craig Avery, Craftsman Book Company, Carlsbad, CA, 1991.

Concrete Manual, International Conference of Building Officials, Whittier, CA, 1992.

Concrete Parking Lots: Eight Steps to Quality Construction, Videotape, Portland Cement Association, Skokie, IL

Concrete Technology, Third Edition, George R. White, Delmar Publishers, Inc., Albany, NY, 1991.

Design and Control of Concrete Mixtures, Thirteenth Edition, Portland Cement Association, Skokie, IL, 1988.

Finishing Concrete Flatwork, Videotape 32:00 minutes, Portland Cement Association, Skokie, IL, 1991.

Guide for Concrete Floor and Slab Construction, ACI 302.1R-96, American Concrete Institute, Farmington Hills, MI, 1997.

The Contractor's Guide to Quality Construction, American Concrete Institute, Farmington Hills, MI, 1997.

ACKNOWLEDGMENTS

Figures 1, 2, 15, 17, 21-23, 25, 27, 29, 31, 42, 43, and 61 courtesy of Associated General Contractors
Figures 3, 33, 39, and 48 courtesy of Roy Jorgensen Associates, Inc.
Figure 4 courtesy of McGraw-Hill Publishing
Figures 5-9, 14, 18-20, 26, 28, 30, 32, 40, 47, 58, and 60 courtesy of Baker Concrete Construction
Figures 10-13, 24, and 25 courtesy of Alberta Ready Mixed Concrete Association
Figures 16, 35, 38, 41, 45, 51-53, and 55-57 courtesy of American Concrete Institute
Figure 28 courtesy of National Association of Homebuilders
Figure 34 courtesy of International Road Federation
Figures 36, 37, 46, 59, and 62 courtesy of National Conference of Building Officials
Figure 44 courtesy of National Crane
Figures 49 and 50 courtesy of SCHWING America, Inc.
Figure 54 courtesy of SOMERO Enterprise, Inc.
Figure 63 courtesy of Federal Highway Administration

SI* (MODERN METRIC) CONVERSION FACTORS

APPROXIMATE CONVERSIONS TO SI UNITS

Symbol	When You Know	Multiply By	To Find	Symbol
		LENGTH		
in	inches	25.4	millimeters	mm
ft	feet	0.305	meters	m
yd	yards	0.914	meters	m
mi	miles	1.61	kilometers	km
		AREA		
in^2	square inches	645.2	square millimeters	mm^2
ft^2	square feet	0.093	square meters	m^2
yd^2	square yards	0.836	square meters	m^2
ac	acres	0.405	hectares	ha
mi^2	square miles	2.59	square kilometers	km^2
		VOLUME		
fl oz	fluid ounces	29.57	milliliters	mL
gal	gallons	3.785	liters	L
ft^3	cubic feet	0.028	cubic meters	m^3
yd^3	cubic yards	0.765	cubic meters	m^3
NOTE: Volumes greater than 1000 l should be shown in m^3.				
		MASS		
oz	ounces	28.35	grams	g
lb	pounds	0.454	kilograms	kg
T	short tons (2000 lb)	0.907	megagrams (or "metric ton")	Mg (or "t")
		TEMPERATURE (exact)		
°F	Fahrenheit temperature	5(F-32)/9 or (F-32)/1.8	Celsius temperature	°C
		FORCE AND PRESSURE OR STRESS		
lbf	poundforce	4.45	newtons	N
lbf/in^2	poundforce per square inch	6.89	kilopascals	kPa

APPROXIMATE CONVERSIONS FROM SI UNITS

Symbol	When You Know	Multiply By	To Find	Symbol
		LENGTH		
mm	millimeters	0.039	inches	in
m	meters	3.28	feet	ft
m	meters	1.09	yards	yd
km	kilometers	0.621	miles	mi
		AREA		
mm^2	square millimeters	0.0016	square inches	in^2
m^2	square meters	10.764	square feet	ft^2
m^2	square meters	1.195	square yards	yd^2
ha	hectares	2.47	acres	ac
km^2	square kilometers	0.386	square miles	mi^2
		VOLUME		
mL	milliliters	0.034	fluid ounces	fl oz
L	liters	0.264	gallons	gal
m^3	cubic meters	35.71	cubic feet	ft^3
m^3	cubic meters	1.307	cubic yards	yd^3
		MASS		
g	grams	0.035	ounces	oz
kg	kilograms	2.202	pounds	lb
Mg (or "t")	megagrams (or "metric ton")	1.103	short tons (2000 lb)	T
		TEMPERATURE (exact)		
°C	Celsius temperature	1.8C + 32	Fahrenheit temperature	°F
		FORCE AND PRESSURE OR STRESS		
N	newtons	0.225	poundforce	lbf
kPa	kilopascals	0.145	poundforce per square inch	lbf/in^2

* SI is the symbol for the International System of Units. Appropriate rounding should be made to comply with Section ASTM E380.

REVIEW/PRACTICE QUESTIONS

1. A highway straightedge is not used _____.
   a. where flatness is critical
   b. instead of a bullfloat
   c. on highways
   d. instead of a rubber float

2. A shovel is used to _____.
   a. carry and spread small amounts of concrete
   b. remove excess mix by dragging it across the surface
   c. consolidate concrete around the edges of a slab form
   d. score and groove contraction joints

3. Placing and finishing hand tools are used to _____.
   a. screed, consolidate, and segregate cement
   b. consolidate, finish, and groove concrete
   c. aggregate, convey, and compact cement
   d. hoist, edge, and groove concrete

4. The walking trowel is different from a bullfloat because it has a _____.
   a. locking adapter to set the angle of the blade
   b. connection to a water hose
   c. swivel to support the flight ends
   d. wide, flat blade

5. Brushes are used in tight places _____.
   a. instead of margin trowels
   b. to seal the surface of a slab
   c. where brooms would not fit
   d. instead of elephant trunks

6. The come-along is sometimes combined with a _____.
   a. bullfloat
   b. fresno tamper
   c. tremie
   d. concrete rake

7. A small, self-propelling cart with a stand for the operator is called a _____.
   a. tremie
   b. bucket
   c. buggy
   d. flight

8. Ready-mix trucks _____.

   a. are truck-mounted batching plants that can move around a site
   b. have rotating blades or paddles on a vertical shaft
   c. have continuous belts powered by chain or hydraulic drives
   d. mix at a high speed and agitate at a lower speed, to keep concrete from setting in transit

9. Continuous-feed pumps and their pipelines can deliver concrete up to _____.

   a. 100 feet vertically and 3,000 feet horizontally
   b. 300 feet vertically and 1,000 feet horizontally
   c. 3,000 feet vertically and 1,000 feet horizontally
   d. 100 feet vertically and 300 feet horizontally

10. Pair edgers have _____.

    a. pole-type handles
    b. a ¾-inch radius at the lips
    c. pointed bits under the plates
    d. the same curve radius at the lips

11. Wheelbarrows usually need to be _____.

    a. washed after every load to prevent contamination
    b. filled from a gravity feed hopper to prevent segregation
    c. run on walkways to prevent segregation
    d. wetted down before filling to prevent contamination

12. Safety features such as brakes, rails, and hand controls are found on _____.

    a. materials hoists
    b. subgrade compactors
    c. personnel hoists
    d. nontilting mixers

13. An exposed aggregate surface is made by _____.

    a. removing mortar with a stiff-bristled push broom and a water hose
    b. deeply embedding aggregates with a jitterbug
    c. sealing the surface after bleed water has dried
    d. pushing a vibrating screed across the form

14. The boom on a tower crane can move in an arc of _____.

    a. 290 degrees
    b. 150 degrees
    c. 95 degrees
    d. 360 degrees

15. Flights, troughed belts, swivels, and chain drives make up a _____.

a. pumping system
b. motorized vibrating screed
c. conveyor
d. lay-down bucket

16. A bullfloat is moved across the surface _____.

a. flat
b. at an angle
c. at right angles
d. after bleed water rises

17. The surface-vibrating screed vibrates and _____.

a. controls the speed of deposit on the surface
b. seals the surface with a textured finish
c. moves forward to strike the surface to a predetermined grade
d. pushes loose aggregates in front of it

18. The finishing trowel is a small _____.

a. steel float with a hand-grip handle
b. stiff trowel with a pointed tip
c. stiff trowel used for patching
d. windowsill edger

19. The subgrade compactor is used to _____.

a. drive air voids out of deposited concrete
b. press large aggregates below the surface
c. evaporate bleed water
d. firm up the soil under a form

20. Edgers are used to _____ of slabs.

a. round the edges
b. joint the centers
c. square off the edges
d. darby the centers

ANSWERS TO REVIEW/PRACTICE QUESTIONS

Answer		Section Reference
1.	d	2.4.3
2.	a	2.5.0
3.	b	2.0.0
4.	a	2.9.2
5.	c	2.12.0
6.	d	2.7.0
7.	c	3.6.3
8.	d	3.2.0
9.	d	3.11.0
10.	d	2.11.3
11.	c	3.6.1
12.	c	3.7.0
13.	a	2.13.3
14.	d	3.8.1
15.	c	3.10.0
16.	a	2.4.1
17.	c	3.12.3
18.	a	2.9.1
19.	d	3.14.0
20.	a	2.11.0

NCCER CRAFT TRAINING USER UPDATES

The NCCER makes every effort to keep these manuals up-to-date and free of technical errors. We appreciate your help in this process. If you have an idea for improving this manual, or if you find an error, a typographical mistake, or an inaccuracy in the NCCER's Craft Training Manuals, please write us, using this form or a photocopy. Be sure to include the exact module number, page number, a description of the problem, and the correction, if possible. Your input will be brought to the attention of the Technical Review Committee. Thank you for your assistance.

Instructors – If you found that additional materials were necessary in order to teach this module effectively, please let us know so that we may include them in the Equipment/Materials list in the Instructor's Guide.

Write: Curriculum and Revision Department
National Center for Construction Education and Research
P.O. Box 141104
Gainesville, FL 32614-1104

Fax: 352-334-0932

Craft ______________ Module Name ______________

Module Number ______________ Page Number(s) ______________

Description of Problem

____

____

____

____

(Optional) Correction of Problem

____

____

____

(Optional) Your Name and Address

____

____

____

notes

Preparing for Placement

Module 23105

Concrete Finishing Trainee Task Module 23105

PREPARING FOR PLACEMENT

NATIONAL
CENTER FOR
CONSTRUCTION
EDUCATION AND
RESEARCH

OBJECTIVES

Upon completion of this module, the trainee will be able to:

1. Describe basic site layout using levels and measuring tools.
2. Properly locate, grade, and build forms for horizontal placement.
3. Perform compaction activities on subgrades.
4. Describe various joints and where to locate them.
5. Describe various reinforcements and how to place them.
6. Describe information needed when ordering concrete.

Prerequisites

Successful completion of the following Task Modules is recommended before beginning study of this Task Module: Core Curricula; Concrete Finishing Level 1, Modules 23101 through 23104.

Required Trainee Materials

1. Trainee Task Module
2. Pencil and paper
3. Appropriate Personal Protective Equipment

COURSE MAP

This course map shows all of the task modules of the Concrete Finishing curricula. The suggested training order begins at the bottom and proceeds up. Skill levels increase as a trainee advances on the course map. The training order may be adjusted by the local Training Program Sponsor.

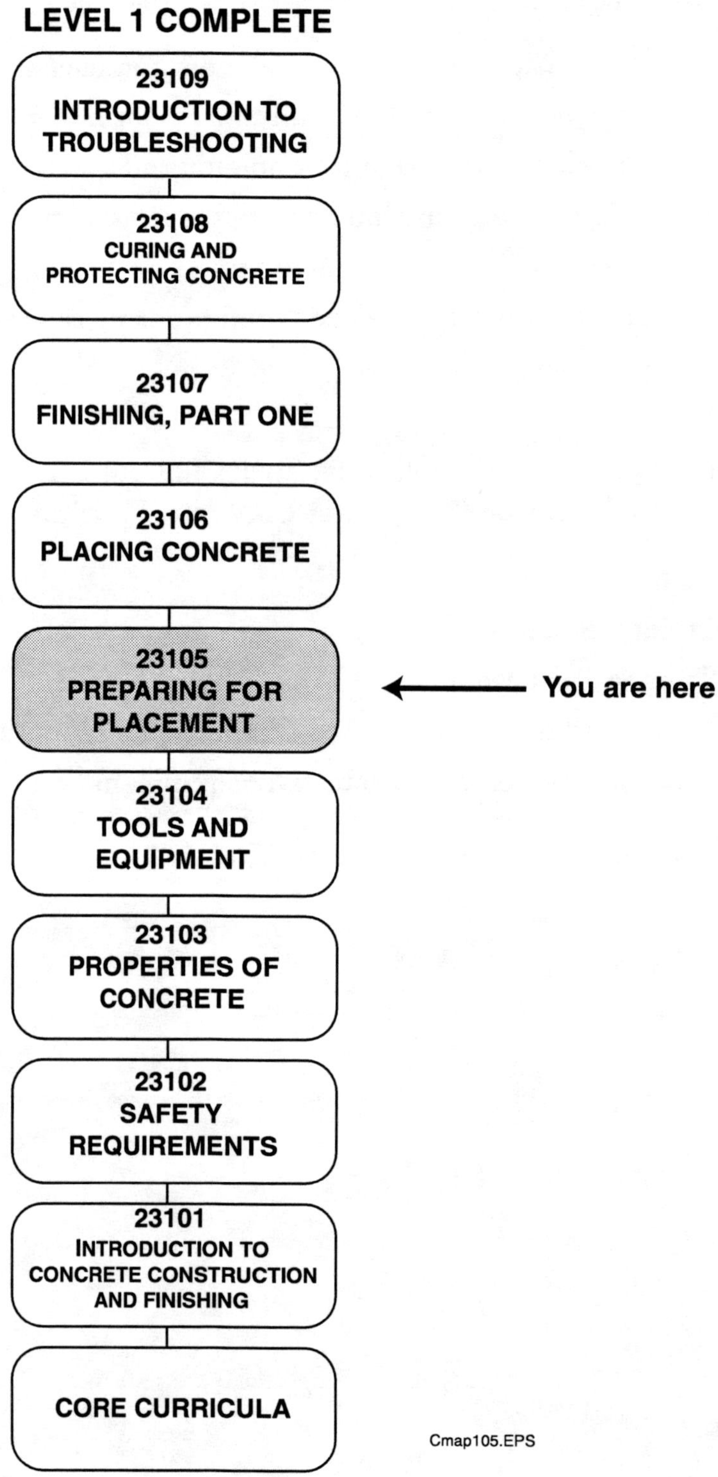

Cmap105.EPS

TABLE OF CONTENTS

TABLE OF CONTENTS (Continued)

Trade Terms Introduced In This Module

Angle of repose: The angle at which a soil will assume a mound and stand without reinforcement.

Backfill: Soil or other material that is compacted into place to correct for over-excavation.

Bulkhead: A partition inserted in either a concrete placement or the formwork to prevent fresh concrete from flowing into another section; a reusable wood or metal form for a construction joint, usually with a keyway.

Clay: A fine, water-retaining soil with plastic properties when moist. Some soils contain clay mixed with other ingredients.

Construction joint: The surface where two successive placements of concrete meet but do not bond; frequently, either keyways or dowels are used as reinforcements across the joint.

Contraction joint: A purposefully formed saw cut or tooled groove in the concrete structure that regulates the location of cracking.

Expansive soil: Soil that is subject to increase in volume with resulting uplift or distortion of concrete members. This is a severe problem in some areas of the United States.

Footing: The part of the foundation placed on grade that transfers the building load to the soil.

Form: A mold of lumber or prefabricated material set up to shape, support, and contain concrete until it has gained sufficient strength to be self-supporting.

Geotextile: A natural or synthetic fabric used to reinforce soils or separate different types of soils or aggregates.

Grade: A prepared surface on which concrete is placed. To prepare a surface to a predetermined elevation.

Honeycomb: Voids left in concrete where cement and sand particles have not filled the spaces between the coarse aggregate particles.

Isolation (expansion) joint: A built-in separation between adjoining elements of a concrete structure to prevent movements of any individual part from causing cracks in the other parts; the location where the bond is interrupted.

Keyway: The part of a bulkhead that shapes a recess or groove in one placement of concrete that is later filled with concrete of the new placement so that the two lock together in a tongue-and-groove configuration.

Lift: A horizontal layer of concrete placed in a form. In a horizontal form, two lifts can be placed to go over and under reinforcement. In a vertical form, a lift is approximately 4 feet high, placed at one time. Large vertical forms are filled by several lifts, depending on height and the ability of formwork to withstand the pressures of the fresh concrete.

Ready-mix concrete: Concrete batched in a concrete plant and mixed in the plant or in a truck mixer that delivers the concrete in a plastic, unhardened state.

Reinforcement: Bars, wires, or strands embedded in concrete in such a manner that the reinforcement and concrete act together to resist forces on the concrete.

Settlement: A change in elevation at times caused by the weight of a structure compressing the supporting soil or subgrade.

Stress: The pressure or load of the internal forces within concrete. Major stresses on concrete are tension, compression, torsion, and shear. Stress acts to deform concrete. Stress is expressed mathematically in terms of force per unit area.

Subbase: A layer in a pavement system between the subgrade and the base course or between the subgrade and a portland cement concrete pavement.

Subgrade: A surface earth or fill material compacted and leveled off to receive a deposit of concrete or the foundation of a structure.

Tolerance: The variation permitted from the dimension given, usually expressed as a number or percent. Tolerance is critical, for example, in the location or alignment of a concrete wall.

Wale: A type of form reinforcement used in vertical forms, cross-braced for rigidity. Also known as walers.

Waterstop: A thin sheet of metal, rubber, plastic, asphalt, or other material inserted in a form across a joint to obstruct the flow of water through the joint.

Welded wire fabric: A mesh made of wires crossing at right angles and welded together. The mesh is used as reinforcement in concrete slabs.

1.0.0 INTRODUCTION

This module describes the work involved in preparing for placing concrete. It includes excavating the site, building foundations, erecting forms, tieing and placing **reinforcement**, site cleanup, and ordering the concrete. This preparatory work is of vital importance and if done properly, makes placing and finishing much easier. It also improves the final quality of the job.

1.1.0 PURPOSE

The purpose of this module is to provide detailed information about the equipment, materials, and activities necessary to prepare a site for placing concrete. In this module, **subgrade** preparation, formwork, reinforcing, joint spacing, embedded fixtures, and placing orders for concrete are discussed. The concrete worker will use the information in this module to get the job started correctly, while avoiding mistakes that could cause problems during placing and finishing.

Much of this material is presented for information only. On a large project, the concrete craftworker may not be on site until much of this work is completed and the ready-mix truck is on its way. On a small project, you may be called to help with some or all of the work, depending on the job.

1.2.0 SAFETY

Good safety practices during this stage call for handling tools carefully and staying clear of any earth-moving equipment. Now is the time to familiarize yourself with company policies and practices about reporting accidents and injuries, marking equipment runways, cleaning up the site, checking work, and wearing Appropriate Personal Protective Equipment. A hard hat and steel-toed work boots are an important part of a craftworkers wardrobe.

Staying aware of the work around you is a habit that should begin now. For a large project, there will be many different tradespeople and craftworkers present, all busy. Every craftworker in every craft is responsible for being aware of what is happening in the immediate area and acting safely in response to it.

Excavating below **grade** may present a safety hazard. If the excavation is over five feet deep, the sides must be shored or the surrounding soil cut back on a slope to prevent cave-ins.

CAUTION: Be alert for any water or soil shifts in excavated areas. They may signal collapse of the excavation. Brace all excavation sides deeper than five feet.

Attention to safety is particularly important in formwork construction. Formwork must support the weight of fresh concrete, so it must be strong, stable, and reliable. Poorly constructed forms can collapse, injuring or killing nearby workers.

2.0.0 GRADE AND SUBGRADE PREPARATION

For large projects, surveyors will establish reference points from which the dimensions and elevations of the building components can be established. On smaller projects, there may be no building plans, or the plans may not specify a finished floor elevation. For these jobs, elevations must be established from nearby buildings or from the site grading and drainage work.

Construction of any slab on grade, foundation, or **footing** begins with subgrade preparation. This work might be as uncomplicated as removing vegetation and topsoil, or as complicated as rock blasting and excavation. After the site is cleared and leveled, the subgrade is prepared. A well-prepared subgrade provides uniform support to the structure. A poorly prepared subgrade can lead to differential **settlement**. Many soils will compact under a heavy load, such as the weight of a building. Differential settlement involves the uneven movement of parts of the footing, foundation, or slab. This often overstresses the concrete, and cracks result. In some cases, the building or wall will lean. The most famous example of differential settlement is the Leaning Tower of Pisa in Italy, as shown in *Figure 1*. Originally built vertically plumb, the tower now leans at an angle of approximately 12 degrees.

105F01.TIF

Figure 1. The Leaning Tower Of Pisa

While most soils consolidate and settle under load, there are soils that will expand when wet, even under load. This swelling and shrinkage can be destructive to footings, walls, and floors if they are not designed to withstand this process. The pressure from the **expansive soil** can cause walls to move, slabs to rise, and footings to be displaced.

2.1.0 PROPER SUBGRADE CONDITIONS

Subgrade conditions can be highly variable due to soil type, density, the **angle of repose**, and physical properties, plus the presence of rock and water. Proper subgrade conditions will depend not only on the site conditions, but also on the requirements of the concrete structure that will be supported by the subgrade. The objective is to provide uniform support for the structure.

A large construction site will be surveyed, mapped, and planned before any construction begins. Soils will be tested. Reference points, benchmarks, and elevations for all subgrade work will be established by a survey crew. On small sites, especially in rural areas, the excavation or concrete crew may be expected to provide site layout for their own work.

After the site is laid out, the first task is to remove all vegetation, large rocks, and topsoil from the construction area. Topsoil has a large amount of organic material that is compressible, so it cannot be used as **backfill**. It should be stored out of the way so that it does not interfere with operations or mix with other excavated material. Next, the site is excavated to the established line and grade. If concrete is to be placed directly on the ground, do not overexcavate, as undisturbed soil has better support capacity than soil that has been disturbed. However, some soil may have to be removed and replaced to get uniform support. For example, soft or mucky spots must be excavated, backfilled with good soil, and compacted. Rocks must be removed and the holes backfilled and compacted. Spots that are considerably firmer than the rest of the excavation should be loosened and recompacted. The objective is to get a base of uniform support. *Figure 2* shows how a slab will react to soft and hard spots that are not removed during subgrade preparation. Because soft spots will not support the slab, it will tend to sink and crack. A hard spot such as a loose boulder may cause upward movement of the slab and force it to crack where the boulder is forcing the slab upward.

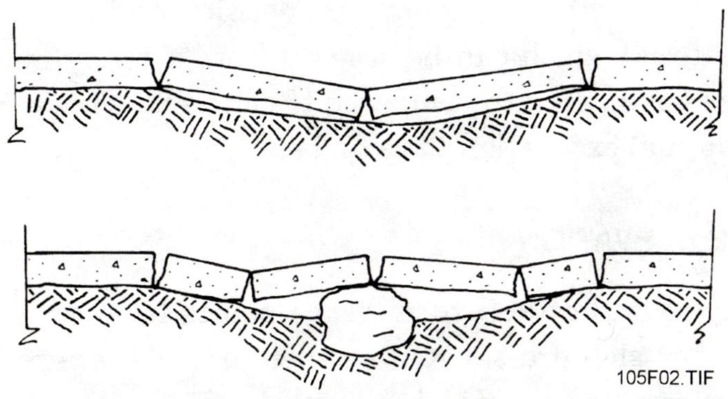

Figure 2. Soft Spots And Hard Spots

2.1.1 Soil Reinforcements

Where problem soils create uneven conditions, improvements must be made. One method is to reinforce the soil with **geotextiles**, engineering fabric, geogrids, or a fiber mesh. These products can substantially increase the load-bearing strength of the soil. Another method is to use a **subbase** of granular material. Granular material for the subbase can be sand, gravel, crushed rock, blast furnace slag, crushed concrete, or a combination of these. Granular material must be trimmable, compactible, and stable.

2.1.2 Flowable Fill

In recent years, a flowable fill or unshrinkable fill has been used in utility cuts and trenches as a backfill. It can also be used as a subbase. *Figure 3* shows flowable fill being placed from a ready-mix truck into a large excavation. The fill is a thick semiliquid that will fill available space quickly.

105F03.TIF

Figure 3. Flowable Fill

Flowable fill is a mixture of fly ash, water, and portland cement. It is a low-strength concrete with a maximum specified strength. The density is specified and controlled. Flowable fill is a standard construction material which can be ordered from a ready-mix plant. It is designed to have a very low compressive strength compared to concrete. This ensures that the fill can be excavated later with normal excavation tools and equipment.

2.2.0 ESTABLISHING ELEVATION

For large projects, surveyors will establish reference points from which the dimensions and elevations of the building or slab are set. On smaller projects where there are no detailed plans or the plans do not require specific elevations, horizontal and vertical control can be established from adjacent buildings, from site drainage, or from other points at the boundaries of the site.

There are several ways to establish dimensions and elevations. If working from a benchmark or other reference point, one corner of the structure is established and the remaining corners are set from that one corner. The benchmark provides a reference elevation for use throughout the project. Elevations can be established using an engineer's level, a laser level, or a hand level and string. Once the main dimensions and elevations are established, other controls such as batter boards can be located at corners and string lines run between them.

2.2.1 Engineer's Level

An engineer's level is used to sight from point to point and establish elevation. The start must be at a point of known elevation, such as a benchmark. The desired elevation points are set in advance. For a foundation slab, for instance, they may mark the corners of the structure.

Figure 4 is a diagram showing the basic concept of setting a level and establishing elevations from point to point. If point B is a known elevation, you can find the elevation of point A and determine how much excavation or filling must be done to bring point A to the same elevation as point B. In this case, you must bring up point A four feet to make it level with point B.

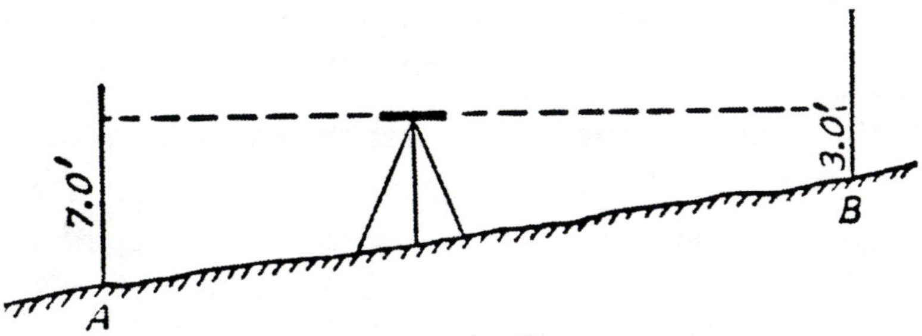

105F04.TIF

Figure 4. Direct Leveling Of Points

To establish the elevation of a point, set the level up between the known point (benchmark) and the unknown point. Make adjustments to the instrument as necessary. Sight on the level rod at the unknown point by looking through the level's telescope and reading the mark on the rod. Rotate the telescope and sight on the level rod held over the unknown point. The difference between the two readings will establish the elevation of the second point.

2.2.2 Laser Level

A laser level is quicker and easier to use than conventional levels because it has many automatic features and only one person is needed to perform the work. To use a laser level, the first step is to locate and level the instrument. Some devices have small spirit levels attached for this purpose, while other devices are self-leveling. Rotate the adjustment wheels at the base of the unit until the level bubbles are centered.

Take the leveling rod with the receiver box to the nearest benchmark. Set the base of the rod on the benchmark. Turn the receiver until it faces the transmitter, then slide the receiver up and down until it intercepts the laser beam. The receiver box will give a steady tone and show a bar when it is on target, as shown in *Figure 5*.

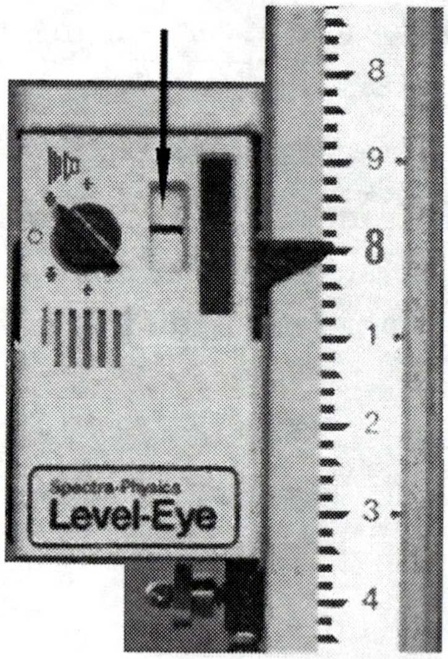

105F05.TIF

Figure 5. Laser Receiver On Target

Once the elevation of the benchmark is known, set the elevation of the benchmark to the beam height on the leveling rod. This is the elevation of the beam and can be used on all grade shots from that point. Most rods are marked in feet and tenths of a foot; some are marked in meters. The tables in Appendix A will allow you to convert these measurements from U.S. Customary to metric measures, and back.

2.2.3 Batter Boards

Batter boards are temporary wooden stakes and horizontal boards erected 3 to 4 feet outside the edge of a structure to mark corners. The horizontal boards are set at a level reference elevation marked to locate the outside dimensions of the foundation or the floor slab. Nylon string is then stretched taut between the batter boards of each corner to indicate the foundation line. The strings also establish a reference elevation from which vertical measurements can be made. To excavate, the strings can be removed and reset after the excavation is finished. *Figure 6* shows one set of batter boards erected at one corner of an excavation. Note the string lines and plumb bob used to obtain the corner point at the bottom of the excavation. In this case, it locates the outside corner of the footing.

Forms are placed using the string lines as a guide. If the slab dimensions are greater than the length of the screed available, then intermediate elevation points must be established inside the slab. This is accomplished by using a laser level or hand level to transfer elevations from the forms to the top of a stake pounded into the subgrade. These stakes are pulled up as the concrete is leveled by the screed.

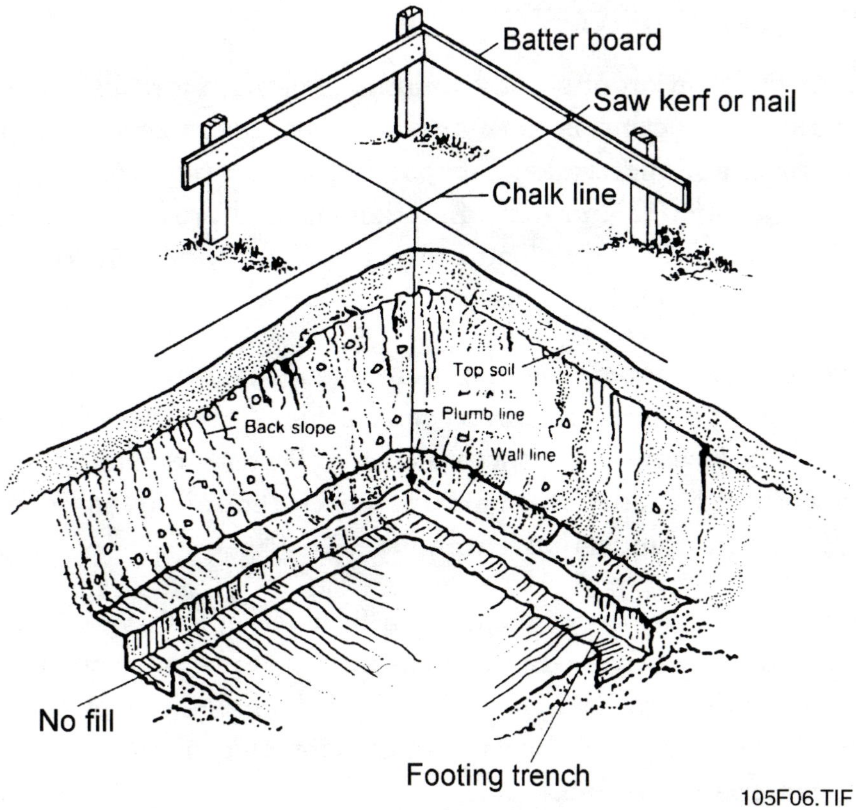

Figure 6. Batter Board

2.2.4 Drainage And Slope Requirements

Positive drainage should always be present on exterior concrete work. In the absence of other specifications, a slope of 1 inch vertically in 4 feet horizontally is adequate. The slope can apply to the longitudinal dimension or the transverse dimension, or to both. For example, sidewalks normally follow the ground contours and may or may not have longitudinal slopes. They should always have transverse slopes that drain the water off to the side of the sidewalk. Sidewalk and ramp slopes should not exceed 5 percent grades, to conform to the access regulations of the Americans with Disabilities Act.

Another requirement involves the slope of the sides of any excavation. In some cases, the foundation must be excavated below the surface of the ground, or below grade, in order to establish proper building elevation. When excavating, the sides of the excavation must be braced or the soil cut back on a slope. If sloping is done, it must be in accordance with OSHA requirements. These requirements are based on the type of soil and the depth of the excavation. Their purpose is to prevent the sides of an excavation from caving in and burying workers.

CAUTION: Keep alert for any movement in the soil when working in excavations and trenches.

2.3.0 COMPACTING REQUIREMENTS

Compacting is the consolidation of soil by tamping, rolling, vibrating, or the application of weight. Some form of compacting is normally required before construction of foundations, slabs, floors, driveways, and sidewalks. Compacting is also required for roads and parking areas, as it reduces the amount and rate of settlement and provides uniform support to the structure above. It also improves the load-bearing capacity, and it increases the stability of the soil by decreasing the amount of voids.

After the site has been excavated, the subgrade should be compacted with hand tampers, portable mechanized compactors, or some type of roller. Any base materials should also be thoroughly compacted before placing concrete on top. When compacting a subgrade or base, it is important to work in a methodical pattern. Check that all areas have been compacted. Recheck all elevations after compaction is complete.

Specific compaction requirements are basically a function of soil type. Two broad categories are defined: granular and cohesive. Granular material includes soils such as sand and gravel that do not cohere or stick together. Vibrating compaction equipment is needed to properly compact this type of soil. *Figure 7* shows a small vibrating plate compactor being used to compact subgrade material for a slab.

105F07.tif

Figure 7. Small Vibrating Plate Compactor

Cohesive soils are mostly composed of silt, **clay**, or other materials that hold water. Tamping and rolling equipment works best for this material. *Figure 8* shows a padded roller used for compacting cohesive soils.

The moisture content of the soil being compacted is important because the moisture acts as a lubricant on the soil particles, allowing them to move closer together. For each soil type there is an optimum moisture content, the point at which the greatest amount of density can be obtained. The optimum moisture content for cohesive soils may be in the range of 5 to 8

105F08.tif

Figure 8. Padded Roller

percent, while granular soils may be in the 10 to 15 percent range. Tests to determine optimum moisture content are normally conducted in a soils laboratory. Because of the criticality of the moisture content, compaction should not be done in the rain or snow.

In fill areas, the thickness of the fill is also important. The thinner the layer, the better the compaction. Depending on the compaction equipment being used, fill thicknesses should always be uniform and installed per specifications.

2.4.0 GROUNDWATER

Whenever construction requires placing concrete below natural grade, the influence of groundwater needs to be considered. The level of groundwater will vary greatly between locations and may also rise and fall with the seasons.

Groundwater close to the surface can hinder subgrade and foundation preparation. In these cases, the groundwater may need to be removed so that construction can proceed in dry conditions. Dewatering is accomplished by drainage or pumping. Drainage involves cutting a channel or installing pipe to drain the water to a lower elevation. Pumps can also be used to lift the water vertically and move it via pipes or hoses to a storage area or drainage channel. Sometimes it is more productive to lower the water table across the whole building site. This is accomplished by sinking a series of well points and extracting the water with pumps. Well point systems use equipment similar to water wells.

After construction, groundwater must be kept away from foundation walls as they are not normally strong enough to withstand this pressure. Drain pipe can be installed outside the foundation wall at its base to drain the water away. *Figure 9* shows the location of the drain pipe as well as any geotextile membrane that may be installed. Make sure the pipe is placed lower than the bottom elevation of the floor.

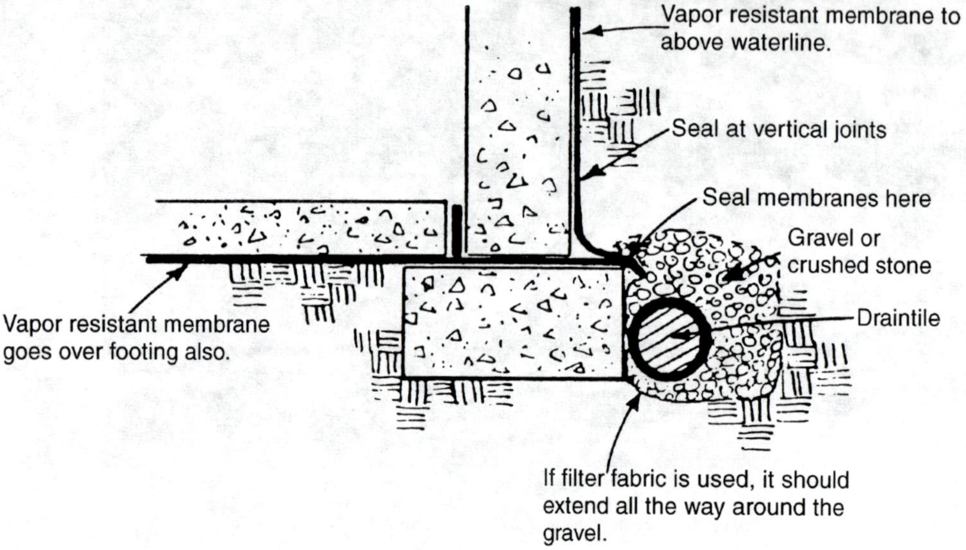

Figure 9. Installation Of Geotextiles And Drainage Pipe

As shown in *Figure 9*, the foundation wall is waterproofed and the perforated drainage pipe is laid at its base. The pipe must be sloped to drain the water. Perforations are necessary in the top of the pipe to allow the water to drain into it. The area around the pipe is backfilled with gravel that is large enough not to get into the pipe and clog it. The gravel acts as a free-draining backfill. Under conditions where the soil backfill erodes, it may be necessary to wrap the drainage pipe with a geotextile so that the soil particles do not pass into the pipe. This provides for clean drainage that can be connected to a storm drain.

2.5.0 FOUNDATIONS

The foundation is the basic support structure for any construction work. It provides a solid and stable platform for the rest of the structure and transfers the load (weight) of all materials above it to the ground. It is important that the foundation be built strong and true. If not, the structure will perform poorly and could even fail.

2.5.1 Foundation Types

The two main classifications for foundations are deep and shallow bearing foundations. The type of foundation to be constructed depends on the structure and the loads it will impose on the soil. Deep foundations consist of piles or caissons and are used for heavy loads. Shallow foundations consist of footings for walls and columns and are used for lighter loads. *Figure 10* illustrates five basic types of footing, seen from above and in side view.

Wall footings (*Figure 10a*) consist of a continuous slab strip at the bottom of a wall with a width larger than the wall thickness. Bearing-wall footings are usually reinforced with steel bars. The footing reinforcement perpendicular to the length of the wall carries tension, while the reinforcement parallel to the length of the wall prevents shrinkage cracking.

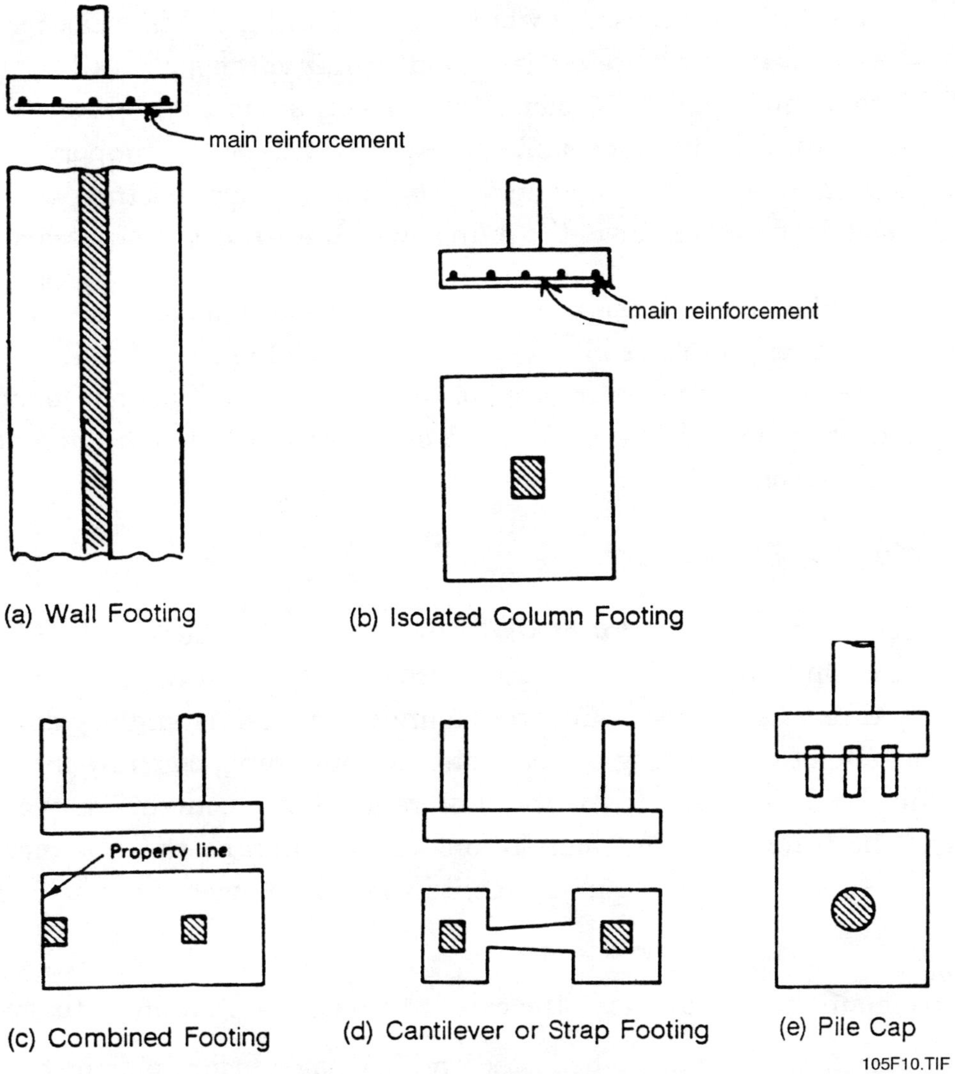

Figure 10. Basic Foundation Types

Independent isolated column footings (*Figure 10b*) are the most common type of spread footing. They consist of a rectangular or square slab of constant thickness. Reinforcing bars are placed near the bottom of the footing and run in both directions. These footing slabs are used to support columns of various material placed on top.

Combined footings (*Figure 10c*) support two or more columns. It is sometimes necessary to support a column within the building line on or near the edge of a property line. The nearest interior column is usually selected, and a combined footing is constructed under both columns. Combined footings are also used when the distance between the adjoining column is relatively small and it becomes more economical to build a combined footing. Cantilever or strap footings (*Figure 10d*) look similar to combined footings. These footings for exterior and interior columns are built independently and joined by a strap or tie beam.

Pile footings (*Figure 10e*) are necessary when the supporting soil is weak for a considerable depth. Wood, steel, or precast concrete piles are driven into the soil until they reach bedrock or a layer of soil that can support the load. Sometimes there is no good support layer and the pile must be driven until it develops sufficient surface friction to support the load. Surface friction is the friction between the pile and the soil pushing against the pile. For this type of foundation, the slab is placed on top of the pilings which provide the necessary support.

Another type of foundation is the raft or floating foundation. This type may be used when the allowable bearing capacity of the soil is very low and piles are not feasible. The foundation substructure is spread under the entire area of the building. Raft or mat foundations may be flat slabs of plain or reinforced concrete, beams or girders with a slab underneath, or beams and girders with a slab on top.

2.5.2 Constructing Foundations

On large projects, concrete foundations will be constructed before the concrete finisher arrives on the site. On small projects, craftworkers may be called to construct foundations, as well as the structures they support. Concrete foundation work is slightly different from that needed for slabs and other building components. For one thing, concrete in foundations does not need finishing, only leveling off. Another difference is that concrete may not need a **form**, only a trench. The **tolerances** for foundations are greater, so there is more leeway (2 to 3 inches) in the size. However, reinforcements must be placed as specified within strict tolerances.

There are three main considerations in the construction of foundation footings:

- The soil bearing capacity should be tested and inspected before placing concrete.
- The concrete must meet the specified strength.
- The footings should be accurately positioned according to the plans.

Since footings are normally below grade, their appearance is not usually important, as long as quality is not sacrificed.

Concrete can be cast against the earth, depending on soil conditions, or in formwork, as shown in *Figure 11*. Old or used material may be used for footing forms provided it is sound. If concrete is cast against the earth, keep material from falling into it while being placed and cured.

3.0.0 FORMWORK

A form is a temporary structure or mold that shapes and supports concrete while it is setting, hardening, and gaining strength to be self-supporting. Formwork is more than just the form. It is the total system of support and molding for freshly placed concrete. It includes the form, form liners, and all supporting members with their hardware and bracing.

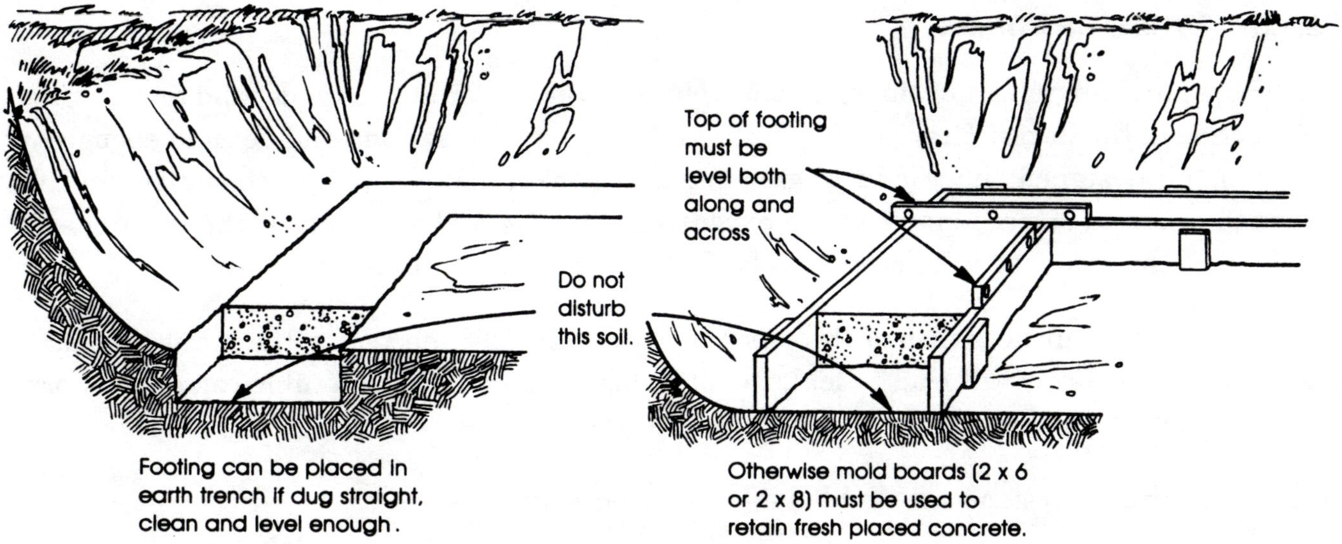

Figure 11. Footing In Trench And Form

As stated in earlier modules, attention to safety is absolutely required in formwork construction. The most effective way to achieve safety is to have good supervision and knowledgeable workers. Most formwork failures can be attributed to human error in design or construction, substandard materials, and poor equipment.

Form safety begins with the planning and management of the project. For projects involving vertical construction, the form must be correctly designed and strong enough to handle the expected loads. This requires following local code requirements as well as OSHA regulations. Any formwork on the job must be constructed exactly as designed, following a safe erection procedure so that no members are temporarily overlooked. Some common construction deficiencies that may lead to form failure include:

- Inadequate diagonal bracing of shores
- Insufficient nailing, bolting, or fastening of bracing
- Failure to provide adequate support for lateral pressure on formwork
- Shoring not plumb
- Locking devices on metal shorings not locked, inoperative, or missing
- Form damage from embankment failure (for below grade construction)
- Inadequate bearing under mudsills
- Failure to comply with manufacturer's recommendations for use and erection

The contractor should have a program of field safety inspection of formwork as the job progresses. All personnel should be aware of proper forming procedures and safety requirements.

3.1.0 TYPES OF FORMS

Size, shape, and alignment of slabs, beams, columns, and other elements depend on accurate construction of the forms. They must be built to the correct dimensions and be rigid enough to maintain the designed shape and alignment. If the form deflects too much, bulges in the hardened concrete surface may require expensive chipping and grinding. If the form moves out of place, the resulting misalignment can destroy the integrity of the structure.

The quality of the surface finish of the concrete is also directly affected by the forms and form material. Sloppy forms will lead to leakage and rough finishes. Considerable corrective work such as grinding, patching, or coating may be required.

Forms and forming systems are divided into three categories.

- *Job-built forms* for one-time use are assembled piece by piece from lumber or other materials. Cutting, nailing, and assembly take place on the job site.
- *Prefabricated job-built forms* are assembled on site and can be reused. Commonly called gang forms, these are usually fabricated in standard sizes. Cutting, nailing, coating, assembly, and storage take place on the job site.
- *Manufactured forms* are purchased or leased from a form manufacturer or supplier. Most steel forms are in this category.

Forms are made from many types of materials. Earth, plywood, lumber, and steel are the most widely used formwork materials. Fiberglass, precast concrete, aluminum, particle board, gypsum board, hardboard, cardboard, rubber, polyvinyl chloride, and polystyrene are also used.

Polystyrene forms are usually supported by wood or steel backing. After the concrete is placed and set, the backing is removed. The polystyrene remains in place as insulation. Fiberglass is used for pans and other void forms. It can be fabricated to fit complex shapes and special applications.

Plywood is widely used for job-built and prefabricated job-built forms. Exterior grade plywood is used because it has waterproof glue. The grade of plywood used will depend on the type and smoothness of the surface required. Lumber or steel is used to support the plywood panels.

Steel is another widely used material for formwork. It is used for all-steel panel systems, framing and bracing of wood and plywood, pans for slab forming, and stay-in-place forms. The latter are left as part of the permanent structure. Steel shoring and structural steel members are also used to frame and support formwork.

3.1.1 Job-Built Single-Use Forms

Job-built single-use forms are built in place by assembling individual components piece by piece. They are almost exclusively made of wood or plywood because these materials can be easily cut to shape. A typical job-built form with components identified is shown in *Figure 12.* Built-in-place forms are used for both large and small jobs. This includes vertical forming as well as most forms for the construction of slabs on grade.

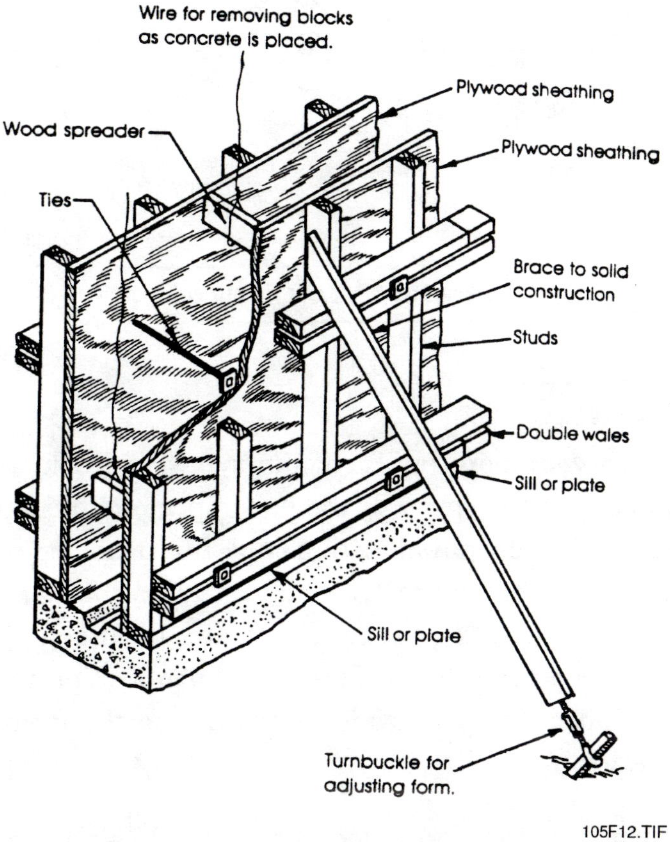

Figure 12. Typical Job-Built Wall Form

Built-in-place forms for vertical construction, as shown in *Figure 12*, are erected over a footing or concrete slab that acts as a platform for the erection. The most common procedure is to fasten a base plate or double base plate to the footing or slab. Wood studs are nailed to the base plate and can be tied together with a temporary ribbon board. The sheathing is nailed to the studs and the **wale** braces are then placed. These pieces can be assembled in advance as a complete panel and then hoisted into place. The inside walls are constructed in the same manner. Spreaders are placed between the inside surface of the forms. Wall ties are inserted through predrilled holes in the sheathing. Finally, the forms are aligned and braced.

Job-built forms for slab or foundation footings are usually single-use and made from dimensional lumber. They are braced by stakes at the corners and along the length. In some cases, they are braced by retaining walls of earth or fill. *Figure 13* shows a job-built form for a low footing wall. Note the braces to the sides of the excavation.

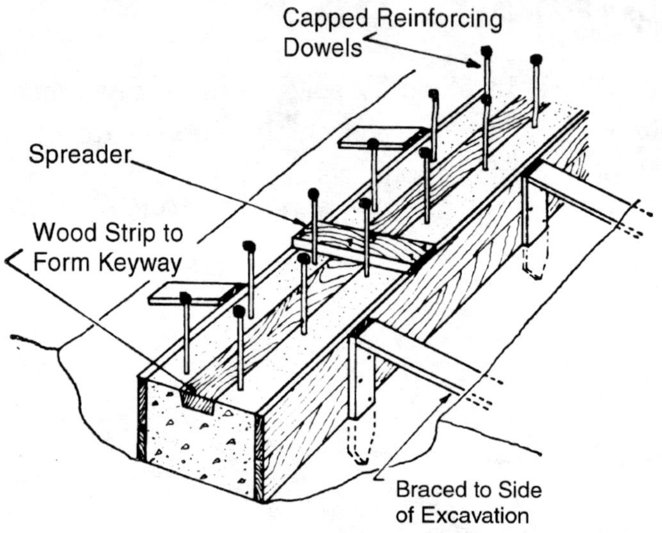

Figure 13. Low Footing Form

3.1.2 Prefabricated Forms

Prefabricated forms are very popular for some types of work because they reduce costs, increase precision, and save time through the use of production-line techniques. Ready-made or contractor-built prefabricated panels are commonly used for wall forming, and for deck forming where multiple floors are being erected. *Figure 14* shows an example of a special type of prefabricated form called a gang form. These standard-size rectangular panels can be placed together to make one large form. The gang form section is lifted into place by crane. When the forms are stripped, the process is reversed, and the forms are moved for use in the next scheduled job location.

Some prefabricated forms of metal and wood are used for slabs on grade. These also include **bulkhead** joints, which are discussed below.

3.1.3 Manufactured Forms

There are a number of form manufacturers specializing in making forms designed to be reused many times. The main types of manufactured forms include pan, internal, panel, column, and stay-in-place. These systems are generally purchased or leased. Each proprietary panel system has its own special ties and other accessories. For some complex systems, workers must be trained and certified in panel construction before they are allowed to work with these forms.

Pan forms are made of metal, fiberglass, or plastic and are used in floor slabs of multistory buildings. The pans block out a portion of the concrete to lighten the dead weight of the floor. The resultant slab is often called a waffle slab because it resembles a waffle on its underside. Internal forms are round or rectangular fiber or cardboard forms that are placed in thick floors to create voids and lighten the dead load.

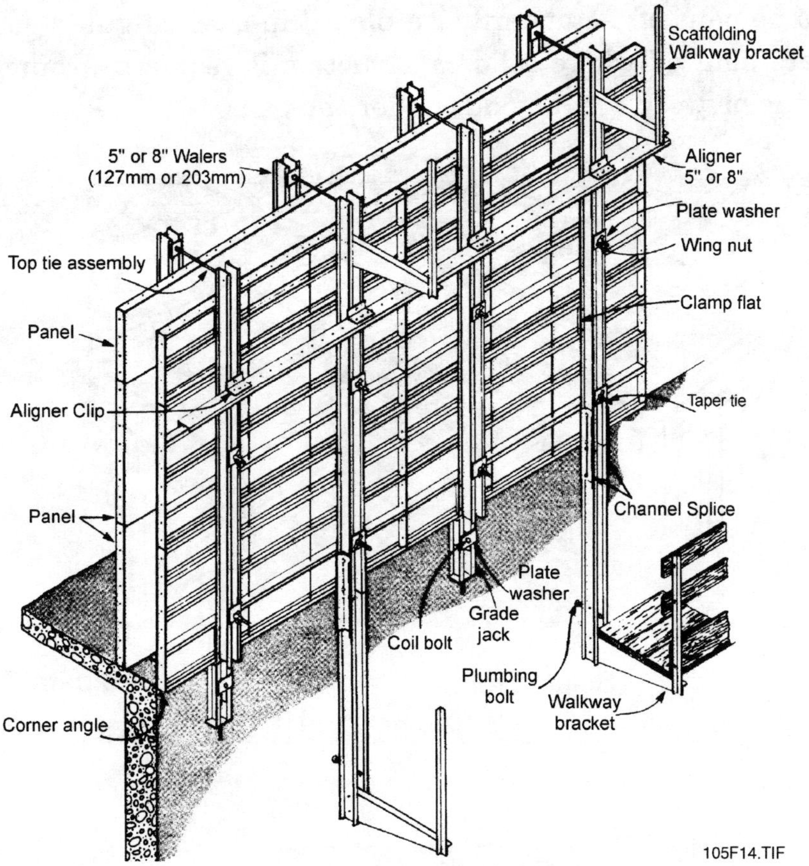

Figure 14. Ganged Forms

Figures 15 and *16* show two examples of pans used on above-grade slabs. The nail-down pans in *Figure 15* are used for joist construction. In this installation there are two different sizes of pan. The end units are tapered, while the pans in the middle are a standard rectangular shape. The alignment of the shapes should be checked before concrete placing begins.

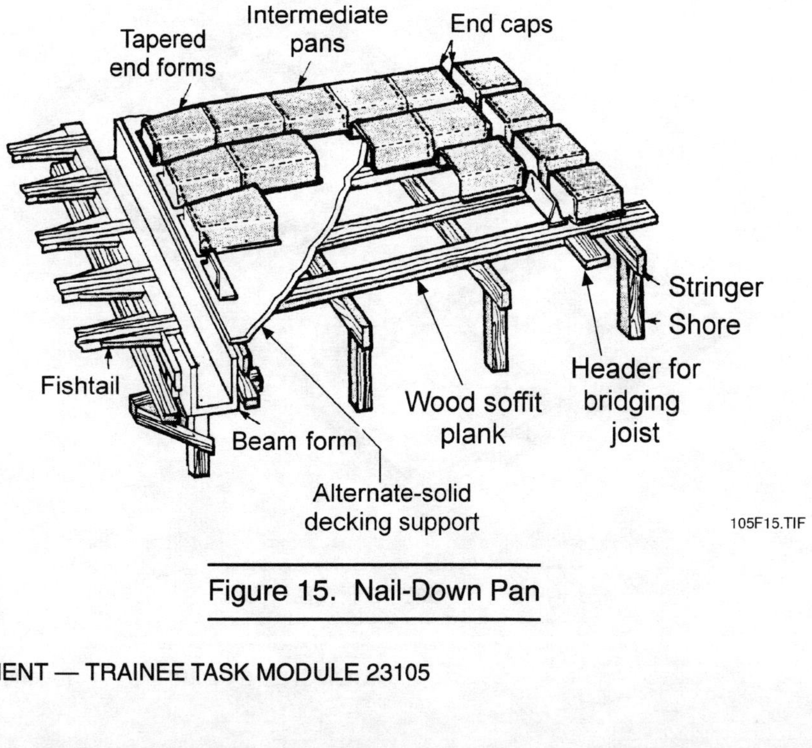

Figure 15. Nail-Down Pan

Figure 16 shows dome pans of a standard size placed in a rectangular pattern to produce a waffle slab. Pans are omitted where solid construction is required around columns. In this figure, you can see the plywood that extends under the pans.

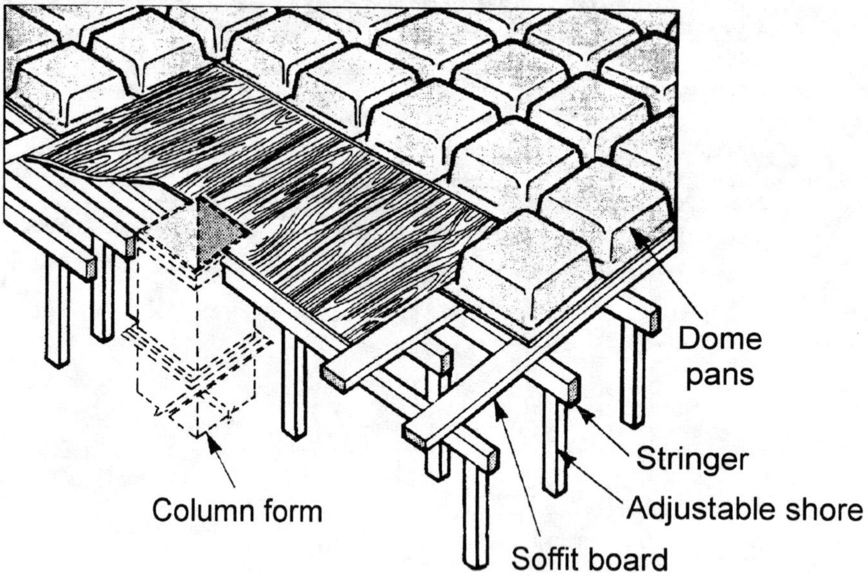

Figure 16. Dome Pan

Panel forms are prefabricated of plywood, aluminum, or steel, and they are fastened together on the job site to form large areas of wall. They offer easy assembly, no need for skilled labor, and many reuses in a variety of large to small assemblies. The ganged forms in *Figure 14* are joined panel forms similar to the single panel form in *Figure 17*.

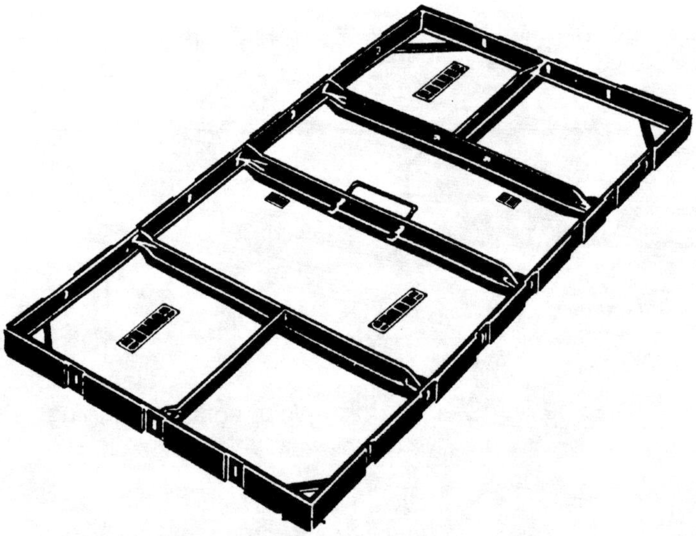

Figure 17. Single Panel Form

Figure 17 shows a metal framed panel with plywood sheathing and cross bracing using metal angles. These panels are fastened together with special metal wedge bolts. The panel edges are notched to accommodate form ties.

Prefabricated panel forms are available in modular sizes from 2 feet × 2 feet to 4 feet × 8 feet. Accessory panels are also available as are small filler and corner units. Square or rectangular column forms are formed from prefabricated panel forms normally used for walls. Round column forms are available in steel, fiberglass, and laminated fiber tubing. Manufactured forms are available in almost any size imaginable.

3.1.4 Prefabricated Forms For Slabs On Grade

Prefabricated forms for construction of slabs on grade are simple compared to formwork for suspended slabs or vertical elements. Usually, these are edge forms placed around the perimeter of the slab. For small jobs, these may be single-use forms made of new or used dimensional lumber. Cross forms, bulkheads, and box-out forms will become isolation joints after the concrete is placed. These are placed around columns and other structural or embedded components that need to be separated from the slab.

Some jobs use a combination of wood and prefabricated metal panel forms, as shown in *Figure 18*. These are economical for large jobs as they can be reused many times. Any form must be stiff enough to take the movement of vibrating and screeding the concrete as it is placed. Note the bracing by stakes which keeps the form rigid so it will not collapse as it is filled.

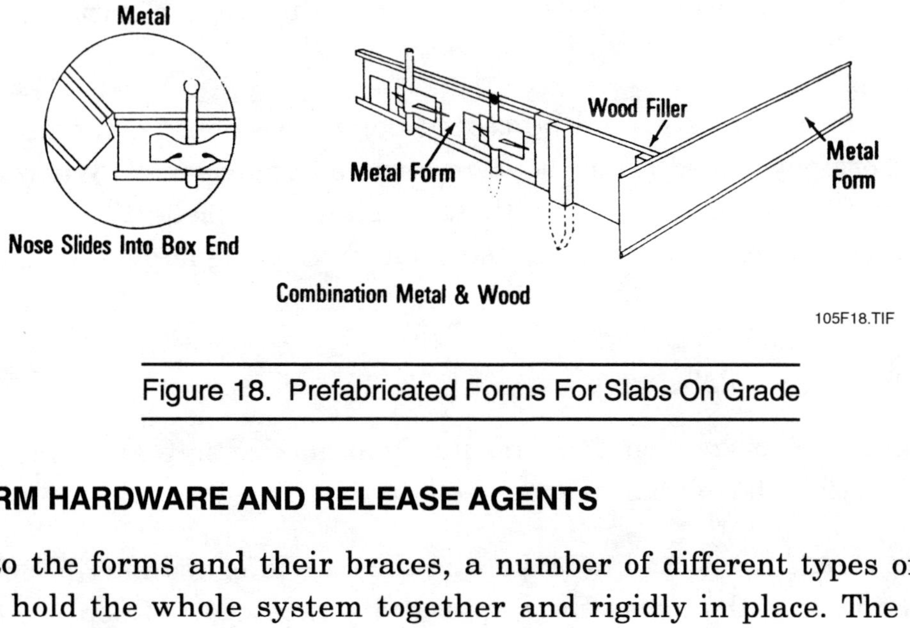

Figure 18. Prefabricated Forms For Slabs On Grade

3.2.0 FORM HARDWARE AND RELEASE AGENTS

In addition to the forms and their braces, a number of different types of hardware are necessary to hold the whole system together and rigidly in place. The major types of hardware are:

- *Ties* keep the form from spreading under the pressure of the concrete mix.
- *Anchors* fasten the form to previously placed concrete.
- *Hangers* attach forms to a structural frame of steel or precast concrete.

- *Spacers* and *chains* hold reinforcing bars the specified distance from the inside surface of the form.
- *Form liners* produce a decorative finish on the concrete surface.

The type and number of accessories and embedments may influence the type of form and hardware used for a specific job. It is necessary to make sure that all the hardware does not interfere with the placing, consolidating, and final finishing of the concrete.

Most forms must be treated with a form oil or form release agent to prevent adhesion of the concrete to the form. The agent should protect the form and prevent absorption of water from the concrete into the form. The agent should not interfere with subsequent curing, painting, plastering, or other surface treatments of the concrete, nor should it stain or discolor the concrete surface.

Form release agents that work on wooden forms may not necessarily work on metal forms. On wood and plywood forms, commercial form oils generally work well. Light-colored petroleum oil and oil emulsions of various types also are used. Other products used on wood and plywood forms include shellac, lacquer, resin-based products, or plastic compounds and epoxy coatings. Compounds used as release agents on metal forms include commercial form oil, paraffin-based oils, and oils blended with synthetic castor oil, silicone, and graphite.

The form surfaces must be clean and smooth. The release agent should be applied evenly by brushing, rolling, spraying, wiping, or dipping, depending on the compound and type of form. The objective is to make the coating complete and uniform while reducing excessive coatings that stain the concrete. On vertical construction, use as thin an application as practical to prevent excessive bugholes.

Whenever possible, form panels should be coated before erection. Forms that are reused continuously are normally coated after they are stripped and cleaned. When coatings are applied after forms are erected, the application should precede the placing of steel reinforcement in order to keep the form release agent from being applied to the reinforcing steel.

3.3.0 SQUARING UP FORMS

Satisfactory alignment of concrete surfaces results from making sure all forms are properly aligned and square within the tolerances specified in the plans.

Vertical wall forms should be checked to make sure they are level and plumb. Spacers placed within the forms will hold the forms the required distance apart. Keep the forms level, plumb, and straight. One of the ways to check for size is by measuring with a steel tape and checking for levelness with a carpenter's level. A plumb bob is also helpful in checking vertical straightness of corners and seams.

On floor or slab forms, squareness may be a problem. A small error in measuring the 90-degree angle in each corner can throw the forms out of alignment. Squareness requires that the diagonal distance between opposite corners be the same. If they are not, then the forms need to be adjusted until the distances are equal. This should be done when the corner points are originally established by survey. Do not wait to adjust the forms after they have been placed.

3.4.0 SETTING FORMS TO LINE AND GRADE

Setting forms to proper line and grade is normally accomplished by using a string line or laser level as explained earlier. When making a level slab or a slab on a slope, use a straightedge and a level to assure accuracy. *Figure 19* shows a grade being set for a slab by leveling up one side of the form.

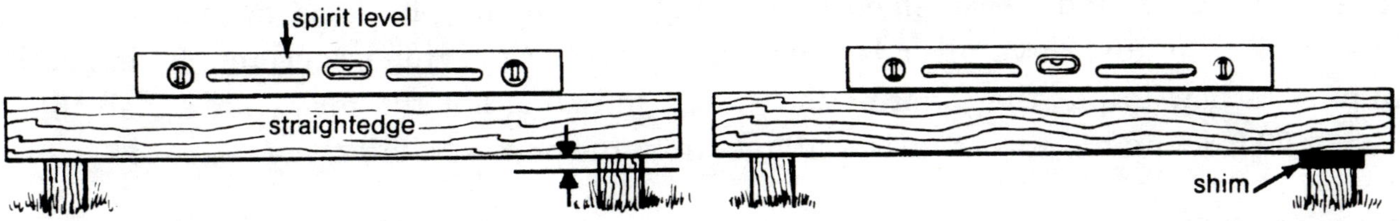

Figure 19. Setting Grade For Slab

The stakes are set level, then one of the stakes is pounded into the ground until the proper slope is set. To continue, level the straightedge again by using a shim, then repeat the process for the remainder of the slab. Once the stakes are set, the form lumber is attached. The forms should be staked and braced to keep them in vertical and horizontal alignment. For ease in placing and finishing, the stakes should be driven or cut off slightly below the top of the form. For easy form removal, use double-headed nails.

3.5.0 INSPECTING FORMS

Before the concrete placing work begins, formwork must be inspected to see that it is in the correct location, at the correct elevation, and built so that it will produce the required finish and dimensions with adequate safety. Any debris, standing water, snow, or other foreign material must be removed from the forms.

Formwork inspection will include a check of alignment, adequacy of bracing and ties, stability of the grade on which the forms are set and braced, and form treatment. The forms are designed for specific pressures. Therefore, lower temperatures, use of retarders, higher slump (especially with superplasticizers), and excess vibration will increase in **lift** heights. Confirm that no changes have occurred in the specifications before you check the forms. All reinforcement has to be properly set and tied within the form and free of rust or scale.

Make sure that all the wall ties and connection hardware are correctly installed. One missed form tie can cause a form to bulge or fail. Remember that the concrete faithfully takes the shape of the form. If the form bulges, the contractor may have to grind it out. One missed connection can be very costly to replace in hardened concrete.

3.6.0 FORMING FOR SLABS ON GRADE

Concrete placed in the wrong location or at the wrong elevation is an expensive error to fix. If the forms are not properly placed, then the concrete that fills them will also be in the wrong place or position. If a survey crew has not done all the work for establishing the location, elevation, and grade of the forms, you may have to perform this work. At a minimum, you should be able to check any work already done.

The following is an example of how to establish the location of the corners of a slab given certain information and measurements that may be on a site plan or drawing. Most of the procedures to follow have already been presented in previous sections. This section summarizes how to put that information to use in laying out the forms for a slab that is 30 feet × 20 feet and has its front edge 110 feet from the property line along the roadway.

3.6.1 Establishing Baseline

Figure 20 is a diagram of a site for construction of a house on a slab. At the top of the figure there is a property line (dashed line) set back slightly from a roadway. Also, on the right and left sides of the figure are dashed lines showing other property lines.

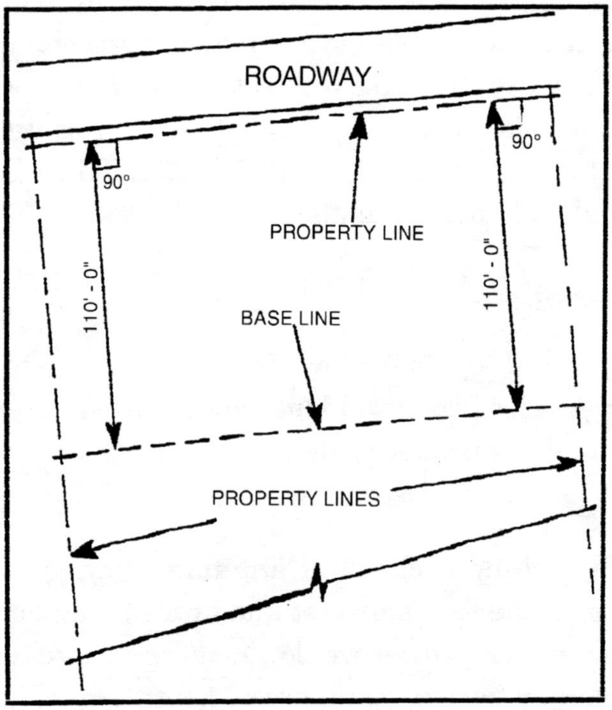

105F20.tif

Figure 20. Finding The Baseline

Start with the corner stakes for the property line next to the roadway. Stretch a string line tightly between these two stakes that represent the front of the property line.

Check the site plan or surveyor's drawing. Using the front property line as a guide, measure back the required distance (110 feet) to establish a baseline for the slab. Use a carpenter's square and tape to measure away from the property line and at right angles to it.

Make at least two measurements at points along the front property line and drive a stake at each measured location on the baseline. Make sure the stakes are evenly pointed or cut. Otherwise, they will twist as they are driven into the ground.

Stretch a string tightly between the two stakes marking the front edge of the slab. This is the baseline, as shown in *Figure 20*.

3.6.2 Establishing Corners

Figure 21 shows the baseline along with two points that establish the first two corners of the slab. Measure along the baseline from one side of the property to the corner of the slab. Drive a stake at this point to mark the first corner which is 100 feet from the left side of the property.

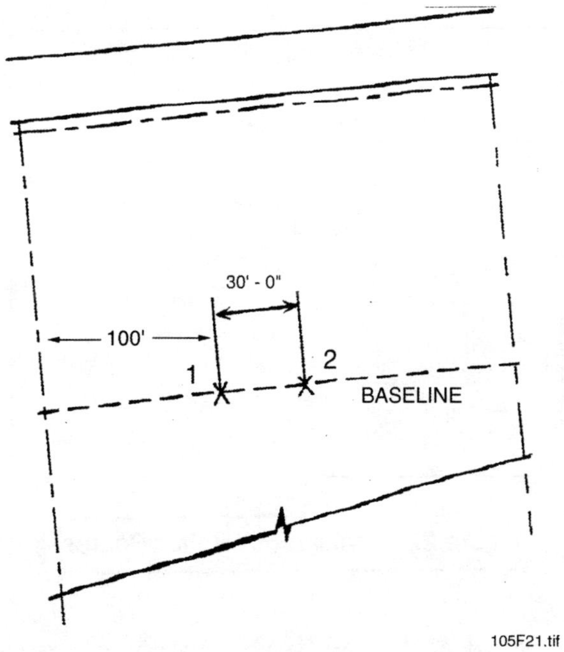

Figure 21. Staking The First Two Corners

Starting at the first corner, measure out the length of the slab (30 feet) along the baseline. Drive a stake here for the second corner.

Check dimensions by measuring the distance from the second corner to the second side of the property line. Check all measurements against the site plan.

Use a carpenter's square and tape to locate the remaining two corners. Once all corners have been marked, check their accuracy by measuring the diagonals. The length of the diagonals should be exactly the same. If they are not, readjust the lines by remeasuring each side.

Corners can also be checked with a 3-4-5 triangle. Mark 3 feet along one leg, 4 feet along the other. The diagonal between them should measure exactly 5 feet. If it does not, the corner is not square and adjustment of the line on one or both sides is required.

When all corners have been accurately squared, drive a nail at the top of each stake to mark the exact point of the corner location.

3.6.3 Setting Batter Boards

Batter boards should be set approximately 10 feet outside the building line so they will not interfere with any excavation.

Drive the first batter board stake directly in line with the diagonal of the building line. Use a string or hand level to sight the diagonal and locate the stake.

Drive the two side stakes at right angles to the first, and parallel to the slab line on each side, as shown in *Figure 22*.

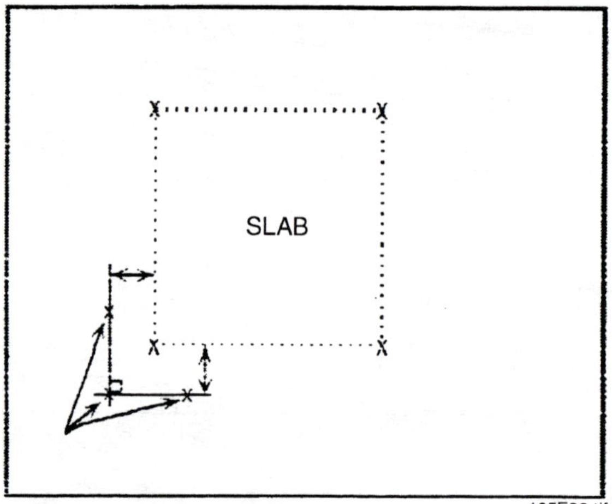

105F22.tif

Figure 22. Staking For Batter Boards

Mark the side of the batter board stakes with the proper elevation using a level and steel tape. Then nail the side boards between the stakes with their tops flush to the elevation marks. Check each side board with a carpenter's level. Repeat for the remaining three corners.

After the batter boards have been erected for each corner, stretch a string line tightly between them so that the lines cross at the point above the building corner stakes. Adjust the lines with a plumb bob until they cross directly above the nail on each corner stake. Check all four corners.

Place a nail in each batter board at the location of the string and anchor the string by tying it to the nail. Recheck that the lines still cross directly above the corner stakes.

3.6.4 Resetting Corners

When everything has been checked, remove the corner building stakes to make room for the excavation work. The subgrade must be excavated to the point where the top of the slab will be at the specified elevation for a given thickness of slab.

After excavation has been completed, reset any string lines that have been moved or damaged. Drop a plumb bob at the intersection of the lines to locate the building corners again and drive corner stakes at the exact locations. Set grade for the corner stakes and check all dimensions. Check diagonals as well. It is easier to move a stake than to move a form.

3.6.5 Setting Form Boards

With the corners set to the proper elevations and dimensions, stretch a string line between the corner stakes at the proper elevation.

Place form boards flush against the outside edge of the string line for positioning. *Figure 23* shows different sizes of dimension lumber that can be used for different slab thicknesses. Use the size appropriate to the slab. Drive stakes against the outside edge of the form, using it as a guide to keep them straight. Space stakes 3 to 5 feet apart and at expected butt joints of the form boards.

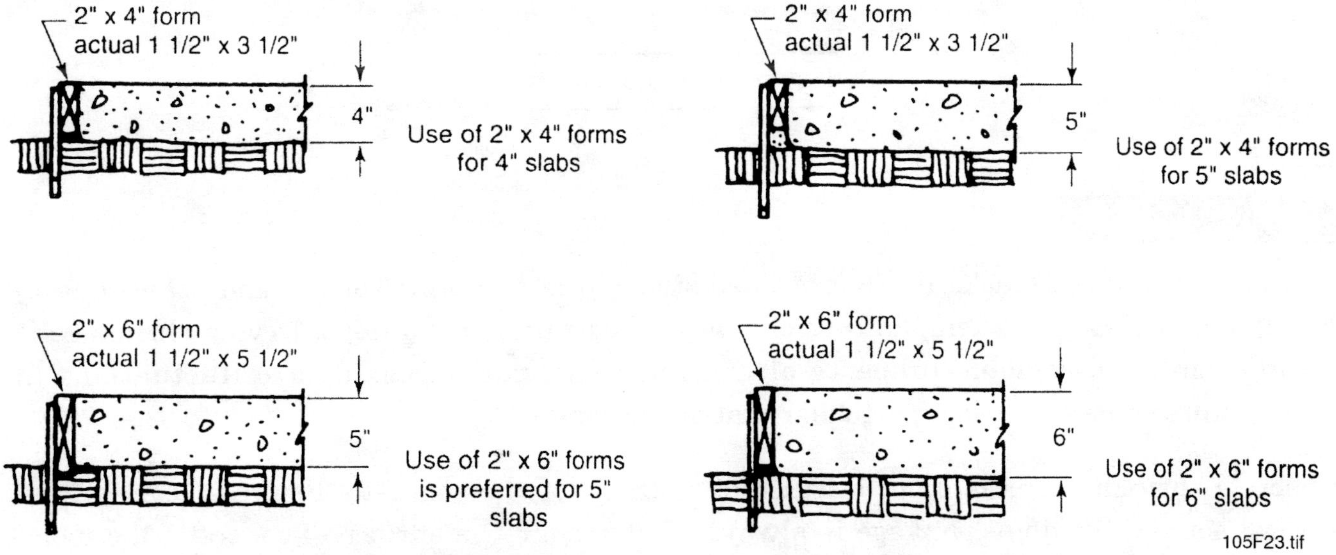

Figure 23. Dimension Lumber Sizes And Slab Thickness

Set the form boards so their tops are level with the string line. If there is not enough room under the string, dig out the subgrade. Butt the ends of the form boards against each other and place stakes at the butt joints. Drive additional stakes as needed to keep the form straight and firm.

Nail form boards to the stakes with double-headed nails. Drive the nails through the stakes and into the form boards. Drive the nails at a slant to avoid splitting the form boards and to give a better grip. Do not cut any form boards unless absolutely necessary.

Drive down any additional support stakes until they are at the level of the top of the forms. Then brace the stakes from the back and nail the braces to the stakes, as shown in *Figure 24*.

Set up any required screed rails at this time.

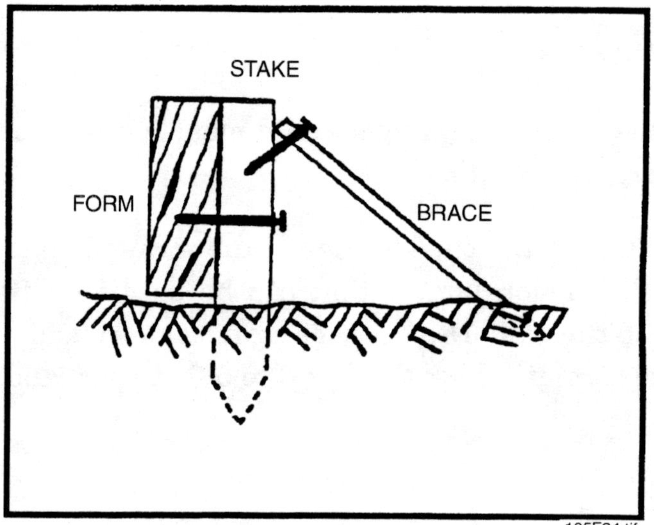

105F24.tif

Figure 24. Brace Stakes

4.0.0 JOINTS

Two basic causes of cracks in concrete are **stress** due to applied loads and stress due to drying and shrinkage or temperature changes in restrained conditions. Drying shrinkage is an inherent, unavoidable property of concrete. Thermal stress due to fluctuations in temperature causes cracking, particularly at an early age.

Concrete shrinkage cracks occur because of restraint of the concrete. Restraint comes from several sources. Drying shrinkage is always greater near the surface of concrete; the moist inner portions restrain the concrete near the surface, which can cause cracking. Other sources of restraint are reinforcing steel embedded in concrete, the interconnected parts of a concrete structure, and the friction of the subgrade on which concrete is placed.

Joints are the most effective method of controlling unsightly cracking. If a sizable expanse of concrete (a wall, slab, or pavement) is not provided with properly spaced joints to accommodate drying shrinkage and temperature contraction, the concrete will crack in a random manner.

- Jointing weakens the slab in places so the cracks occur at the joint cuts, not randomly. This type of joint is called a control or **contraction joint**.
- Jointing isolates or separates the concrete to let it shift vertically and horizontally without moving other structural parts. This type of joint is called an expansion or **isolation joint**.
- Jointing also marks the end of a day's placing activity or an interruption of the work for any unforeseen reason. This type of joint is called a **construction joint**.

Figure 25 shows a typical joint layout for a floor on grade. The isolation joints run between the wall and the slab, and around the columns. The contraction or control joints run across the slab in a regular pattern. The joint at the bottom is a construction joint set up for the next day's work.

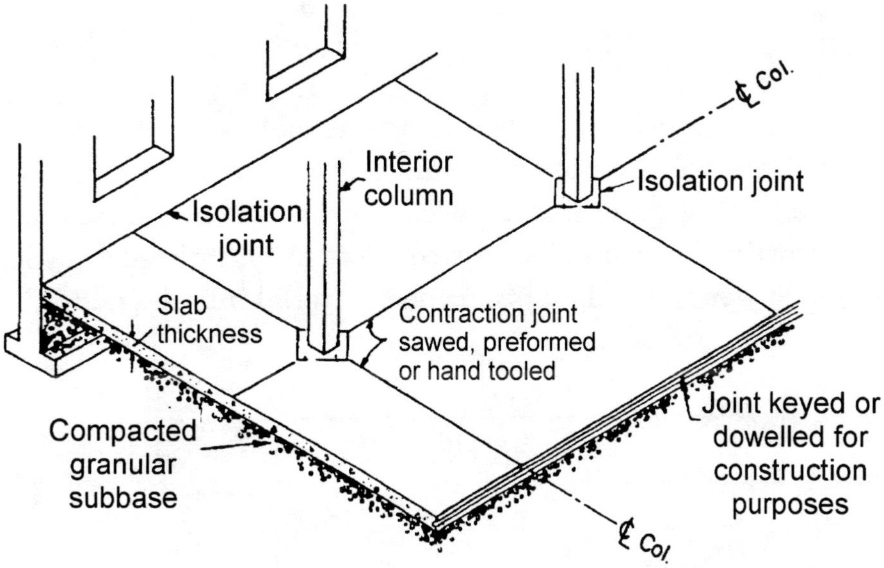

Figure 25. Three Types Of Joints On A Floor

The layout of joints and joint details is usually provided by the designer or architect. If none are provided, the contractor must draw up a set of joint plans for approval by the engineer. Each type of joint uses slightly different materials and techniques. *Figure 26* shows typical joint locations on exterior slabs at a residential site.

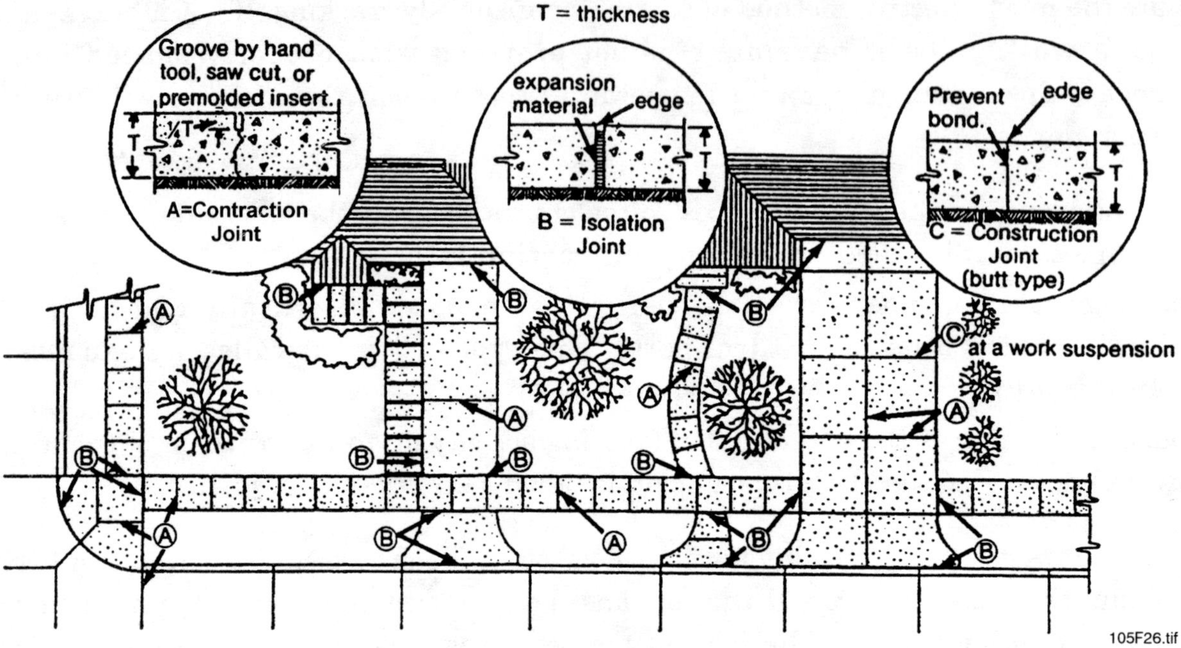

105F26.tif

Figure 26. Outside Locations Of Joints

4.1.0 CONTRACTION OR CONTROL JOINTS

Contraction joints control cracking by weakening the slab and focusing the stress at that point. Cracks should therefore occur at the weakened area instead of randomly. Contraction joints are made with a hand groover in preset locations while the concrete is still fresh. They are also made by sawcutting the joint after the concrete has hardened or by inserting strips of plastic, pressure-treated wood, metal, or preformed material into the unhardened concrete, as shown in *Figure 27*.

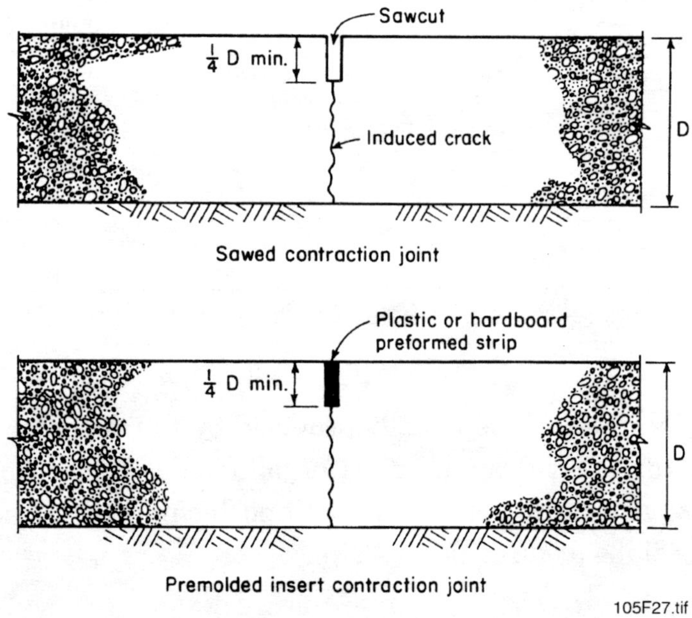

105F27.tif

Figure 27. Sawcut And Preformed Contraction Joints

When joints are cut by hand, it should be done right after or during edging. Contraction joints can also be made by sawing after the concrete is finished. Wide-cut joints usually need a sealant of some sort. If a floor is to be covered in tile, no contraction joints are used because random cracking is preferable to tooled joints under tile.

Joint locations are usually identified in the project drawings. While the design and location of contraction joints are the responsibility of the designer or specifier, the proper installation of joints in the structure is the responsibility of the contractor. The depth and maximum distance between these joints is a factor of slab thickness. If there are no drawings, these rules will help you locate contraction joints:

- Joint depth equals approximately ¼ of the slab depth, as you can see in *Figure 27*.
- Joint spacing in feet is usually two to three times the slab thickness in inches.

The panels formed by these joints should be as nearly square as possible. Avoid elongated and ell-shaped panels, as they will crack. Never make the long side more than 1½ times as long as the short side. Make these joints continuous, not offset. Locate contraction joints at all re-entrant corners, as shown in *Figure 28*. This will avoid random cracking at these locations.

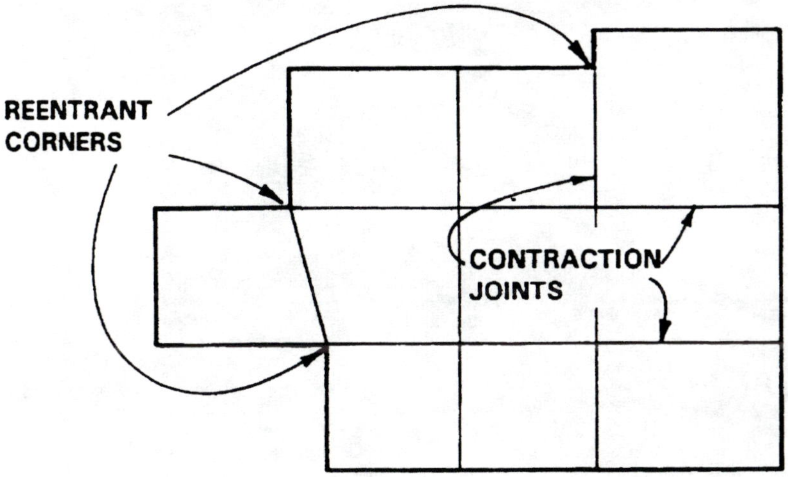

Plan of Slab on Grade

105F28.tif

Figure 28. Joints At Re-Entrant Corners

For driveways and sidewalks, contraction joints are spaced at distances about equal to slab width. If the slab is wider than 10 to 12 feet, it should also have a joint lengthwise down the center.

For slabs on grade, mark the location of contraction joints on the edges of formwork. Use a marker or grease pencil, as chalk will fade. Measure locations accurately before marking.

4.2.0 ISOLATION JOINTS

Isolation joints allow a slab to move vertically and horizontally by freeing it from adjoining parts of a structure. *Figure 25* shows isolation joints between the slab and the walls, and around the columns. The rule for placing isolation joints is to separate the concrete from anything that should not be moved. These joints are not cut, but made of preformed materials placed before the concrete is deposited.

Usually, joint material is from ¼ to ½ an inch thick and made of preformed asphalt-impregnated sheeting, treated wood, various types of plastic, or similar materials. It is important that the joint extend the full depth of the slab to fully separate the concrete. It is also important that the joint not stick up above the top of the slab so the slab can be finished properly. *Figure 29* shows an isolation joint placed in a slab form to mark a planned change in slab depth. The asphalt-impregnated sheeting is used on the back side of the 2 × 4 which will be removed after the fresh concrete is placed on either side of the joint. Isolation joints also mark changes in the thickness in the same structural member, as the different thicknesses will expand at different rates.

105F29.tif

Figure 29. Slab Isolation Joint

Columns on separate footings should be isolated from the floor slab with either circular or square isolation joints, as shown in *Figure 30*. Note that square isolation joints must be set with their corners meeting contraction joints, as shown in *Figure 30b*. If the corners of an isolation joint do not meet contraction joints, it will cause stress cracking at re-entrant corners like those shown in *Figure 30c*.

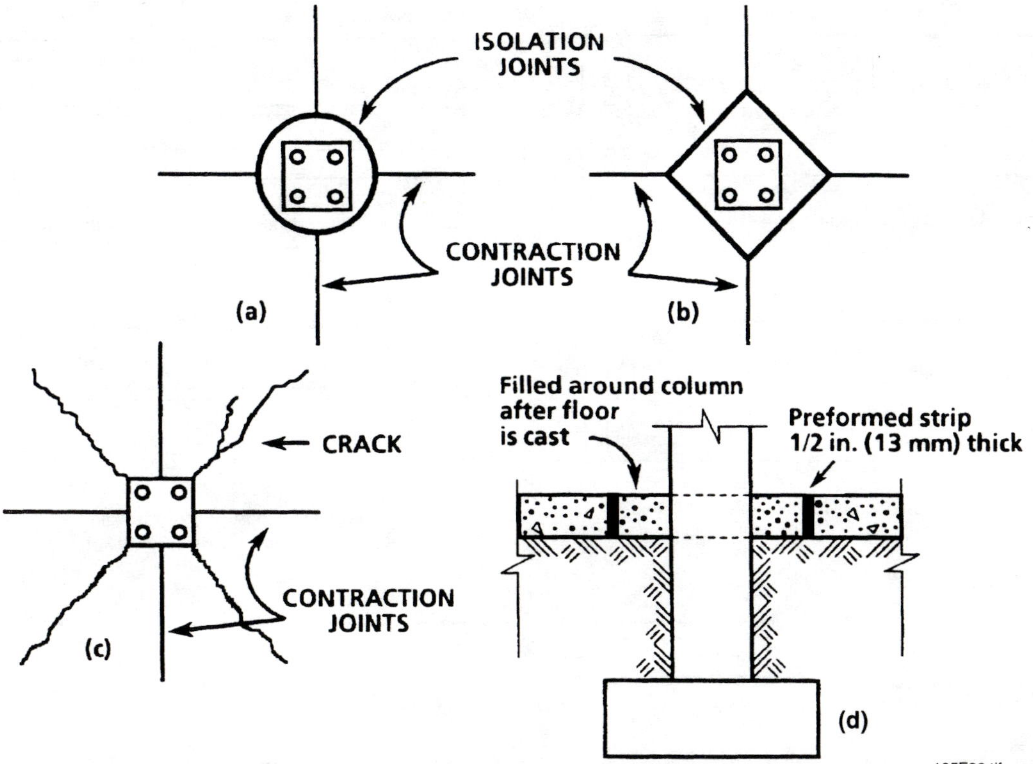

Figure 30. Isolation Joints Around Columns

The isolation joint box-outs are formed before the slab is placed. Again, location is critical. Check dimensions for forming isolation joints as you build the forms.

After the form is removed, put the isolation joint material at the formed surface and around the column base. Then place concrete to fill the gap in the slab. Delay this fill-in as long as possible to minimize grade differences if the slab corners settle.

4.3.0 CONSTRUCTION JOINTS

Construction joints mark the end of placing for the work period. Or, they mark an interruption long enough for placed concrete to harden. They should be spaced 5 feet or more from any parallel joint. Except for butt joints, they bond new to existing concrete. They permit little or no movement and transfer loads and stresses. Deformed bars, as shown in *Figure 31d*, are often used in construction joints to restrict movement and transfer the load from one slab to the other. In thick, heavily loaded floors, coated doweled construction joints (*Figure 31c*) are common. For thin slabs, flat-faced butt joints (*Figure 31a*) are enough.

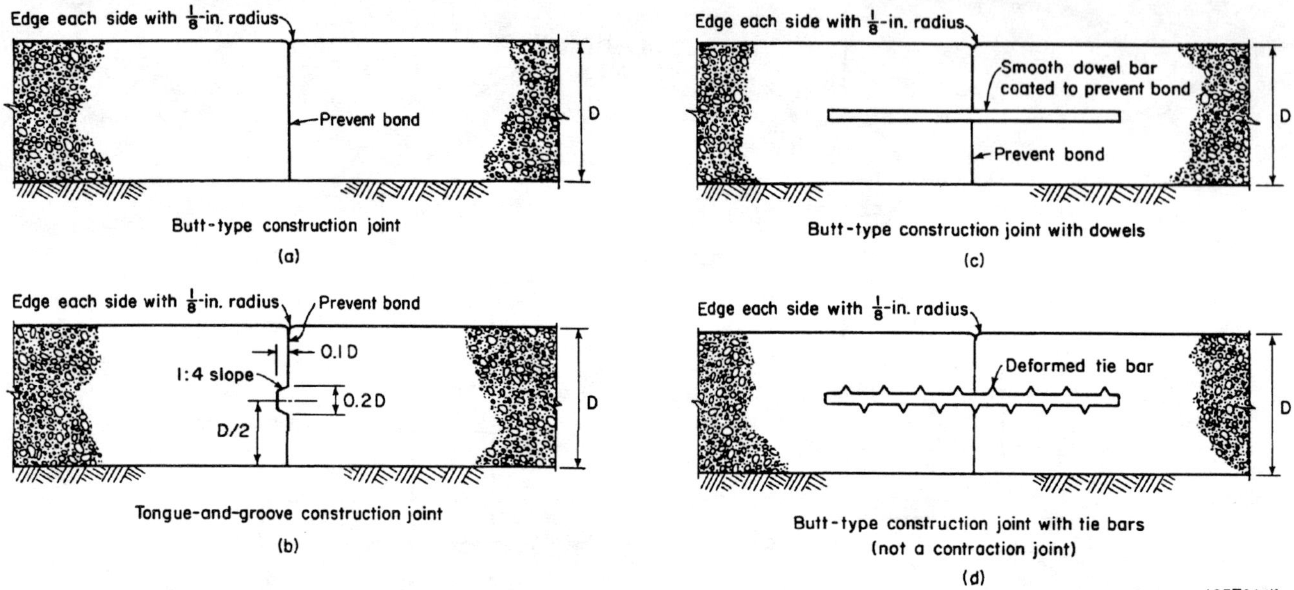

105F31.tif

Figure 31. Types Of Construction Joints

Because they are complicated, planned construction joints are sometimes designed and built to act as contraction joints. In these cases, they are deliberately made unbonded. Sometimes construction joints occur where no joint is wanted. In these cases, tie bars or **welded wire fabric** is buried at mid-depth in the slab, across the joint, as in *Figure 31d*. The first three joints in *Figure 31* can be used as contraction joints. As shown, break the bond between the old and the new deposits by coating the end of the old slab with oil, paint, a curing compound, or a form-release agent.

Details on the construction plans usually show how construction joints are fabricated. They must be placed within the middle third of the span. This location avoids weakening slab integrity because shear forces in the middle of a slab are smaller. If large forces must be transferred through the construction joint, tie bars must be used in accordance with the design by the structural engineer.

4.3.1 Bulkheads And Keys

Tongue-and-groove construction joints use forms called bulkheads. For slabs thicker than 6 inches, the bulkheads have keys, as shown in *Figure 32*. The keys give slabs the tongue-and-groove fit after concrete has been placed on both sides of the bulkhead. If the bulkheads are made of lumber, a beveled 1 × 2 strip will form the key for slabs 7 inches thick. The 1 × 2 needs a lengthwise saw cut to allow it to swell without cracking the concrete. Metal and premolded joint formers with **keyways** are also used. Install these according to the manufacturer's directions.

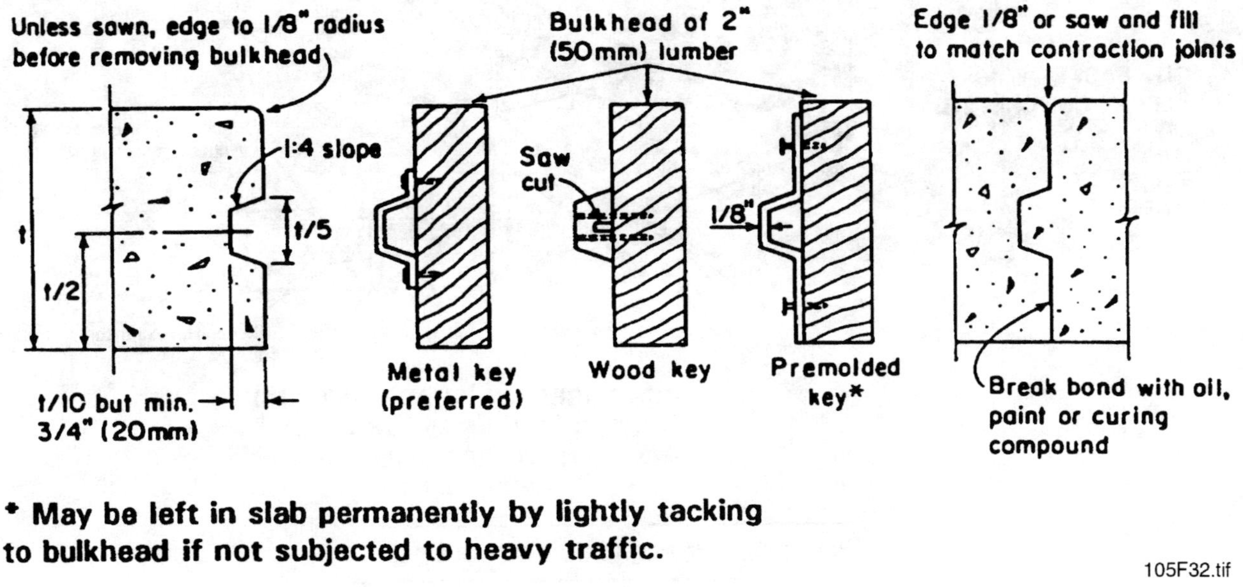

Figure 32. Bulkheads And Keys

Before placing concrete on the other side of the joint, clean the old surface thoroughly, removing all debris and loose paste. Then wet the old surface to minimize absorption of water from the new concrete.

4.3.2 Dowel Baskets

Keyed joints are not recommended for slabs with heavy, hard-wheeled traffic. Doweled joints are used for these slabs. Doweled construction joints are common in roads and industrial floors. The dowels do not stop horizontal movement but limit vertical movement. They are supposed to hold the two sides of the joint at the same elevation when loads move across from one slab to the other.

The dowels must be laid level and parallel to the length of the slab. They come prefabricated in assemblies called dowel baskets, as shown in *Figure 33*. Although most dowels are round, studies are underway on using square dowels. The dowels need a bond-breaker applied on their whole length or on half of their length.

Position the dowels so that the slab joint crosses the midpoint of the dowels. Before placing concrete, cut any basket wire that parallels the dowels. Check that the dowels are set parallel to each other and to the slab length, or they will not work properly.

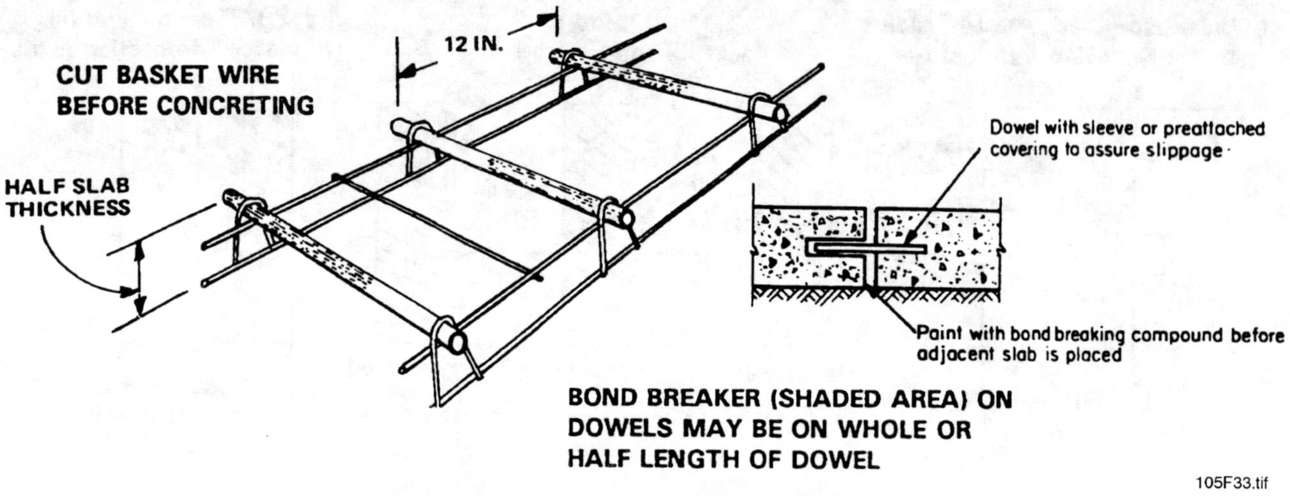

Figure 33. Prefabricated Dowel Basket

5.0.0 REINFORCEMENT

One of the significant functions of reinforcement in concrete construction is to resist the stresses caused by bending, shrinking, or temperature variations. Reinforcement in concrete is used to:

- Reduce the size of concrete members needed to resist compressive and bending stresses
- Resist internal pressure in pressure vessels, tanks, and pipes
- Resist cracking

To be effective, reinforcement must be the right kind, the right size, and in the right location.

The most commonly used reinforcement is steel. Concrete and steel have about the same coefficient of expansion and contraction from temperature changes. Steel and concrete bond together, and expand and contract together. The alkaline nature of concrete keeps steel reinforcement from rusting in many circumstances. In states where chloride-based deicing chemicals are used on highways, the chlorine ions from these chemicals do cause the steel to rust, creating serious problems.

Other types of reinforcement have been used. These include bamboo, fiberglass, and resin-reinforced fiber rods, as well as fibers of steel, plastic, glass, and other materials. Fibers help prevent cracking but cannot replace steel bars or wire mesh fabric in structural applications. Other metals such as aluminum have coefficients of expansion three or more times as great as steel. Aluminum corrodes in the presence of chlorine ions and will not expand or contract at the same rate as concrete.

5.1.0 REINFORCING STEEL BARS AND WIRE

Reinforcing steel for concrete is available as plain steel bars, deformed steel bars, welded wire fabric, deformed welded wire fabric, prestressing tendons, and post-stressing cables. The most commonly used steel reinforcements are deformed bars or post-tensioning cables.

Deformed bars are round steel bars with deformations resembling lugs that are rolled into the surface of the reinforcing bar. The purpose of the deformations is to provide additional bonding surface between the concrete and the reinforcing steel. *Figure 34* shows three standard types of deformed reinforcing bars. Each one has a different deformation pattern on the outer surface.

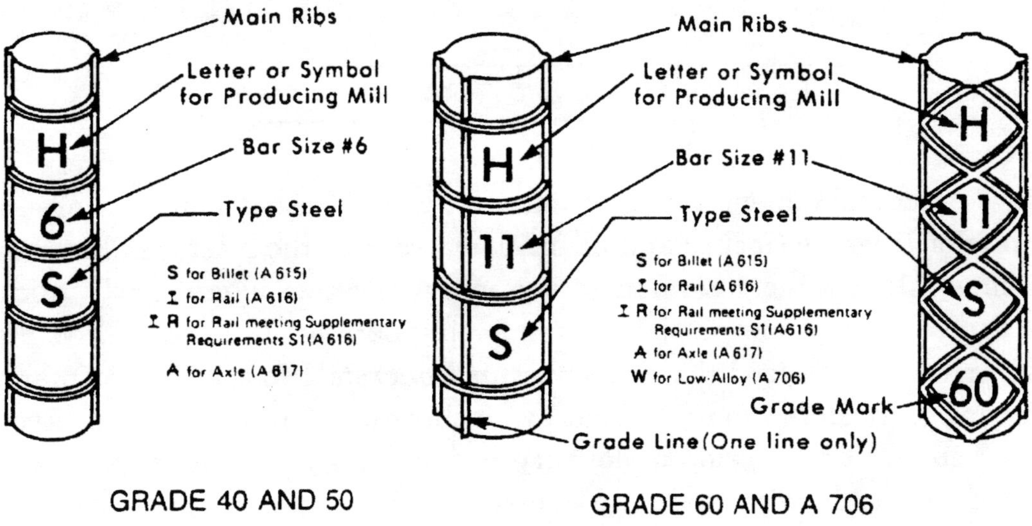

105F34.tif

Figure 34. Standard Deformed Reinforcing Bars

Deformed bars come in different sizes and strengths. This information is stamped on each bar using a series of codes like those shown in *Figure 34*. The main reference number in the code you should become familiar with is the bar size which refers to the diameter of the bar.

Wire fabric properly designed by an engineer may be used as another type of steel reinforcement. It is used primarily in flat slabs and consists of a square or rectangular wire mesh pattern where the intersections of the wires are factory-welded. *Figure 35* shows welded wire fabric placed in an excavation for a basement slab. The fabric comes in sheets or rolls with a variety of wire spacings. The use of rolled wire mesh is not recommended because it can never be properly located within the slab.

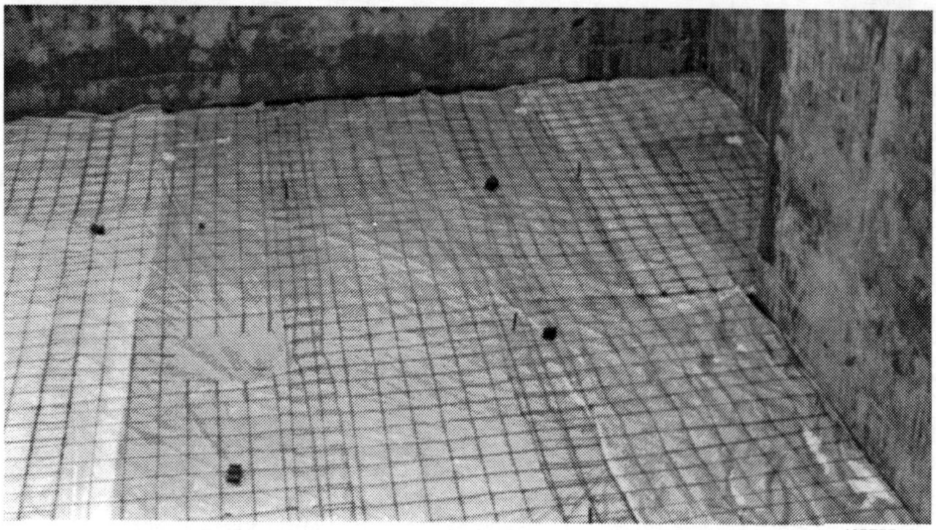

105F35.tif

Figure 35. Welded Wire Fabric Laid In Place

Figure 36 shows nominal dimensions for reinforcing bars. These dimensions are in the U.S. customary units of measure (inches and pounds) and remain the most familiar reference for reinforcing bars, although conversion to the metric system is currently taking place. Examples of the equivalent metric sizes of reinforcing bars used in the United States were given in an earlier course module, *Properties of Concrete*. Most bars are manufactured straight and bent to fit. Bars should be bent by machine in a fabrication shop. Even straight bars should be cut to length prior to delivery to the job site. For some applications, the reinforcing steel is coated with a zinc-galvanizing compound or epoxy coating to improve corrosion resistance. The bars should be cut or bent to fit the forms before coating, if possible. Cuts or nicks in the coating will allow moisture to contact the steel surface causing rust or corrosion.

Standard Reinforcing Bars

Bar #	Nominal weight (lb/ft)	Nominal dimensions		
		Diameter (inches)	Cross-sectional area (sq. in.)	Perimeter (inches)
3	0.376	0.375	0.11	1.178
4	0.668	0.500	0.20	1.571
5	1.043	0.625	0.31	1.963
6	1.502	0.750	0.44	2.356
7	2.044	0.875	0.60	2.749
8	2.670	1.000	0.79	3.142
9	3.400	1.128	1.00	3.544
10	4.303	1.270	1.27	3.990
11	5.313	1.410	1.56	4.430
14	7.65	1.693	2.25	5.32
18	13.60	2.257	4.00	7.09

105F36.tif

Figure 36. Nominal Dimensions For Reinforcing Bars

5.2.0 POSITIONING REINFORCEMENT

Reinforcement needs to be embedded in the concrete to protect it from damage and corrosion, to develop strength of bond between the reinforcement and the concrete, and to protect the steel from damage by fire.

Reinforcement is put into the forms as soon as they are constructed. The reinforcement must be properly tied, aligned, and located prior to the concrete being placed. Care should be taken to bend the ends of the tie wire so that they do not snag clothing. Never attach the tie wire to the form or allow it to penetrate the slab surface. *Figure 37* shows five widely used patterns for tying reinforcement steel bars together.

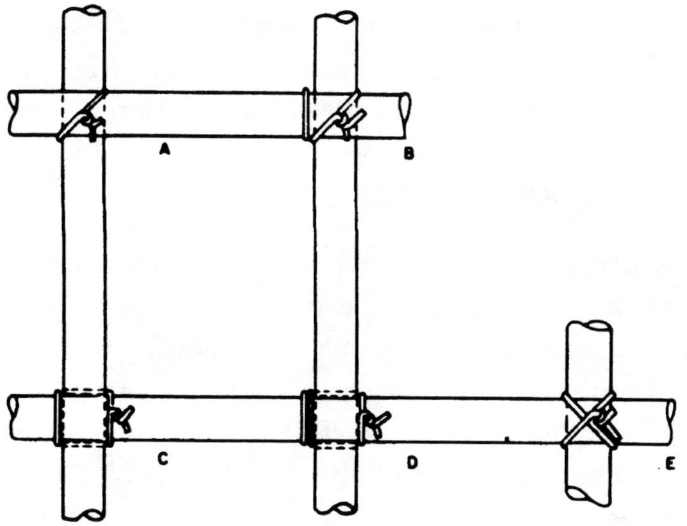

Figure 37. Common Wire Ties

Bar supports or spacers are necessary to keep vertical and horizontal reinforcement placed in the form at the required depth and cover. Bar supports are made from steel wire, plastic, or precast concrete blocks. In wall construction, side form spacers maintain the desired distance between the vertical face of the form and the reinforcement.

All reinforcement should be inspected after positioning and again before the concrete is placed. When placing or checking the placing of reinforcement in forms, remember these tips:

- Wire mesh reinforcement must be overlapped and tied at least one full stay, or column of rectangles, plus 2 inches.
- Steel bars must be overlapped and tied at least 24 bar diameters and never less than 12 inches.
- Horizontal reinforcement must be secured in the specified location, usually never closer than 1 inch to the bottom or 1½ inches from the top of a slab.
- Tie wires should not protrude above or below the plane of the reinforcement.

- Reinforcement must go through planned contraction joints and not go through planned isolation joints. Check form edges for marks indicating joint location.
- Reinforcement should stop at least 2 inches from the vertical edge of a slab.

6.0.0 EMBEDMENTS AND WATERSTOPS

Often it is necessary to place fixtures, accessories, and other materials such as **waterstops** in placed concrete. Also, items such as electrical conduits, pipes, sleeves, and blockouts must be embedded in concrete slabs, walls, and columns for efficient construction.

As long as the items are not harmful to the concrete, this practice is accepted. Incompatible metals can lead to problems. For example, copper piping that touches steel reinforcement can cause corrosion on the steel that may lead to cracking in the concrete and eventual disintegration. Uncoated aluminum reacts with concrete, especially if chloride salts are present. The corrosion produced will cause deterioration similar to that described for steel.

6.1.0 LOCATING EMBEDDED ITEMS

Embedments, fixtures, and anchor bolts are usually attached to, or suspended from, the forms. Sometimes they can be attached to the reinforcing steel to keep them in place. Care must be taken not to dislodge these items when placing concrete. When placing concrete close to an embedded item, you must properly vibrate and consolidate the concrete so that the embedded item is completely surrounded. If this is not done, **honeycomb** or voids under the embedded item may be left, causing failure of the items. Embedments need to be checked for alignment during placement.

Anchor bolts are used for external attachments when significant holding power is required. Anchor bolts must be placed accurately and firmly supported to keep them from moving during placing. *Figure 38* shows two methods of securing anchor bolts and sleeves.

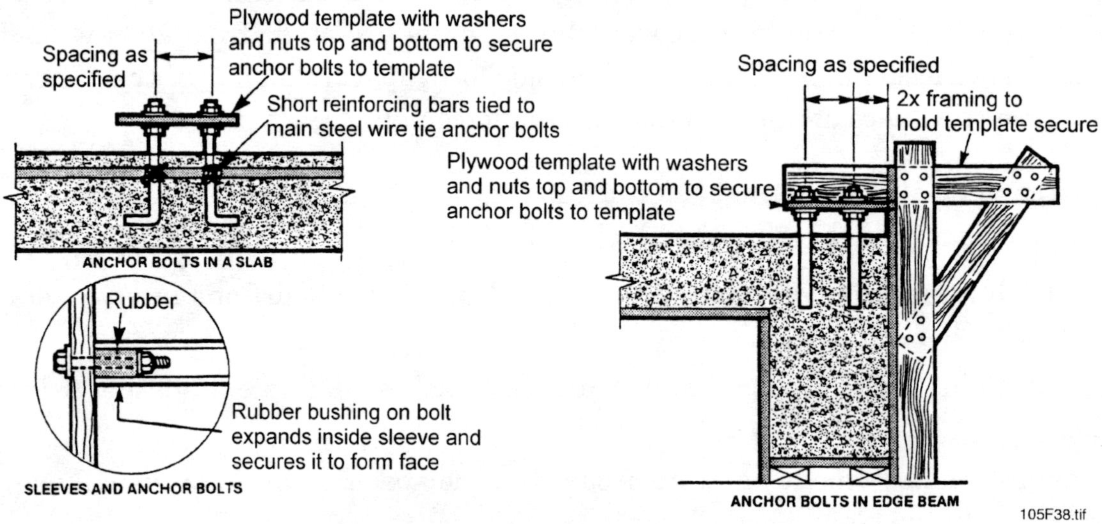

Figure 38. Securing Anchor Bolts

If located on the top surface, anchor bolts should be secured above the concrete surface by means of a template to allow vibration and finishing. Anchor bolts are also sometimes wired to the reinforcing steel to hold them secure. If not wired, then a nut is placed on both sides of the template to hold the anchor bolt secure.

6.2.0 WATERSTOPS

Waterstops are strips of material placed across a joint to restrict or stop the flow of water. They are made of metal, rubber, plastic, or betonite which swells in the presence of moisture. Polyvinyl chloride (PVC) is the most commonly used waterstop material. Waterstops are used both horizontally and vertically. Typically, they are used at the bottom of a wall where it meets a footing, at a horizontal construction or control joint, and in vertical construction or control joints in walls. *Figure 39* shows several types of waterstops.

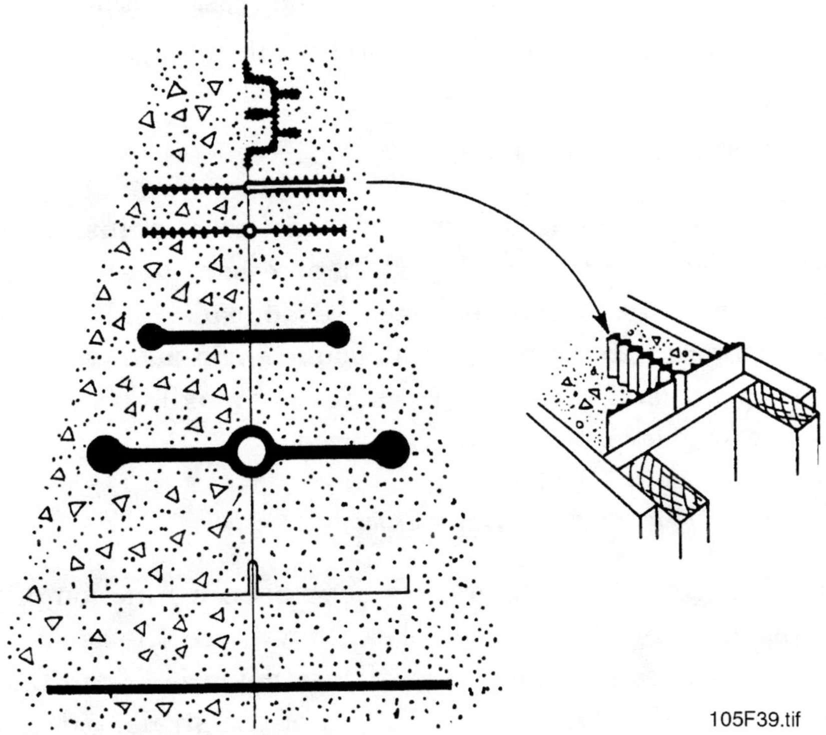

Figure 39. Waterstops

The waterstop must be held securely in place so that it will not be displaced or distorted during placement. Care must be exercised in consolidating and vibrating the concrete next to the waterstop. Honeycomb would defeat the purpose. Waterstops may need splicing at ends or intersections. Depending on the type of waterstop, prefabricated junction pieces, cement, or heat may be used to make the splices.

7.0.0 TOLERANCES

No structure is exactly level, plumb, straight, and true. Tolerances denote acceptable variations in dimension, location, levelness, plumb, flatness, and other parameters for various components of a structure. Tolerances are normally specified by the designer based on design assumptions and project needs. The designer may create project-specific tolerances or use tolerances that are standard to the industry. One source of such standards is the American Concrete Institute publication 117-90, *Standard Specifications for Tolerances for Concrete Construction and Materials*. An example of the kind of detail provided is shown in *Figure 40*. This publication provides tolerances for several hundred items, including:

- Reinforcing steel fabrication and placing
- Placing embedded items
- Concrete batching
- Vertical, lateral, and level alignments, and cross-sectional dimensions of foundations
- Vertical, lateral, and level alignment; floor flatness; and finish for cast-in-place concrete
- Openings through placed concrete
- Precast concrete fabrication

Uniform concrete batching and mixing, placement adjustments, and consistent finishing practices are always important. It may be necessary to set formwork elevations or camber to accommodate deflections from settlement, dead loads, and live loads. There are many reasons why the formwork will deflect. These include shrinkage of lumber, crushing of lumber, poor subgrade support, and water intrusion. Setting maximum acceptable values for this process ensures that some level of quality is achieved.

7.1.0 SLAB AND FLOOR TOLERANCES

Traditionally, slab surface tolerances have been a function of the maximum gap allowed under a ten-foot long straightedge placed anywhere on the slab. The straightedge normally rests on two high points and the maximum deviation from flatness would be the greatest distance between the floor and the bottom of the straightedge. This distance should not be more than the tolerance specified.

A new tolerancing system based on F-numbers is rapidly replacing the traditional straightedge. The system establishes F-numbers for flatness and levelness. F_F is a measure of flatness and F_L is a measure of levelness. F_F, the flatness parameter, defines the floor's bumpiness over a 12-inch length. F_L, the levelness number, defines conformity to a horizontal plane measured over a 10-foot distance. The F-numbers are typically written together as F_F/F_L. Values are determined from sampling procedures using raw data from surface elevation measurements at 12-inch centers. Once the measurements are made, the data and the F_F/F_L numbers are calculated using a computer program. The higher the number, the more level or flat the surface. The lower portion of the right-hand column of *Figure 40* addresses floor finish tolerances using the F-number system. Note the three different tolerance levels: conventional, flat, and very flat.

3.4 — Cross-sectional dimensions

3.4.1 *Footings*

3.4.1.1 *Horizontal dimension of formed members* .... + 2 in.
.... – ½ in.

3.4.1.2 *Horizontal dimension of unformed members cast against soil*

2 ft. or less .... + 3 in.
.... – ½ in.

Greater than 2 ft. but less than 6 ft. .... + 6 in.
..... – ½ in.

Over 6 ft. .... + 12 in.
.... – ½ in.

3.4.1.3 *Vertical dimension (thickness)* – 5 percent

3.5 — Relative alignment

3.5.1 Footing side and top surfaces may slope with respect to the specified plane at a rate not to exceed the following amounts in 10 ft. .... 1 in.

SECTION 4 — CAST-IN-PLACE CONCRETE FOR BUILDINGS

4.1 — Vertical alignment

4.1.1 *For heights 100 ft or less*

Lines, surfaces, and arrises .... 1 in.

Outside corner of exposed corner columns and control joint grooves in concrete exposed to view .... ½ in.

4.1.2 *For heights greater than 100 ft*

Lines, surfaces, and arrises, 1/1000 times the height but not more than .... 6 in.

Outside corner of exposed corner columns and control joint grooves in concrete, 1/2000 times the height but not more than .... 3 in.

4.2 — Lateral alignment

4.2.1 *Members* .... 1 in.

4.2.2 *In slabs, centerline location of openings 12 in. or smaller and edge location of larger openings* .. ½ in.

4.2.3 *Sawcuts, joints, and weakened plane embedments in slabs* .... ¾ in.

4.3 — Level alignment

4.3.1 *Top of slabs:*

4.3.1.1 Elevation of slabs-on-grade .... ¾ in.

4.3.1.2 Elevation of top surfaces of formed slabs before removal of supporting shores .... ¾ in.

4.3.2 *Elevation of formed surfaces before removal of shores* .... ¾ in.

4.3.3 *Lintels, sills, parapets, horizontal grooves, and other lines exposed to view* .... ½ in.

4.4 — Cross-sectional dimensions

4.4.1 *Members, such as columns, beams, piers, walls (thickness only), and slabs (thickness only)*

12 in. dimension or less .... + ⅜ in.
.... – ¼ in.

More than 12 in. dimension but not over 3 ft dimension .... + ½ in.
.... – ⅜ in.

Over 3 ft dimension .... + 1 in.
.... – ¾ in.

4.5 — Relative alignment

4.5.1 *Stairs*

Difference in height between adjacent risers .... ⅛ in.

Difference in width between adjacent trends .... ¼ in.

4.5.2 *Grooves*

Specified width 2 in. or less .... ⅛ in.

Specified width more than 2 in. but not more than 12 in. .... ¼ in.

4.5.3 *Formed surfaces may slope with respect to the specified plane at a rate not to exceed the following amounts in 10 ft*

4.5.3.1 Vertical alignment of outside corner of exposed corner columns and control joint grooves in concrete exposed to view .... ¼ in.

4.5.3.2 All other conditions .... ⅜ in.

4.5.4. *The offset between adjacent pieces of formwork facing material shall not exceed:*

Class of surface:

Class A .... ⅛ in.
Class B .... ¼ in.
Class C .... ½ in.
Class D .... 1 in.

4.5.5 *Floor finish tolerances shall meet the requirements of either Section 4.5.6 or 4.5.7, as set forth by the specifier.*

4.5.6 *Floor finish tolerances as measured in accordance with ASTM E 1155-87 Standard Test Method for Determining Floor Flatness and Levelness Using the F-Number System (Inch-Pound Units)*

Floor profile quality classification	Minimum F_F F_L number required			
	Test area		Minimum local F number	
	Flatness F_F	Level F_L	Flatness F_F	Level F_L
Conventional				
Bullfloated	15	13	13	10
Straightedged	20	15	15	10
Flat	30	20	15	10
Very flat	50	30	25	15

4.5.6.1 The F_L levelness tolerance shall not apply to slabs placed on unshored form surfaces and/or shored form surfaces after the removal of shores. F_L levelness tolerances shall not apply to cambered or inclined surfaces and shall be measured within 72 hr after slab concrete placement.

105F40.tif

Figure 40. Example ACI Standard Specifications for Tolerances

7.2.0 OTHER TOLERANCES

As shown in *Figure 40*, ACI 117-90 lists the deviations allowed for different types of structural members. Item 4.3 requires that alignment for tops of slabs on grade be within ¾ of an inch of the specified elevation. ACI specifications are agreed upon by committee, over a long period of work and negotiation, and are generally accepted as industry standards. For large jobs, they are specifically referenced in the contracts.

Some specifications appear difficult to meet. Highway surfaces, for instance, call for levelness within ⅛ of an inch of a specified depth. Contractors routinely meet this ⅛-inch specification when building highways. One reason for doing so is economic. If the contractor places more concrete than specified, that results in paying for the extra concrete out of the profits. If the contractors place less than is specified, they are usually penalized under the terms of the contract. This means they may get only 75 or 80 percent of the agreed-upon payment for not meeting the specifications.

A second reason for meeting specifications is for safety. Tolerances are critical for calculating stresses, loading, weightbearing, and strength. Any part of a structure that does not meet the required tolerances may not be able to support its designed load. As a result, it may fail, causing the entire structure to collapse.

8.0.0 ORDERING CONCRETE

The most convenient way to obtain concrete is to get it from a local ready-mix supplier. The ingredients will be accurately batched, mixed, and delivered to the job site at the schedule agreed upon.

Local practices in ordering concrete will vary throughout the country. Usually, you can specify the performance needed or the ingredients by recipe or prescription. When the contractor specifies the performance, the 28-day strength desired is the governing factor. Contractors should advise the ready-mix supplier of the intended use of the concrete and the desired slump, maximum size of the aggregate, how much is needed, the batch size, and any admixtures. The supplier will often already have mix designs available to meet the various strength and needs requirements. When a strength level is specified, the ready-mix supplier is responsible to provide the properly proportioned mix. If the contractor chooses to specify the mix proportions for aggregate, cement, water, and admixtures, then they are responsible for the results. If the strength does not meet the minimum required, the contractor, not the supplier, is responsible.

At one time, concrete was ordered by the number of sacks of cement in a one-cubic-yard mix. For example, a five-bag mix was used for footings and foundations; a six-bag mix for floors; and a seven-bag mix for beams and columns. The bag-ordering method is not a very accurate way to specify strength and performance, particularly now that the composition of cement can include so many different materials.

8.1.0 ESTIMATING QUANTITIES

The amount of concrete needed for a placement is the volume of the forms plus about 5 percent for overage. As you will recall, the volume is the length × width × depth. For example, the volume of concrete needed for a floor 20 feet wide × 50 feet long × 6 inches deep is calculated as follows:

	V = (20 ft. × 50 ft. × 0.5 ft.)	=	500 cubic feet
	500/27 (cu. ft./cu.yd.)	=	18.518 cubic yards (volume)
	18.518 cubic yards × .05	=	.926 cubic yards (overage)
Therefore:	18.518 + .926 = 19.44 cubic yards (required)		
	Order 19.5 or 20 cubic yards		

This estimate provides extra concrete to account for waste, spillage, or underyield. A good builder will keep records of the amount used and the amount ordered. By comparing the amount ordered with the amount delivered and used, the builder can determine, over time, how much to increase or decrease the 5 percent overage ordering amount.

8.2.0 SITE ACCESS

Many building sites are accessible to ready-mix trucks while others are not. **Ready-mix concrete** trucks are heavy and need a firm, dry site to park and discharge their loads. If good access to the placement site is not available, then placement may be difficult because the concrete could require transport to the forms by wheelbarrow or buggy. If other placing equipment such as a conveyor or pump is used, a clear, level staging area is required for this operation. Because of their weight, ready-mix trucks cannot go over curbs and sidewalks without causing damage. They will also damage most residential driveways. Therefore, plan the access path to the point of placement with care.

8.3.0 DELIVERY SCHEDULE

The delivery schedule must be carefully planned. The ready-mix supplier needs to schedule the batch plant and ready-mix trucks; the contractor needs to make everything ready to receive the concrete and have the placing and finishing crews on site with their equipment ready to work when trucks arrive; and inspectors and other quality control personnel also need to be on site. Everyone involved needs to coordinate with the delivery schedule.

8.4.0 FINAL INSPECTION

Job site preparation may be done quickly or over a very long period of time, depending on the size and difficulty of the project. It is important to make sure the site is ready because it will be too late to perform any major rework while the concrete is being delivered. Allow enough time for a thorough inspection of formwork, reinforcement, subgrade, and placement equipment, without the pressure of trucks on the way or waiting. Before the concrete arrives, the wise contractor will schedule one last inspection.

The general rule for final inspection is that everything that was inspected previously needs to be inspected again. Everything is usually inspected three times, with the last time being just before the ready-mix truck leaves the plant. The following items summarize the final checks that should be done.

The subgrade should be checked for:

- Uniform compaction
- Accurate grading
- Proper use and placement of vapor retarders and geotextiles
- Proper location of all drainage items

Formwork inspection includes a check of:

- Alignment and elevation referenced back to the survey control or benchmark
- Adequacy of all bracing and ties
- Stability of the grade on which the forms are set and braced
- Form release treatment

Remember that formwork is designed to withstand specific pressures. Be aware of factors such as lower temperatures that may increase pressure.

A check of all reinforcement will make sure that it is:

- Properly set and tied
- Supported as required to maintain proper depth and cover
- Free of rust and scale

All materials and tools needed for placing should be on the job and accessible. This includes jointing materials, reinforcing steel, and curing materials. For large slabs, have polyethylene sheeting on hand to cover the slab quickly in case of rain.

Before the first truck shows up, check the access roadway in and out from the point of discharge. Access is an important factor as it affects the speed of placing and the quality of the final product.

With all the preparatory work finished and the last inspection complete, the site is ready for the concrete to arrive and placing to begin.

SUMMARY

There is an old saying to the effect that prior preparation prevents poor performance. The saying is very applicable to concrete work. This module starts with preparation of the subgrade and continues through the preplacement preparations for layout of formwork, joints, and reinforcement, as well as placement of embedded items and ordering the concrete. These preplacement activities are necessary for quality placement and finishing of the concrete.

Subgrades should provide uniform support to the slab or footing. Compaction of the subgrade lessens the amount of settlement and the rate at which settlement occurs. There are two different classes of soil to compact: granular and cohesive. Vibrating compaction equipment works best on granular soils. Tamping and rolling equipment works best with cohesive soils.

Foundations are the base upon which a building or structure is constructed. Wall and column footings are considered to be shallow foundations. Pile foundations are deep foundations. Sometimes, the water table is above the subgrade level and hinders subgrade preparation. In such instances, the water level will need to be lowered through pumping or drainage ditches.

Once the subgrade preparation is complete, the forms can be built and erected. Forms may be built for one-time use or multiple uses. They can be built on the job or manufactured in a factory. Earth, plywood, lumber, and steel are the most widely used form materials.

There are three kinds of joints for slab-on-grade construction. Isolation joints provide for movement between separate structural members. Contraction joints induce cracking at preselected locations. Construction joints provide stopping places where placing or construction is interrupted or stopped. Construction joints should be planned so that they can also be used as contraction joints when the placement continues. In order to provide load transfer across contraction and construction joints, steel dowels or reinforcement are used, as well as other designs such as tongue-and-groove.

Inspection of all preplacement work is very important. It may be too late to correct a mistake after the concrete has arrived and placing has begun. As a general rule, everything that was inspected when it was installed needs to be inspected again before the concrete is ordered. This includes the subgrade, formwork, reinforcement, joints, and site access.

Reference

For advanced study of topics covered in this task module, the following materials are suggested:

Concrete Construction and Estimating, Craig Avery, Craftsman Book Company, Carlsbad, CA, 1991.

Concrete Manual, International Conference of Building Officials, Whittier, CA, 1992.

Concrete Parking Lots: Eight Steps to Quality Construction, Videotape, Portland Cement Association, Skokie, IL.

Concrete Technology, Third Edition, George R. White, Delmar Publishers, Inc., Albany, NY, 1991.

Design and Control of Concrete Mixtures, Thirteenth Edition, Portland Cement Association, Skokie, IL, 1988.

Finishing Concrete Flatwork, Videotape 32:00 minutes, Portland Cement Association, Skokie, IL, 1991.

Guide for Concrete Floor and Slab Construction, ACI 302.1R-96, American Concrete Institute, Farmington Hills, MI, 1997.

The Contractor's Guide to Quality Construction, American Concrete Institute, Farmington Hills, MI, 1997.

ACKNOWLEDGMENTS

Figures 2, 23, 25-27, and 31 courtesy of Portland Cement Association

Figures 3, 4, 7, 8, 20-22, 24, 29, and 35 courtesy of Roy Jorgensen Associates, Inc.

Figure 5 courtesy of Spetra Physics, Inc.

Figures 6 and 9 courtesy of National Association of Homebuilders

Figures 10-19, 28, 30, 32-34, and 36-40 courtesy of American Concrete Institute

SI* (MODERN METRIC) CONVERSION FACTORS

APPROXIMATE CONVERSIONS TO SI UNITS

Symbol	When You Know	Multiply By	To Find	Symbol
		LENGTH		
in	inches	25.4	millimeters	mm
ft	feet	0.305	meters	m
yd	yards	0.914	meters	m
mi	miles	1.61	kilometers	km
		AREA		
in^2	square inches	645.2	square millimeters	mm^2
ft^2	square feet	0.093	square meters	m^2
yd^2	square yards	0.836	square meters	m^2
ac	acres	0.405	hectares	ha
mi^2	square miles	2.59	square kilometers	km^2
		VOLUME		
fl oz	fluid ounces	29.57	milliliters	mL
gal	gallons	3.785	liters	L
ft^3	cubic feet	0.028	cubic meters	m^3
yd^3	cubic yards	0.765	cubic meters	m^3
NOTE: Volumes greater than 1000 l should be shown in m^3.				
		MASS		
oz	ounces	28.35	grams	g
lb	pounds	0.454	kilograms	kg
T	short tons (2000 lb)	0.907	megagrams (or "metric ton")	Mg (or "t")
		TEMPERATURE (exact)		
°F	Fahrenheit temperature	5(F-32)/9 or (F-32)/1.8	Celsius temperature	°C
		FORCE AND PRESSURE OR STRESS		
lbf	poundforce	4.45	newtons	N
lbf/in^2	poundforce per square inch	6.89	kilopascals	kPa

APPROXIMATE CONVERSIONS FROM SI UNITS

Symbol	When You Know	Multiply By	To Find	Symbol
		LENGTH		
mm	millimeters	0.039	inches	in
m	meters	3.28	feet	ft
m	meters	1.09	yards	yd
km	kilometers	0.621	miles	mi
		AREA		
mm^2	square millimeters	0.0016	square inches	in^2
m^2	square meters	10.764	square feet	ft^2
m^2	square meters	1.195	square yards	yd^2
ha	hectares	2.47	acres	ac
km^2	square kilometers	0.386	square miles	mi^2
		VOLUME		
mL	milliliters	0.034	fluid ounces	fl oz
L	liters	0.264	gallons	gal
m^3	cubic meters	35.71	cubic feet	ft^3
m^3	cubic meters	1.307	cubic yards	yd^3
		MASS		
g	grams	0.035	ounces	oz
kg	kilograms	2.202	pounds	lb
Mg (or "t")	megagrams (or "metric ton")	1.103	short tons (2000 lb)	T
		TEMPERATURE (exact)		
°C	Celsius temperature	1.8C + 32	Fahrenheit temperature	°F
		FORCE AND PRESSURE OR STRESS		
N	newtons	0.225	poundforce	lbf
kPa	kilopascals	0.145	poundforce per square inch	lbf/in^2

* SI is the symbol for the International System of Units. Appropriate rounding should be made to comply with Section ASTM E380.

REVIEW/PRACTICE QUESTIONS

1. Poorly prepared subgrades often result in _____.
   a. uniform non-support
   b. honeycomb and lattice
   c. settlement and cracks
   d. high groundwater tables

2. The Leaning Tower of Pisa is a serious case of _____.
   a. differential settlement
   b. poor concrete foundation
   c. clay particles slipping
   d. unbalanced foundation load

3. In preparing a site, the first task is to _____.
   a. remove large vegetation
   b. blast all rocks
   c. excavate
   d. remove all vegetation

4. Which of the following is not a corrective measure for problem soils?
   a. Engineering fabric
   b. Geogrids
   c. Fiber mesh
   d. Slip forming

5. Flowable fill is used ____.
   a. mainly as backfill
   b. as a top covering of footings
   c. for non-structural columns
   d. for decorative walkways

6. Sidewalks should have _____ to drain water off to the side.
   a. weep holes
   b. aggregates
   c. transverse slopes
   d. joints

7. Cohesive soils contain mostly _____.
   a. silt particles
   b. aggregates
   c. silt and clay particles
   d. clay and sand particles

8. Shallow foundations consist of _____ for walls and columns.

   a. footings
   b. walls and footings
   c. pile piers
   d. caissons

9. Deep foundations usually require _____.

   a. spread footings
   b. large aggregates
   c. caissons or piles
   d. drilled friction walls

10. The water table is the _____ of groundwater.

   a. mid-level
   b. upper surface
   c. bottom surface
   d. gradient

11. Well point systems are used to _____.

   a. moisten backfill for compaction
   b. pump water for concrete
   c. filter groundwater
   d. dewater building sites

12. Formwork safety is absolutely required. The most effective way to achieve safety is to have _____.

   a. good supervision and trained workers
   b. a comprehensive safety manual
   c. safety awards and bonuses
   d. automatic alarm systems

13. Historically, _____ have been the most widely used form materials.

   a. plastic, plywood, and lumber
   b. precast concrete, lumber, and aluminum
   c. earth, lumber, hardboard, and sheet metal
   d. earth, plywood, lumber, and steel

14. Exterior grade plywood has _____.

   a. water resistant glue
   b. waterproof glue
   c. waterproof plastic coating
   d. painted surfaces

15. An internal form creates ____ in thick floors and decreases the dead load.

    a. hidden raceways
    b. voids
    c. permanent supports
    d. storage areas

16. Panelized prefabricated forms are available in modular sizes from _____.

    a. 2 feet × 2 feet to 4 feet × 8 feet
    b. 2 feet × 2 feet to 2 feet × 8 feet
    c. 2 feet × 2 feet to 3 feet × 5 feet
    d. 2 feet × 3 feet to 6 feet × 9 feet

17. Line and grade for forms are set using _____.

    a. benchmarks
    b. plumb bobs
    c. string lines
    d. offsets

18. Rectangular forms can be checked for squareness by ensuring that the _____.

    a. opposite sides are the same length
    b. diagonals are the same length
    c. interior angles add up to 360 degrees
    d. adjacent sides meet at the corner stake

19. Isolation joints in a floor allow movements between the floor and _____.

    a. doorway
    b. subbase
    c. embedded items
    d. walls and columns

20. Control joints induce _____ at preselected locations.

    a. load transfer
    b. water flow
    c. cracking
    d. torque

notes

ANSWERS TO REVIEW/PRACTICE QUESTIONS

Answer	Section Reference
1. c	2.0.0
2. a	2.0.0
3. d	2.1.0
4. d	2.1.1
5. a	2.1.2
6. c	2.2.4
7. c	2.3.0
8. a	2.5.1
9. c	2.5.1
10. b	2.4.0
11. d	2.4.0
12. a	3.0.0
13. d	3.1.0
14. b	3.1.0
15. b	3.1.3
16. a	3.1.3
17. c	2.2.0/3.4.0
18. b	3.3.0
19. d	4.2.0
20. c	4.1.0

NCCER CRAFT TRAINING USER UPDATES

The NCCER makes every effort to keep these manuals up-to-date and free of technical errors. We appreciate your help in this process. If you have an idea for improving this manual, or if you find an error, a typographical mistake, or an inaccuracy in the NCCER's Craft Training Manuals, please write us, using this form or a photocopy. Be sure to include the exact module number, page number, a description of the problem, and the correction, if possible. Your input will be brought to the attention of the Technical Review Committee. Thank you for your assistance.

Instructors – If you found that additional materials were necessary in order to teach this module effectively, please let us know so that we may include them in the Equipment/Materials list in the Instructor's Guide.

Write: Curriculum and Revision Department
National Center for Construction Education and Research
P.O. Box 141104
Gainesville, FL 32614-1104

Fax: 352-334-0932

Craft ____________________ Module Name ____________________

Module Number ____________________ Page Number(s) ____________________

Description of Problem

____

____

____

____

(Optional) Correction of Problem

____

____

____

(Optional) Your Name and Address

____

____

____

notes

Placing Concrete

Module 23106

Concrete Finishing Trainee Task Module 23106

NATIONAL
CENTER FOR
CONSTRUCTION
EDUCATION AND
RESEARCH

PLACING CONCRETE

OBJECTIVES

Upon completion of this module, the trainee will be able to:

1. Describe how concrete is conveyed and placed.
2. Draw up a pre-placement checklist.
3. Demonstrate the use of equipment and tools for placing concrete.
4. Demonstrate the process of depositing, spreading, consolidating, and striking off concrete in a form.
5. Associate trade terms with the appropriate processes and equipment.

Prerequisites

Successful completion of the following Task Modules is recommended before beginning study of this Task Module: Core Curricula; Concrete Finishing Level 1, Modules 23101 through 23105.

Required Trainee Materials

1. Trainee Task Module
2. Pencil and paper
3. Appropriate Personal Protective Equipment

COURSE MAP

This course map shows all of the task modules in the first level of the Concrete Finishing curricula. The suggested training order begins at the bottom and proceeds up. Skill levels increase as a trainee advances on the course map. The training order may be adjusted by the local Training Program Sponsor.

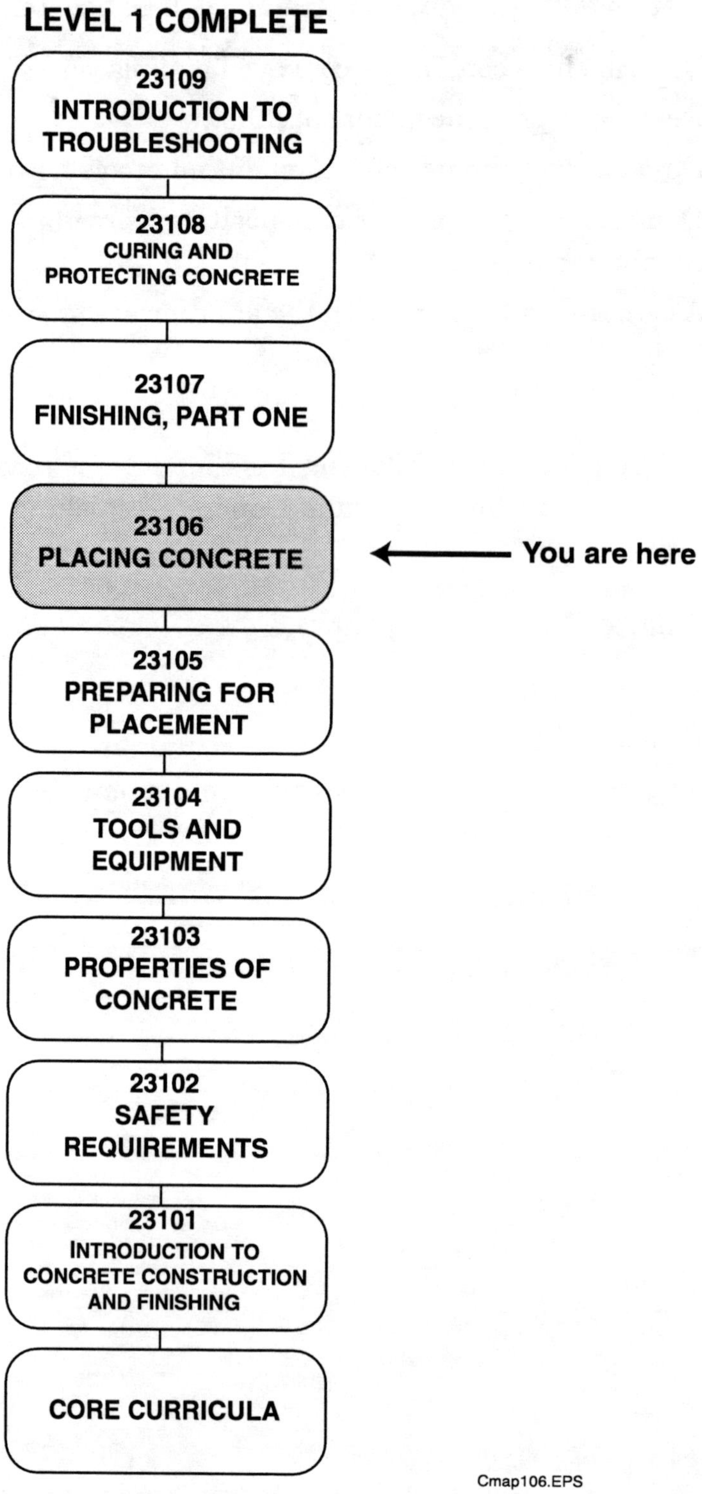

Cmap106.EPS

TABLE OF CONTENTS

TABLE OF CONTENTS (Continued)

Trade Terms Introduced In This Module

Bleed water: Water that rises to the surface of concrete after it is placed; caused by the settlement of cement paste.

Bridging: The poor placement of concrete over reinforcement, so that an air pocket is created underneath the reinforcement. This can be corrected by revibration.

Bughole: Air voids or small cavities on formed concrete surfaces.

Cold joint: A joint (usually visible) in a concrete wall or floor where the fresh concrete has bonded imperfectly or not at all to the previously placed concrete because too much time has elapsed between placements.

Consolidating: Removing the entrapped air voids from concrete by pressing, rolling, rodding, vibrating, spading, or tamping.

Heel: The edge of a concrete deposit against which the next increment of concrete is to be placed.

Honeycomb: Voids left in concrete where cement and sand particles have not filled the spaces between the coarse aggregate particles.

Lift: A horizontal layer of concrete placed in a form. In a horizontal form, two lifts can be placed to go under and over reinforcement. In a vertical form, a lift is approximately 4 feet high, placed at one time. Large vertical forms are filled by several lifts, depending on height and the ability of the formwork to withstand the pressures of the fresh concrete.

Revibration: Vibrating concrete for a second time to reconsolidate, usually to avoid a cold joint.

Surcharge: The rolling wave (1 to 2 inches high) of concrete that is pushed ahead of a screed or straightedge; the surcharge fills in hollows as the screed passes.

1.0.0 INTRODUCTION

This module introduces the subject of placing concrete. Placing is the process of depositing the wet concrete mix into the form, **consolidating** it, striking it off to the predetermined grade, and smoothing it. Placing does not include finishing, which is the subject of the next module, *Finishing, Part One*.

This module concentrates on horizontal placing. Vertical placing is covered in Level 2 of Concrete Finishing.

1.1.0 PURPOSE

This module introduces the checklists, equipment, and procedures you will need to:

- Place concrete into the form
- Consolidate and level the concrete
- Strike off the concrete
- Smooth and float the concrete

These tasks are grouped together because they all occur between the time the concrete is mixed and the time **bleed water** appears on the surface. This is a short period of time in which many tasks need to be completed. By the end of this module, you will know how to perform these tasks.

1.2.0 SAFETY

General safety procedures while placing concrete begin with checking tools and equipment, setting procedures for materials movement, and wearing appropriate personal protective equipment.

1.2.1 Tools And Equipment

As noted in the previous module, *Tools and Equipment*, all gear should be kept clean and inspected before use. It is especially important to wash down equipment after it is used for placing. Hardened concrete can contaminate the new batch as well as ruin ready-mix trucks, chutes, and other equipment. Ready-mix trucks in particular need to be washed out with pressurized hoses after each batch is deposited. Hardened waste concrete must be disposed of in accordance with Environmental Protection Agency requirements.

Hand tools become useless when contaminated with hardened concrete. Wash tools thoroughly with water and a wire brush after each use. Be sure that all concrete is completely removed. Check tools to see if they need repair or oiling.

1.2.2 Movement Safety

Placing when concrete is deposited by lift or bucket calls for setting up safe materials movement procedures. Personnel safety becomes important when concrete is moved overhead by bucket, conveyor, lift, or chute. Heavy materials movement, especially by crane, poses a safety hazard to all workers on the job. Good safety practices call for being aware of materials movement at all times. Work safety while moving concrete calls for:

- Establishing clear pathways for concrete movement
- Using a consistent set of signals to alert workers to concrete movement
- Assigning individual responsibilities for each worker during concrete movement
- Staying alert to what is going on around you

During placing, craftworkers also interact with heavy equipment on the ground. This calls for a different kind of movement safety. One craftworker is designated to signal the operator of the heavy equipment being used, which may be a crane, a ready-mix truck, or a concrete pump. For example, this craftworker will direct the operator of a ready-mix truck to:

- Move the equipment as required
- Start the concrete flow
- Stop the concrete flow

When working with large moving equipment, it is important that only one person give signals. Confusion and accidents result when more than one person tries to give directions.

1.2.3 Form Safety

One of the most potentially dangerous times on a concrete job is when concrete is being placed in a form. All workers must be especially aware of bulges, shifts, or movement in the form as it fills. Any changes in form condition can warn of form collapse. Crew members should be designated as form watchers who must be alert to any changes in the forms as they are filled.

CAUTION: Formwork must be continuously watched during and after placing. Precautions must be set up beforehand to protect formwork watchers and to maintain a safe area and exit route during placement.

Telltales such as string or wire lines enable form watchers to constantly check form elevation, camber, and plumb. Watchers should continue checking until concrete has been floated and telltales show that deflection, or movement of forms, has stopped. A form often gives warning of failure by gradually tilting.

If there are any signs of form movement or unfamiliar noises during placing, work must stop while formwork is checked. The affected area can be damaged by the pause in placing or by the removal of any placed concrete to allow form repair. In such situations you may have to rip out all the concrete, reform the area, and place new concrete.

1.2.4 Personal Safety

Concrete is very abrasive, and fresh concrete is very caustic. If not handled properly, it can cause skin burns or cement dermatitis. A few simple precautions can protect you from injury or skin irritation.

CAUTION: Workers handling and placing concrete need to wear appropriate personal protective equipment such as safety glasses, rubber boots, gloves, and clothing that covers the legs and arms.

If skin irritation occurs, wash the area thoroughly with soap and water. If the irritation continues, see a doctor. If your eyes are affected, wash them with clean water and consult a doctor immediately. Do not rub your eyes, and do not use any other treatment without consulting a doctor.

2.0.0 PREPARE CHECKLIST

A ready-mix truck does not just show up one morning and start discharging concrete. Many things need to be planned, constructed, inspected, approved, and checked off before the first load of concrete is placed. Even the smallest concrete job needs a checklist.

2.1.0 PLANNING

Appendix A at the end of this module shows a typical pre-placement checklist with many questions to be answered. Most companies use some version of this checklist; become familiar with the checklist as it is used by your company. *Figure 1* shows a part of this checklist. As you can see, much planning takes place before the first truck arrives.

As a concrete finisher, you may not be responsible for determining what questions to ask. However, the answers to the questions will affect how you do your job.

The following questions directly influence how you place concrete on the job site:

- How will the concrete come to the site?
- How will the concrete get from the truck or portable mixer to the form?
- Has the subgrade been dampened?
- Are the forms built? Braced? Inspected? Approved?
- Are the forms cleaned out? Are they wetted down?
- Has a form release agent been specified? Has it been applied?
- Are the screed rails in place?
- Is the reinforcement clean and properly placed?
- Are all embedded items and drains properly set?
- Is the expansion joint material in place?

Pre-Construction Meeting Checklist

1. **Scheduling:** Schedule at least two weeks prior to placing floor slabs with a special finish. If shrinkage compensating concrete is involved, are all necessary tests completed and sufficient data available to ensure predictable results? See special checklist.

2. **Location:** A conference shall be held at the project site.

3. **Attendance:** A responsible representative of the following shall attend:

   a. Owner's representative
   b. Architect/engineer
   c. Construction manager
   d. General contractor
   e. Concrete contractor and a superintendent
   f. Rebar or reinforcing contractor (if separate company)
   g. Ready-mix supplier
   h. Rebar or reinforcing supplier
   i. Other subs affected by slab on grade industrial floors
   j. Dry shake floor hardener representative
   k. Testing laboratory
   l. If expansive concrete is involved, Type K or component manufacturer

4. **Purpose:**

   a. To establish close coordination of all parties by not only reviewing in detail all plans, specifications, bid documents, and understanding, but by determining and specifically agreeing on the procedures for installing the slab under the apparent job conditions.

106F01.eps

Figure 1. Example Checklist Questions

- Are the finishing tools and equipment handy to the site?
- Are the curing materials handy to the site?
- What sequence will be used to place the concrete into the form?

Many companies use specific checklists for this part of the process, as it is very complicated. For instance, weather conditions can require special materials such as plastic sheeting to be on hand to protect the placement site. *Figure 2* shows a more specific placement checklist used by a different company.

The checklist holds signoffs by the inspectors, the contractor, and the client. It also has space to note weather conditions so that precautions can be taken to protect the concrete and the protective materials can be on hand.

Concrete Placement Checklist

Job Name: ______________________ **Job No.:** __________

Contractor: ______________________ **Contract No.:** __________

Pour location: ______________________________________

Pour date: __________ Estm CY: __________ Estm SF: __________

Type of structure: ______________________________________

DATE	QC REP	SUPT/ENG	COMPANY	ITEM OF WORK
				Layout – lane dimensions
				Subbase/fine grade – elevation – stability
				Mechanical/plumbing/fire protection – storm drain
				Electrical (ground cable, plates) – elevation – location
				Formwork (screed rails) – location – alignment – rigidity/bracing – elevation
				Embedments (guard posts, anchor bolts) – elevation – location
				Reinforcement (resteel, dowels) – spacing, elevation, clearance – lap location and length – rigidity of chairs, resteel
				Waterproofing
				Preparation for pour – clean – pumps – equipment ready – machines – screeds – vibrators
				Other

Pour start	Concrete mix	Temp Lo	Temp Hi
Pour finish	Concrete supplier	Wind	
Finish	Testing agency	Precipitation	
Curing		Sky	

Signature Concrete Contractor Carpenter Superintendent:	Date:
Signature Concrete Contractor Superintendent:	Date:
Signature Concrete Contractor QC/QA Representative:	Date:

106F02.TIF

Figure 2. Example Placement Checklist

2.2.0 METHOD OF PLACING

The contractor must have the equipment and workforce available to place the concrete without segregation. Placing equipment transports the concrete from the delivery source to its final position. Once placing begins, the work must continue without interruption because concrete will start to stiffen as soon as it is mixed. However, concrete that is agitated generally can be placed and compacted within 1½ hours of mixing.

The placing equipment needs to be large enough that one load of concrete does not set before the next load is placed next to it. Partially hardened concrete cannot be deposited into forms.

For these reasons, the method of transporting and placing concrete is very important. *Figure 3* shows some different methods of transporting concrete. Concrete gets carried from the batch plant or mixer to the job site, typically in some type of truck. At the job site, concrete is moved from the truck to the deposit site. This second transport can be accomplished with many different types of equipment. Concrete from a mixer on the job site moves directly to the point of placement.

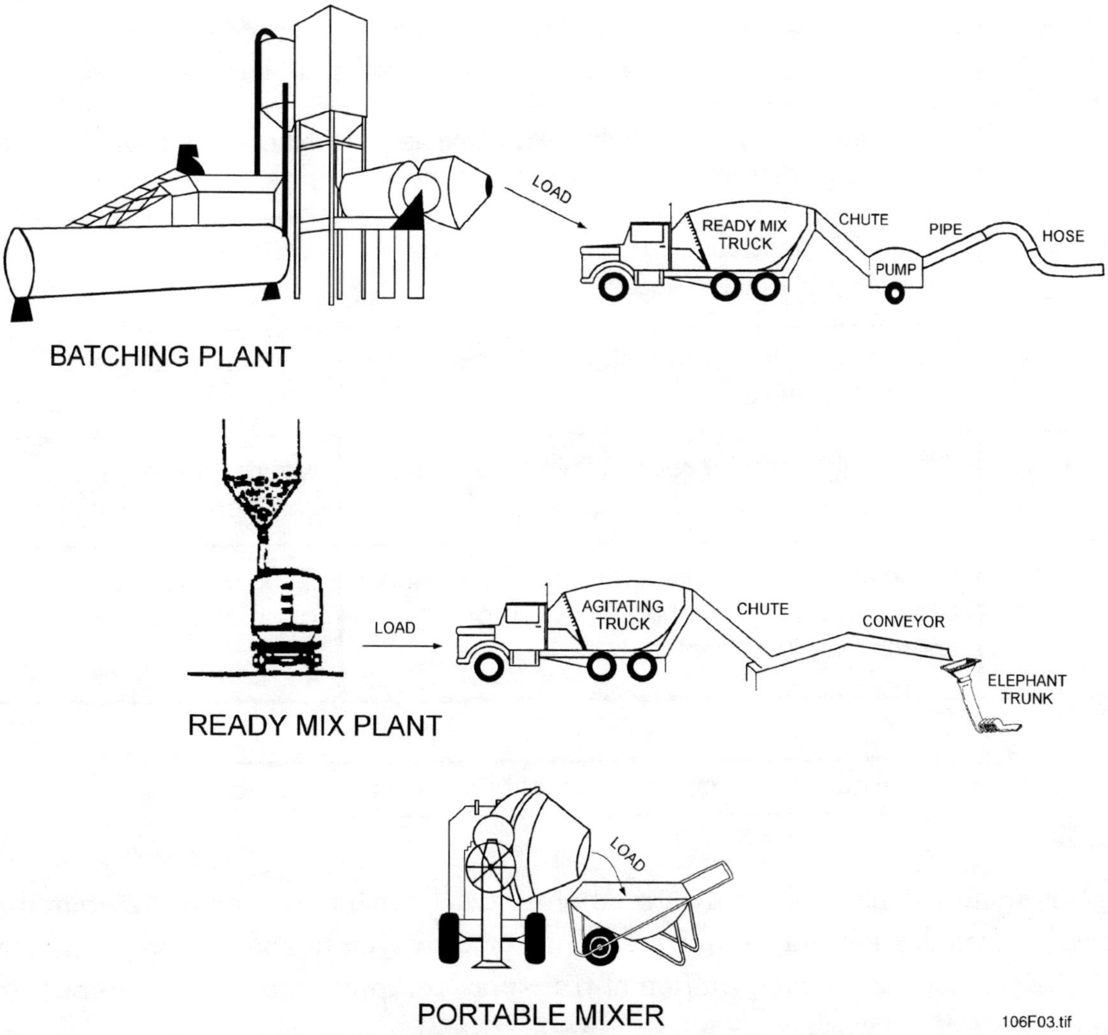

Figure 3. Methods Of Transporting Concrete

Figure 4 lists some common equipment for moving concrete from the mixer to the form. Each kind of equipment has advantages and disadvantages. The contractor selects the equipment for each job according to size, cost, availability, mix characteristics, and other considerations.

EQUIPMENT	ADVANTAGES	DISADVANTAGES
Wheelbarrow	Versatile for small loads and short hauls where conditions are always changing	Labor intensive; may need runways to avoid segregation
Buggy	Good for short hauls where placing conditions are always changing	Labor intensive; may need runways to avoid segregation
Chute	Gravity-powered; good for shorter distances and below-grade deposit	Must be supported, may need drop chute
Conveyor	Carry horizontally up to 3,000 feet, vertically about 100 feet; adjustable	Needs drop chute to prevent segregation
Bucket and Lifter	Moved by cranes or hoists over any distance	Needs cranes or hoists for lifting
Pump	Can deliver in continuous flow with flexible pipes over obstructions and around corners	Needs continuous input of fresh, plastic mix
Cableway	Commonly used for dams and other large-scale projects taking years to build	Very expensive
Hopper	Gives on-site temporary storage; transfers to other equipment	Complicates deposit
Dump truck	Commonly used for road building; can have end, side, or bottom dump gate	Needs mixer nearby; can carry only 30 minutes
Ready-mix truck	Combines mix and transport; most commonly used; has chute attached for reach of approximately 12 feet	Time deliveries to pace of on-site work

106F04.eps

Figure 4. Common Equipment For Moving Concrete

The simplest small job may need only a ready-mix truck with a one- or two-section discharge chute. A large, complex job may require a pump or conveyor to span a long distance to the placing site. Some jobs use a combination of transport equipment, or use different equipment at different stages of the work.

The equipment chosen will affect how the concrete is placed, and the type of mixture will affect the equipment chosen. For example, large aggregate may not flow through a concrete pumping system. Loose concrete with a high slump may not move well on a steeply-inclined conveyor.

2.3.0 JOB SITE CONDITIONS

Whatever the size of the job, the placing site must be accessible to the placing equipment. If trucks will be used, there must be a truckway set up. Lighting, power, and shelter from the elements are also critical for a successful placing.

Concrete should be placed under the following conditions to meet ACI specifications:

- Concrete cannot be placed during rain, sleet, or snow unless protection from these elements is provided.
- Concrete cannot be placed on a frozen subgrade.
- Preventive measures should be taken during cold weather to keep the internal temperature of concrete warm.
- At placing temperatures below 40°F, the internal temperature of concrete must be 55°F for slabs of 12 inches or less, and 50°F for slabs thicker than 12 inches.
- Internal temperature should not exceed 90°F at the time of placing. Preventive cooling measures are required in hot weather.
- If the temperature of steel forms or reinforcements exceeds 120°F, they must be sprayed with water just before the concrete is placed.

Mixing and placing need to be carefully coordinated with finishing. Concrete should not be placed at a faster rate than it can be spread, leveled, and floated, since these activities are done before bleed water begins to surface.

2.4.0 CREW SIZE AND MAKEUP

The size of the finishing crew will depend on the size of the job, the weather, the concrete temperature, the set time, and the placement rate needed to finish the job on schedule. More workers will be needed to spread and level the same amount of concrete in hot weather than in cold weather.

Teamwork safety involves being aware of what your fellow workers are doing. Teamwork also calls for working quickly and efficiently, especially in warm weather. One member of the team could be guiding the end of a discharge chute while others are leveling the discharged concrete with square-end shovels and come-alongs. Someone further away could be consolidating while a mechanized screed advances slowly. Other workers can be moving at the same time. Be alert at all times to what is happening around you.

Concrete must not harden while it is being deposited, so pacing of the work is essential. Remember that concrete starts to set soon after mixing.

2.5.0 PRE-PLACEMENT PROCEDURES

The job site needs to be readied before the concrete is ordered. Checks must be made of the subgrade, formwork, reinforcements, and materials on hand.

2.5.1 Subgrade Check

For flatwork, the subgrade must be accurately graded and uniformly compacted. Concrete workers should check the elevation of the subgrade by measuring down from a straightedge spanning between the side forms. High spots should be removed and low spots filled in and compacted. If the surface of the subgrade is 1 inch too low, it will take an extra cubic yard of concrete to fill an 18 × 18-foot section.

In hot weather, the subgrade may need to be moistened to prevent uneven absorption of moisture from the concrete. Semiporous subgrades should be dampened, with no standing water; porous subgrades are sealed. If a vapor retarder is used, compact 2 to 3 inches of granular fill on top of it. As shown in *Figure 5*, dampen the fill only if severe drying conditions threaten plastic shrinkage.

106F05.tif

Figure 5. Dampen Fill As Needed

There should be no muddy or soft spots or standing water on the subgrade when concrete is placed. Also, if the temperature is cold enough, there should be no frost on the subgrade.

2.5.2 Formwork Check

Formwork must be in place, set to grade, and correctly positioned. If slabs are to be screeded by mechanized screeds, check that the screed rails are in place. Sometimes, the tops of the forms also serve as the screed rails.

In cold weather, any snow, ice, or water in the forms must be removed, and the forms must be dampened. In hot weather, wet down forms immediately before placing. Apply any form treatment or form release agents before placing concrete into the forms.

2.5.3 Reinforcement Check

Reinforcement has to be properly installed. It should be free of dirt and rust prior to concrete placement. If you are using bare rebar, make sure all rust and scale has been removed. If you are using epoxy-coated bars (green bar), make sure the coating is not cut or broken.

Is the reinforcement secured in place? Is the reinforcement the proper distance from the subgrade? Is it set on chairs or bricks or other bracing materials? *Figure 6* shows a concrete worker checking the bracing and tying of reinforcements. Are expansion joint materials, anchors, and other embedded items positioned securely? Have they been inspected and accepted?

106F06.TIF

Figure 6. Checking Reinforcement

2.5.4 Materials Check

All tools and materials needed for placing the concrete should be at the site and accessible. This includes jointing materials, reinforcing steel, placing tools, finishing tools, and curing supplies. The workers should be on the job and ready to go. Valuable time can be wasted if equipment, tools, and workers are not ready to perform the necessary activities.

Weather conditions during placing can require special materials. Slabs may require polyethylene sheeting as an emergency cover if rain is expected. Reinforcements may need to be cooled down in very hot weather. Wind shields may be needed. These situations call for additional supplies.

To be sure everything is on hand, a daily list is a good idea. The daily list can include items needed for work as well as items for possible emergencies such as a sudden storm, a significant drop or rise in temperature, a power breakdown, or fire.

On some jobs, the foreman prepares a daily list and assigns a worker to be sure that all of the items are on hand. On other jobs, the foreman makes each worker responsible for specific items. The purpose is to avoid delays once placing starts.

2.6.0 PLACING SEQUENCE

Concrete must be placed continuously or in **lifts** of such thickness that no part of it hardens during the placing. If this happens, the concrete forms seams, **cold joints**, or layer lines.

Always deposit new concrete on the **heel** of the fresh concrete already in place. Place quickly so that the fresh and newly deposited concrete are both still plastic. Concrete that is partially hardened or contaminated by foreign materials should not be placed.

If a section cannot be placed continuously, construction joints are used. The placing pattern and the location of construction, contraction, and isolation joints should be mapped out in advance.

Learn the placing pattern for your job before the concrete arrives, and be prepared to place the concrete according to the preset placing pattern.

2.6.1 Large-Slab Pattern

The most efficient way to place concrete in large slabs is in long strips or lanes, as shown in *Figure 7*. Strip placements allow easier access to sections being placed. Intermediate contraction joints are set at specific distances across the strips as shown.

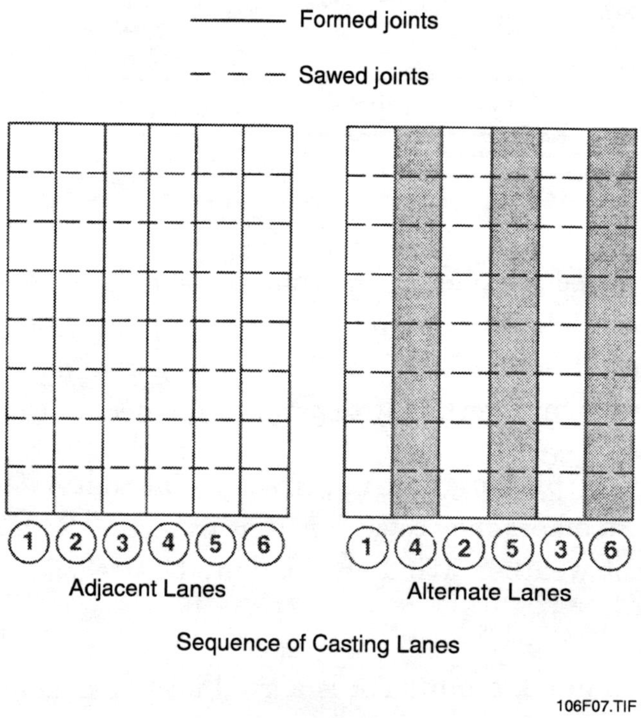

Figure 7. Placing Patterns For A Large Slab

When using this pattern, the first lane or strip is usually next to a wall. This means only one line of side forms is needed. For the second lane, the side forms are removed from the first lane and moved over for the second lane. Intermediate contraction joints are installed or sawed at measured intervals across the lanes. If the strips are wide, they may need longitudinal joints as well. Construction joints are used at the end of the day's placement.

Sometimes, the strips are deposited alternately instead of next to each other. In this case, forms are needed for both sides of each strip. This method gives more time to put in joints while still continuing to place new concrete.

2.6.2 Small-Slab Pattern

For small slabs, begin placing against the form edge farthest away from the concrete truck, as shown in *Figure 8*. Work back toward the supply and place concrete back toward the starting point as you go. If possible, work the shortest way across the slab to prevent cold joints.

Be sure to deposit the concrete as close to its final location in the slab as possible. If the slab is on a slope, start at the lower end and work uphill.

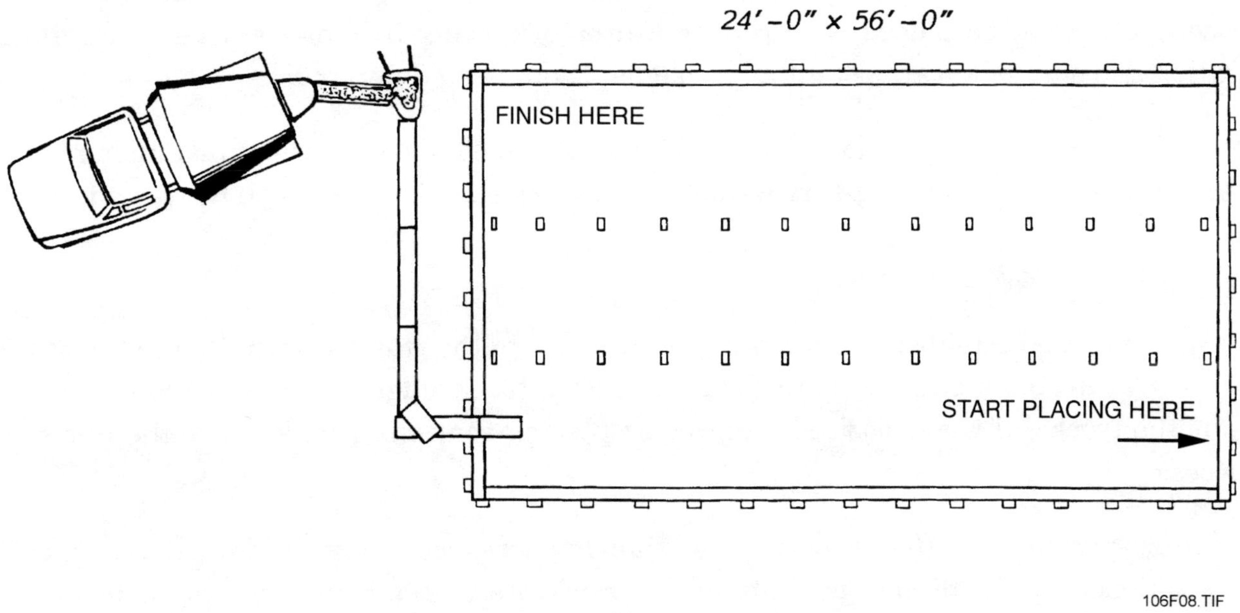

Figure 8. Placing Pattern For A Small Slab

2.7.0 SITE ACCESS

Site access can get very complicated. The first concern for concrete craftworkers is the pathway between the concrete and the form. Access to the form must be planned according to how the concrete will arrive. If the concrete comes by truck, a truck path must be set up to the form in a place that agrees with the placing pattern. A ramp may be needed if the deposit site is far from ground level. If the form is on a slope, the access should be at the top of the slope.

If the concrete travels by wheelbarrow or buggy, runways must be planned with separate paths for loaded and empty travel. Loaded buggies or wheelbarrows must have the right-of-way on their runways. All runways and equipment paths must be clearly marked so that workers can keep clear. Staying aware of traffic is an important part of job site safety.

On a large project, other materials will be arriving as well. Do not use equipment paths for temporary materials storage or leave tools in runways where they can be hit and damaged.

In rainy weather, inspect paths and lay planking in low spots. Sometimes, you can move paths to avoid drainage problems and puddles. If runways and paths are moved, be sure to move markings and inform workers of new locations.

3.0.0 DELIVERY AT SITE

Transporting concrete from the batch plant or mixer to the job site occurs most often in a ready-mix truck. Sometimes, agitating trucks carry premixed concrete to the deposit site. They are used when minimal mixing is needed to retain plasticity. Agitating trucks are used on large construction projects with some distance between the mixing plant and the deposit site. Dump trucks are often used on large road-building projects. They carry short distances, as the concrete must be placed within 30 minutes of mixing if it has not been agitated. Segregation and slump loss are potential problems when using dump trucks.

Before any truck arrives, equipment, tools, and workers need to be on site and standing by in a convenient place. Otherwise expensive time, and concrete setting time, can be wasted.

3.1.0 POSITIONING TRUCKS

A truck path must be established from the site entrance to the staging area. Before the truck arrives, check that truck pathways are clear and that a truck staging area has been set up to wash out the trucks. If the deposit site moves as placing continues, make sure there is still clear access.

Keep a clear space around the deposit site so that hoppers, conveyors, and other equipment can be positioned by the discharge chute of the truck. As trucks come and go, be aware of movement. Stay out of the truck path.

One person should be assigned to spot the truck by giving directions to the driver. It is important, when working with large moving equipment, that only one person give signals. Confusion and accidents result when more than one person tries to give signals.

3.2.0 TRUCK OPERATOR RESPONSIBILITY

The truck operator is responsible for delivering the batch of concrete to the proper site within the 90-minute, or 300-revolution deadline. A batch ticket should accompany every load of mix.

The operator drives or backs the truck to and from the specified discharge site according to signals from the designated spotter.

The truck operator is responsible for:

- Controlling the discharge and gate mechanisms on the truck
- Wetting the chute before concrete is released
- Starting the discharge of concrete on signals from the designated craftworker
- Stopping the discharge of concrete on signals from the designated craftworker

The truck operator washes out the mixer and chute so the next batch does not get contaminated by hardened concrete.

The operator of a dump truck is responsible for dumping the load of concrete according to signals from the designated concrete craftworker. After the concrete is discharged, the truck operator moves the empty truck to a designated area for washing.

The concrete craftworker is responsible for unloading the concrete from the chute and staying out of the way of the truck as it is positioned for discharging the concrete.

3.3.0 BATCH OR DELIVERY TICKET

Each load of concrete is accompanied by a batch or delivery ticket. A sample ticket is shown in *Figure 9*. At a minimum, the ticket should contain:

- Name of ready-mix batch plant
- Serial number of ticket
- Date and truck number
- Name of contractor
- Name and location of job
- Class of concrete and specification
- Amount of concrete in cubic yards
- Time loaded or first mixing of cement and aggregates
- Starting number on the revolution counter before the truck left the plant
- Amount of water that can be added at site
- Mix designation or batch weight of concrete mix design

Make sure you check the batch ticket *before* you start discharging and placing the concrete. Note that the ticket must be signed and a copy returned to the plant. Another copy goes to the job office.

Rev. 8-25-75

NAME OF CONCRETE COMPANY

Del. Ticket Serial No. 10210

Plant No. 86-C10

Date 8-6 1975

Delivered To San Marco Construction

Address Road 860 at 50th Street

Truck No.	Time Loaded	Arrived	Discharged	Cubic Yards This Load
26	1:55 PM	*	*	8
DOT Class	DOT Mix No.			Cubic Yards Total Today
II	2001			8

Allowable Jobsite Water Addition	Mixing Revolutions: At Plant
1½ gals./lbs.	Mixing Revolutions: At Plant 75 At Jobsite * (or Central Mixing Time:)

FILL OUT ON FIRST DELIVERY AND ON EACH CHANGE OF AGG. WEIGHTS	DOT Job No. 86925 -3503
Air Darex (Brand) 5.5 (Amount) oz	Retarder WRDA 79 (Brand) 45 (Amount) oz
Cement I (Type) Fla.Ptld. (Brand) 564# (Amount)	Batch Water 22 (Amount) gals./lbs.
Coarse Agg. 1773# (Amount) 2.0 (%Moisture)	Fine Agg. 1217# (Amount) 4.0 (%Moisture)
Flyash — (Source) (Amount)	

Issuance of this ticket constitutes certification to the accuracy of the above recorded information.

Tom Burgess
Signature of Plant Operator or Company Rep.

Water Added by Receiver of Concrete: * gals./lbs.

Received By: ______

106F09.TIF

Figure 9. Load Or Batch Delivery Ticket

3.4.0 CHECK BATCH/DELIVERY TICKET

The truck operator is responsible for filling out and presenting the ticket at the point of delivery. The designated concrete craftworker is responsible for reading the ticket to make sure this is the right truckload. If the truck is in the wrong place, do not let it unload.

Check the batch ticket and the revolution counter. Make sure the mixer has turned less than 300 revolutions. Check the time to make sure the concrete was mixed less than 90 minutes earlier. If the concrete has turned more than 300 revolutions or has mixed longer than 90 minutes, report it to your supervisor.

3.5.0 VISUAL INSPECTION

After checking the batch ticket, the revolution counter, and the time, inspect the concrete. If the operator has to add water at the job site, let it mix 30 revolutions after the water is added, then reinspect it.

Ask the truck operator to release a test batch to look at. Catch it in a wheelbarrow or on a board. Look for segregation, dry or hardened lumps, or water ponding on the surface. If you see any of these problems, report it to your supervisor. After you have visually inspected the concrete, and determined that it is the right truck, you can accept the batch. The next step is attaching the discharge equipment, usually extensions to the chute, and lining up the placing tools.

4.0.0 DISCHARGING CONCRETE

The important fact to remember in placing is that concrete begins to set soon after it is mixed. Whatever discharge equipment you use, the time limit is still the same.

Most concrete comes in a ready-mix truck that has several detachable chutes. Job conditions determine if the truck chute gets attached to another chute, a hopper, a conveyor, or other equipment. When moving concrete from the truck to the point of placement, you must prevent segregation and contamination. Check the conveying equipment before connecting it to make sure there is no old concrete or foreign matter inside.

The next sections describe specific discharge and placing equipment.

4.1.0 CHUTES

Chutes are the most commonly used method of discharging concrete into a form. The simplest discharge chute comes off the back of a ready-mix truck. Chutes can be joined to deliver over distance. The limitation on chutes is that they work by gravity, so they can only move the concrete downhill.

Chutes can cause problems of segregation, slump loss, and reduction of plasticity. Adding drop chutes will prevent segregation. Check that the slope of the chute can move the concrete at the slump specified. Long chutes in hot or windy weather cause the mix to lose slump through evaporation, so they must be covered. For short distances, chutes are the easiest and quickest method for placing.

Make sure the chute is wet before the concrete starts moving down. Guide the end of the chute as close as possible to the deposit site and keep the end down close to the subgrade. Deposit new concrete against the heel of concrete that is already there. *Figure 10* shows a ready-mix truck using two sections of chute to reach the deposit area.

To stop the flow from a chute, the designated worker must signal the truck operator. Remember that some concrete will still be moving down the chute after the flow is stopped.

106F10.TIF

Figure 10. Placing With Joined Discharge Chutes

4.2.0 WHEELBARROWS

Wheelbarrows can be filled directly from the chute at the ready-mix truck or directly from a portable mixer.

Figure 11 shows a worker loading a wheelbarrow from a small portable mixer. Notice the plywood walkway under the wheelbarrow. Use a square-end shovel to control the concrete. Do not overfill the wheelbarrow.

106F11.TIF

Figure 11. Loading A Wheelbarrow

The designated craftworker must signal to stop discharge from a chute before the wheelbarrow is full. This is because concrete will be moving in the chute after the gate is closed. Even with this delay in stopping the supply of concrete, chutes are easier to stop than other discharge equipment.

When handling a wheelbarrow, remember these tips:

- As you pick up the wheelbarrow, remember to bend your knees.
- Stay on the walkways. If there are no walkways, be sure the path is clear.
- Remember the wheelbarrow has momentum and give yourself time to stop.
- Wheel the load as close as possible to previously placed concrete.
- Dump the concrete on the heel of previously placed concrete.

4.3.0 BUGGIES

Figure 12 shows a power buggy carrying concrete across a wooden walkway. Buggies can be filled directly from a ready-mix truck's discharge chute or other outlet. The designated craftworker must signal to stop the concrete before the buggy is full, allowing for the concrete moving in the chute after the gate closes.

106F12.tif

Figure 12. Power Buggy

Hand-powered or motorized buggies are limited to transporting small quantities over short distances. Because of their weight, buggies need an experienced operator and a firm runway or surface.

- Familiarize yourself with buggy motor controls before driving.
- In transit, stay on the buggyway.
- If there is no set path, make sure your route is clear.
- Remember the buggy has momentum and give yourself time to stop.
- Dump the concrete on the heel of previously placed concrete.

4.4.0 CONVEYORS

Conveyors can be set up to continually move concrete to the discharge point. The discharge end of the conveyor is movable so that concrete can be placed over an area without restaging the conveyor.

Concrete with a slump of between 2 to 4½ inches travels most efficiently on a conveyor. The concrete travels in a ribbon centered on the trough-shaped belt. A border of empty space on either side keeps the concrete from spilling and keeps large aggregates from separating. Usually, conveyors have a drop chute or elephant trunk at the discharge end to bring the concrete closer to the form. *Figure 13* shows a truck-mounted portable conveyor discharging across an inaccessible area that prevents using chutes from a ready-mix truck.

106F13.tif

Figure 13. Portable Conveyor Discharging

The speed of the belt must be fast enough to keep the concrete plastic and unsegregated when it gets to the discharge point. Belts move in the range of 250 to 750 feet per minute, depending on the slump and the angle of the conveyor. Smooth belts can operate at vertical angles between 20 to 25 degrees, corrugated belts at greater vertical angles, between 30 to 35 degrees. A large conveyor can transport 40 to 50 cubic yards of concrete per hour.

Concrete travels on several different types of conveyors.

- *Portable conveyors* have their own gasoline or diesel power units and use a hydraulic motor drive. The maximum length of a portable conveyor is usually around 60 feet. Set-up time is very quick.
- *Series or feeder conveyors* have more than one flight linked by swivels. They should automatically start in sequence with the discharge flight starting first and the others starting back to the charging point. This gets each flight operating at the proper speed before the concrete hits the first belt.
- *Straight line spreaders* are side-discharge conveyors that have a traveling plow or diverter. The diverter uses a wiper-type blade to remove the concrete from the belt. These are often used on road and bridge projects.
- *Radial spreaders* have a support that swings the discharge end through a radius of up to 360 degrees. The support also extends and retracts the conveyor.

When working with conveyors, remember that they are fast moving.

- Make sure the conveyor is clear of foreign material before it starts moving.
- Check that links and swivels are securely set.
- Check and clean the scraper that removes paste from the discharge end.
- Shield or cover conveyors during windy or hot weather to avoid evaporation across the length of travel.

When working with a drop chute or elephant trunk, guide the discharge so that new concrete falls against the heel of concrete previously placed. When working with a radial spreader, stay out of the way of the moving end of the conveyor.

The designated concrete craftworker signals the conveyor operator to start concrete moving. If the need arises, it is easier to slow down the conveyor than to stop it. The designated worker signals to slow down. If the conveyor has to stop, time becomes critical. If the interruption is long, the belts must be cleaned off before they can be restarted.

4.5.0 BUCKETS

Buckets can be filled by discharge chute, hopper, or other conveying equipment. Once filled, they can be moved by crane, hoist, or cableway. For safety, only one person should be allowed to signal the lifting operator. This is especially important when working with a crane or derrick because of the free travel of the bucket. Safety requires that each craftworker learn the crane hand signals so they will know what to expect. These crane hand signals can be found in the Core Curricula module, *Basic Rigging*.

Cranes may generate hazards. In high-rise construction, materials movement by crane poses the greatest safety hazard to workers on the job. Work safety calls for:

- Establishing clear pathways for materials movement
- Using a consistent set of signals to alert workers to movement
- Assigning individual responsibilities for each worker during materials movement

Buckets can be dumped into an intermediate hopper or directly into the forms. *Figure 14* shows a bucket discharging into a form. Note the lever gate control operated by the worker on the right. Before working with a bucket, familiarize yourself with the gate controls or other discharge mechanism. You should be able to open and close the discharge gate at any time during discharge.

106F14.TIF

Figure 14. Bucket Discharging Into Form

Guide the bucket close to the discharge spot. Do not discharge the bucket too close to the lift surface, and do not discharge it while it is moving. Either of these actions will cause segregation. Use the gate to control the discharge so that you get a slow, steady stream of mix coming out of the chute. Aim the discharge against the heel of the concrete already in place.

To prevent contamination, do not shovel spilled concrete back into the bucket. Do not move the bucket across freshly finished concrete. Check the bucket from time to time to make sure there is no buildup of semi-hardened concrete inside.

Do not discharge the concrete too quickly, or the suddenly lighter bucket will jerk upward into the air. This could be dangerous to nearby workers and to the lifting equipment. When the bucket is empty, the designated worker signals for its removal.

You need to be alert and take extra precautions when using a bucket to lift and place large amounts of concrete. Follow these guidelines to ensure safe operation:

- Make sure the capacity of the crane is sufficient to handle the weight of the concrete bucket.
- Never allow yourself to get between a bucket and a stationary object.
- Never allow yourself to get between two buckets.
- Never open the bucket until the discharge gate is less than knee high.
- Always watch the swing of the bucket.
- Do not discharge the concrete too quickly or from too great a height, as this may damage or collapse the formwork.

4.6.0 PUMPS

Pumps work well with medium consistency concrete without large size aggregate. Aggregate size should be smaller than ⅓ of the inside diameter of the pump hose. Pumps can change the plastic properties of the concrete. The temperature will rise, air content will fall, and workability will drop. You should perform air, slump, and temperature tests at the beginning of the pump before the concrete is pumped, as well as at the discharge end of the pump.

Pumps require a steady stream of consistent concrete. They work with pressures ranging from 250 to over 1,000 psi. They require rigid steel or plastic pipes coupled to reinforced hose at the far end. Hose can be used interchangeably with rigid pipe sections as long as pipe is used next to the pump. Lay out the coupled pipes and hoses so they are firmly supported. Check the couplings and gate valves before pumping starts. *Figure 15* shows mix being placed with a hose connector to a concrete pump. Note that the worker is guiding the discharge end of the pump hose with a steel guide bar. Use a harness or a guide bar to control the end of the pump hose. It is under pressure and has a strong force, so do not use your hands directly.

Position the discharge end of the assembly as near to the placing spot as possible. A designated craftworker maintains communication with the pump operator to make sure that the pump line stays full and that concrete is moving steadily through the line.

106F15.TIF

Figure 15. Discharge End Of Pump Hose

Monitor the discharge carefully to prevent segregation. Aim the discharge to prevent displacing reinforcement or damaging the form. Avoid work stoppages once pumping begins. It is best to slow the pump down if a delay is needed. Once the pump stops, time starts running out. The entire assembly may need to be cleaned out before starting again.

Prime the pump with a batch of mortar or cement without aggregates. Special pump priming materials are also available. Discharge this batch into a waste pile, not inside the form.

CAUTION: Be sure the line is clear as it is assembled. For safety reasons:

- Do not stand on the line.
- Do not look into the line.
- Do not hold the end of the line with your hands.

4.7.0 DUMP TRUCKS

Dump trucks do not have mixers or agitators. They are typically used on large projects, such as road construction, where large amounts of concrete are spread over clear areas. They can haul large amounts of concrete for short distances more economically. Usually, dump trucks are limited to 30-minute hauls because there is no provision for agitating. *Figure 16* shows a dump truck depositing mix into a concrete paving machine that is paving an airport runway.

106F16.tif

Figure 16. Dump Truck With Concrete Paver

Segregation and slump loss can be problems when dump trucks are used. Minimize these problems with a tarpaulin to cover the load and a smooth pathway to keep the mix from bouncing up and down.

The designated concrete craftworker gives the truck operator signals to position the truck. Dump trucks have end, side, or bottom dump gates. The truck operator controls the gates based on signals from the designated worker.

4.8.0 COLLECTION HOPPERS

Hoppers collect concrete from the truck or portable mixer, hold it, and transfer it to handling equipment in an uninterrupted stream. Most hoppers are loaded from the top and empty from the bottom. Hoppers provide a vertical drop at the discharge gate. The gate is usually nonjamming with positive action, spring-loaded linkages. Familiarize yourself with the controls for the gate before using them. *Figure 17* diagrams the correct and incorrect methods of using a collection hopper to discharge into a buggy.

Collection hoppers come in three designs.

- One side open to receive concrete from power buggies
- Four-sided for charging by bucket with normal-slump concrete
- Extra steep sided with a large discharge opening to handle extremely low-slump concrete

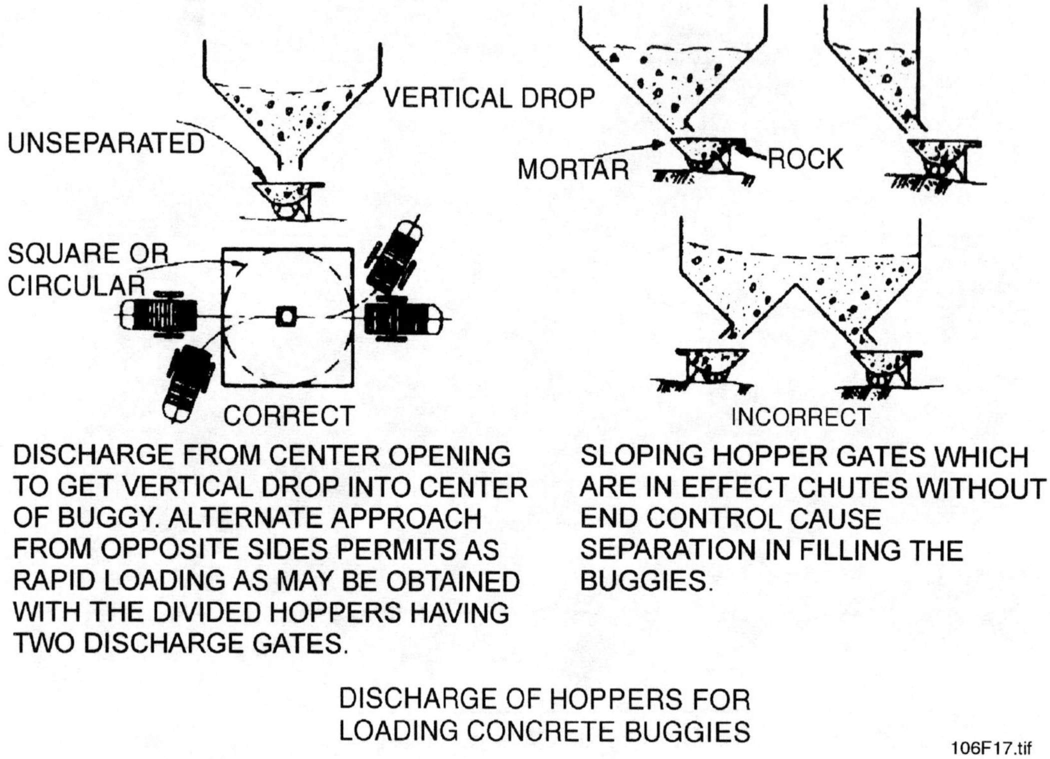

Figure 17. Collection Hopper

5.0.0 DEPOSITING, SPREADING, AND CONSOLIDATING

This section deals with placing work inside the form. It begins with depositing the mix in the form and works through spreading, leveling, and consolidating.

5.1.0 DEPOSITING

Beyond the rules for working with depositing tools and equipment, the following paragraphs give some general principles for placing concrete mix inside formed areas.

During discharging, make sure each deposit is placed against the heel of the previously deposited concrete. Do not dump the concrete in small, separate piles and then try to work them together. Also, never dump the concrete in large piles and try to move it horizontally into position. Never use a vibrator to move concrete horizontally. All of these practices result in segregation, because mortar tends to flow ahead of the coarser material.

Follow the placement pattern for your job. Start placing at the proper end of the form and work according to the lane direction. If concrete is started at both ends and placed toward the center, air and water can get trapped in the center. This will result in a **honeycomb** which causes structural weakness. Do not let water collect at ends, in corners, or along form faces.

5.1.1 Lifts

For many jobs, concrete is placed in lifts, or horizontal layers of uniform thickness. Each lift is leveled and consolidated before the next is placed. Leveling and consolidating a lift needs to happen quickly enough so that it has not yet set when a new lift is placed on top of it. This will avoid flow lines, seams, and cold joints.

Placing in lifts is handy for reinforced slabs. This technique allows laying one lift, consolidating it, then placing the reinforcement on top of that lift. The second lift is laid over the top of the reinforcement. This process makes it easier to get the reinforcement at the proper level. The first lift will usually be about ⅔ of the depth of the slab. The second lift must be at least 2 inches deep to position the reinforcement properly. *Figure 18* shows a cross-section of a two-lift slab with reinforcements.

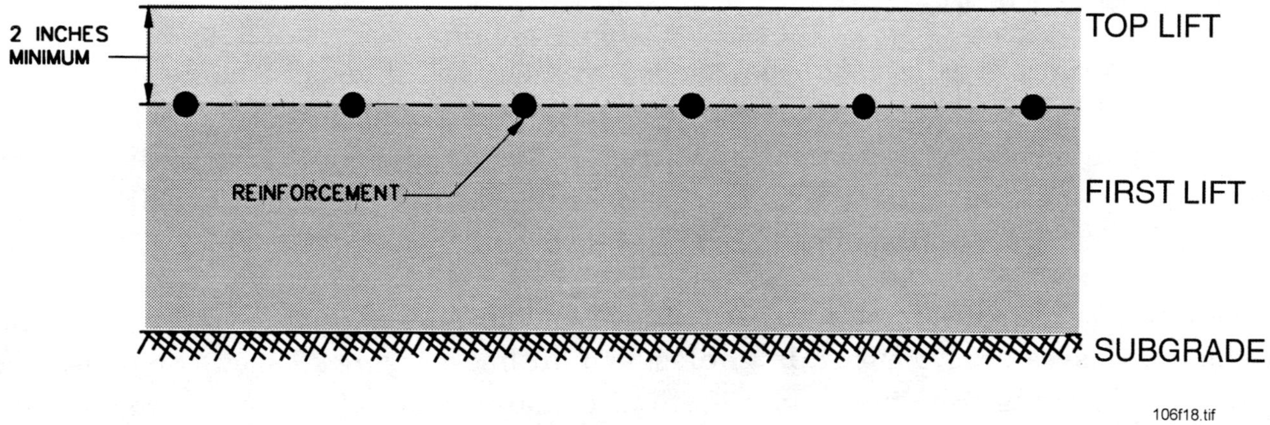

Figure 18. Two-Lift Slab

5.1.2 Bridging

Do not allow the concrete to separate over rods, spacers, reinforcement bars, or other embedded items. This leads to a condition called **bridging**, where the concrete mounds over the reinforcement bar and creates a void underneath. Further placing and consolidating may not fill that void.

To avoid bridging, fill the space underneath the obstruction first, then fill on top. Sometimes you must rod the concrete before finishing the lift to make sure there is no bridging. If bridging does occur, using a vibrator can correct it.

5.2.0 SPREADING

After the concrete is placed in the form, begin spreading it. Start leveling as the concrete moves out of the deposit equipment onto the subgrade. Use square-end shovels, concrete rakes, and come-alongs. Follow these tips for spreading concrete:

- Move the concrete horizontally as little as possible to avoid segregation.
- Working quickly, even out the ridges and hollows left by the discharge equipment.
- Move the spreading tools horizontally; do not press downward into the concrete. The objective is to smooth out the lane.
- Work across the lane, backing toward the most recent placement.
- Do not try to level the concrete with the edges of the form. Leave the surface higher so that strikeoff will produce a true grade.

5.3.0 CONSOLIDATING

Freshly placed concrete is usually honeycombed with entrapped air. If it hardens like this, the concrete will be weak, porous, and poorly bonded to its reinforcement. Consolidating or compacting gets rid of most of the voids in concrete and brings the aggregates closer together. During compacting, the concrete gets solidly packed around reinforcements, embedments, and into form corners. The paste should be evenly distributed.

Consolidating is a necessary step to get the uniform surfaces needed for most jobs. Poor consolidating leaves honeycombing, **bugholes**, and lift lines. *Figure 19* shows bugholes and honeycomb resulting from poor consolidation. Under-consolidating is a more common problem than over-consolidating; so if there is any doubt, continue the consolidating effort longer.

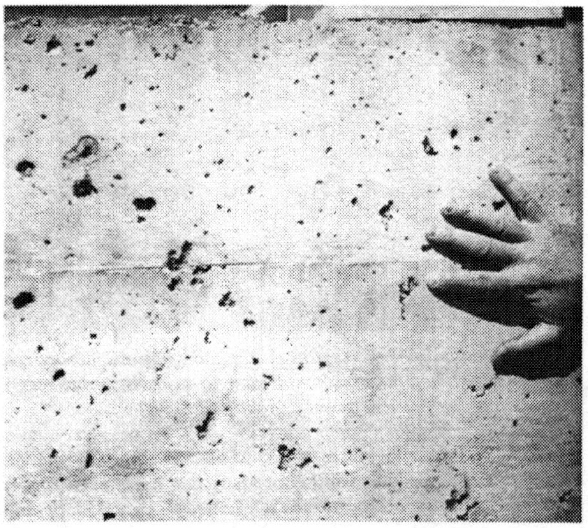

Bugholes

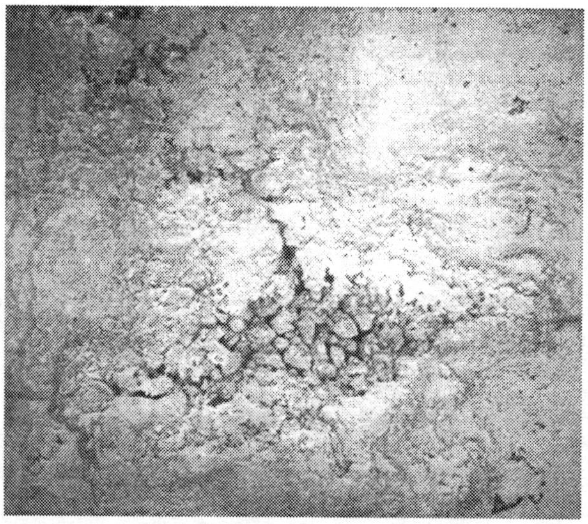

Honeycomb

106F19.TIF

Figure 19. Bugholes And Honeycomb

When consolidating, work the concrete around the reinforcement and embedded fixtures and into the corners of the forms. Common methods are by rodding, spading, or vibration. Consolidating cannot be done with spreading tools, and spreading should not be done with consolidating tools. Do not move concrete horizontally with vibrators.

5.3.1 Hand Consolidating

Hand consolidating includes rodding, spading, and tamping. On some jobs, manual consolidating using a rod or pole is all that is necessary.

Rodding, or hand consolidation, is using the hands to drive a metal or wooden rod into the concrete and withdrawing it slowly. The thrusts of the rod must be fairly close together. This pumping action produces movement of the concrete within several inches of the rod and makes the concrete flow into voids.

Spading is done by driving a flat spade into the concrete at the edge of the form with the spade back against the form, as shown in *Figure 20*. Shift the spade laterally a few times, then slowly withdraw it. Spading along the edges of a form forces larger coarse aggregates away from the form. It also reduces the air bubbles on the concrete surface. Reinsert the spade next to where it was withdrawn so there will be continuous spading along the edges of the form. Spading can improve the surface appearance of concrete that has been mechanically vibrated.

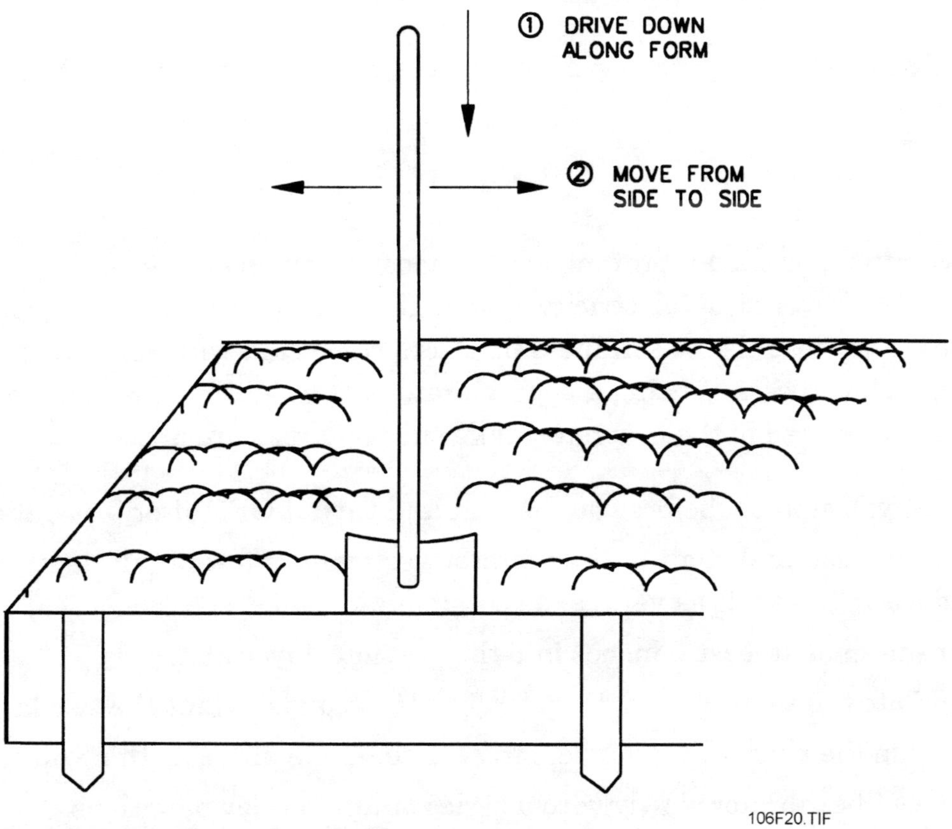

Figure 20. Spading

For very dry, hand-tamped concrete, the surface is rammed with heavy, flat-faced mechanical tampers until a very thin film of mortar appears at the surface. This indicates that the voids inside have filled.

5.3.2 Mechanical Vibrators

Mechanical vibration is the most common method of consolidating concrete. The mechanically-produced rapid vibrations cause the concrete to behave, temporarily, like a liquid. The liquid settles in the form and entrapped air rises easily to the surface. Mechanical vibrators are of three types: form, surface, and internal.

Form vibrators are attached to the outside of the form. They vibrate the form, which vibrates the concrete. These vibrators are used for precast shapes. They are useful for places where it is difficult or impossible to insert an internal vibrator.

Surface vibrators include plate or grid vibrating tampers such as the jitterbug, vibrating roller screeds, and vibrating screeds. Plate or grid vibrating tampers work best on very stiff concrete. Move hand-held tampers slowly in small arcs to push coarse aggregate slightly below the surface. Do not let the tamper sit in one place.

The vibrating roller screed strikes off and consolidates. It has one or more rollers and works well on plastic mixes. It knocks down, screeds, and provides vibration in one slow pass. This screed will run on rails or wheels just outside the slab form.

Some straightedge screeds have vibrators attached and provide surface vibration as they strike off.

5.3.3 Internal Vibrators

Internal vibrators, or spud vibrators, are the most common equipment used for consolidating fresh concrete. Powered by electricity, gas engine, or compressed air, they are suitable for most mixes. They lose their usefulness in places congested with reinforcement or in hard-to-reach areas. The process of using a spud vibrator is straightforward. Operate it continuously so no holes are created in the concrete. Make sure no areas are missed.

- Turn the vibrator ON before you insert it and turn it OFF after it is out of the concrete.
- Lower the head and shaft of the vibrator vertically into the full depth of the layer being placed. Keep the shaft as vertical as possible.
- Lower the spud at least 6 inches into the previous layer also.
- Manipulate the shaft up and down for 5 to 15 seconds to knit the two layers together.
- Do not run the vibrator for any length of time in one place, as this will cause segregation.
- Withdraw the vibrator slowly with a series of up-and-down motions.

- Reinsert the vibrator so that it overlaps the area visibly affected by the previous insertion. Look carefully to be sure of the overlap. If you do not overlap the vibrating field, honeycombing will result.
- Do not let vibrators come in contact with the form, as they will damage form faces. This will cause surface defects on the concrete.

Vibrators can be hazardous. To avoid accidents, remember these rules:

- Keep hoses and electrical cords from snagging on forms and reinforcements.
- Keep electrical wires, connectors, and equipment out of water, oil, or concrete.
- Safety-wire or chain all connectors on an air line before turning on the air.
- When disconnecting an air vibrator, plug the air inlet with a clean cloth to protect it from dirt.
- Check the line oiler for an air vibrator to make sure it contains oil at all times. Do not allow dirt to enter the oiler.
- Clean all equipment thoroughly when you are finished.

Figure 21 shows the correct and incorrect ways of inserting the vibrator into fresh concrete. Note that the vibrator goes into the previous lift. The concrete cools the vibrator, so keep as much of it covered as possible.

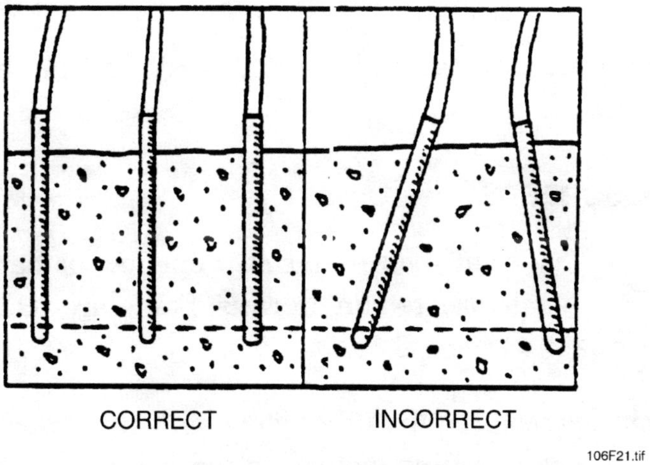

Figure 21. Vibrator Placement

For thin slabs, set the vibrator into the concrete at an angle. Drag the vibrator through the concrete at an angle, keeping the vibrator immersed as much as possible. You will know that vibration is complete when:

- The concrete levels off and settles into place
- Air bubbles rise to the surface
- The surface takes on a sheen, from a thin layer of mortar over the aggregate
- The sound of the vibrator changes
- The feel of the vibrator changes

5.4.0 REVIBRATION

Revibration is the vibration of concrete which has already been placed and consolidated. Revibration occurs unintentionally when vibration on an upper lift penetrates the partially hardened lift beneath.

Deliberate revibration is done when placing is interrupted. The leading edge of the concrete may be partially set when work resumes. It can be revibrated back to plasticity to avoid a cold joint.

Revibrating is done while the running vibrator will still sink into the concrete from its own weight. Revibration releases trapped water underneath horizontal reinforcing bars. It can be done after any length of delay, provided that the mix can be brought back to a plastic state. Mix and weather conditions determine the actual allowable delay.

6.0.0 STRIKE OFF AND SMOOTHING METHODS

After the concrete has been placed and consolidated, it is struck off by a mechanically or manually powered screed or straightedge. Striking off can be done before and/or after compacting, depending on the job. After strike off is complete, floating smooths the surface. All of these activities take place before any bleed water comes to the surface of the placed concrete.

The next sections discuss mechanical and manual strike off methods as well as floating techniques.

6.1.0 MECHANICAL SCREEDS

Mechanical screeds can be gas- or air-powered, self-propelled, or hand-operated. Some move along the tops of the forms, while others run on rails. Some are pulled along the tops of the forms, as shown in *Figure 22*.

Working with a mechanical screed requires that you check the grade and/or tracks before the screed starts its pass. The best screeding happens close to where consolidating is finished, so the screed travels a few feet behind the consolidators.

A small (1 to 2 inch) rolling wave, or **surcharge**, of concrete should travel in front of the screed to fill in any low spots. Check to see that the surcharge does not become too big, which could cause the concrete to surge up behind the screed and create a wavy surface.

Mechanical screeds are used on highways, runways, and other large slabs. Some of them have vibrators, so they also consolidate. The latest models have laser-guided strike off and combine spreading and vibrating with strike off. Internal vibration or a combination of internal and surface vibration is recommended for reinforced slabs. *Figure 23* shows a laser-guided screed that includes vibration.

Figure 22. Mechanized Screed

Figure 23. Laser-Guided Screed

If the screed has a vibrator, be sure to turn off the vibrator when the screed is not moving forward. This will avoid bringing excess fines to the surface if the screed is left stationary.

6.2.0 STRAIGHTEDGE

A straightedge, or manual screed, can be as simple as a 2 × 4 piece of lumber (*Figure 24*) or a square-edged magnesium or steel bar. The requirements for a manual screed are that it be straight and that it extend past the edges of the form. The maximum recommended size for manual strike off is 12 feet. Larger spans will develop rollback or sag with manual strike off.

106F24.TIF

Figure 24. Manual Screeding

For narrow forms, such as walkways, one person can operate a straightedge. For larger forms, two people must work together, one at each end of the straightedge. When straightedging with a partner, it is important that both of you move at the same rate.

Strike off the concrete in two passes soon after it is placed. Follow these guidelines:

- First, pull the straightedge forward in a side-to-side motion with its top edge tilted 15 to 20 degrees forward. The tilt keeps the edge from tearing out large aggregate.
- Keep the bottom edge of the screed in contact with the top of the form as much as possible.
- For the second pass, tilt the straightedge in the opposite direction.
- Pull the straightedge forward while sliding it back and forth across the tops of the forms.
- Keep a low (½ to 1 inch) wave of concrete in front of the straightedge to fill in low spots.

If there is too much surcharge, the surface will tear. Remove the excess with a shovel, then continue screeding.

6.3.0 FLOATING

Floating is the compaction and consolidation of the unformed concrete surface. Floating operations take place at two separate times during the concrete finishing process.

The first floating is by hand and takes place immediately after screeding. Initial floating should be completed before any excess moisture or bleed water is present on the surface. Any finishing operation performed while there is excess moisture or bleed water on the surface will cause dusting or scaling.

The first floating operation is performed using a bull float, darby, or highway straightedge. The second floating operation takes place after evaporation of most of the bleed water.

6.3.1 Highway Straightedges

Highway straightedges are special floats used to create and maintain a flat surface during the plastic finishing process. They vary in width from 4 to 16 feet with a pole handle of equal length. Straightedges are generally rectangular in cross-section and attached to a handle with an adjustable head connector. The highway straightedge can be set with a camber in it to match the camber required for the road. Straightedges are also used on floor slabs, as shown in *Figure 25*.

106F25.TIF

Figure 25. Highway Straightedge

When using a straightedge, remember these tips:

- Use the highway straightedge at a 90-degree angle to the direction of the screeding.
- The highway straightedge should be pulled toward you in a continuous motion.
- Keep the handle low.
- Each pass should overlap the previous pass by ⅓ to ½ its width.
- If the surface has many irregularities, the highway straightedge can be moved in a back-and-forth scraping motion to level out the high spots.

For superflat floor slabs, several steps are needed to ensure flatness. The following checklist is one company's way of assuring a superflat floor by using a straightedge. Different procedures can be used on different jobs.

- Set edge forms using optical or laser levels to ensure proper grade.
- Place concrete uniformly in front of the screed.
- Use a vibrating screed for initial strike off. Maintain uniform surcharge in front of the screed. Check the grade behind the screed frequently and adjust the screed as required.
- Turn off the screed vibrator when the screed is not moving to avoid bringing excess fines to the surface.
- Bullfloat or use a highway straightedge before bleed water rises to the surface.
- After bleed water evaporates, power float across strip width.
- Restraighten finely with a 10-foot highway straightedge. Use the straightedge in at least two directions, preferably at a 45-degree angle to the strip length.
- Make multiple passes with a power trowel.
- Remeasure flatness.

6.3.2 Bullfloats

The bullfloat is used to eliminate ridges and to fill voids left by screeding operations. Bullfloating should embed the coarse aggregate only slightly. This process prepares the surface for subsequent edging, jointing, floating and troweling.

A traditional bullfloat that is 4 to 5 feet wide can be used to smooth and consolidate the concrete surface after screeding. The use of this width of bullfloat, however, can make achievement of higher flatness extremely difficult. An 8-foot to 10-foot bullfloat is better at removing surface irregularities early in the finishing process for an eventual superflat floor. For air-entrained mix, magnesium or aluminum bullfloats are best, as they will not tear out embedded aggregate. Wood bullfloats are rougher and best used on low-slump mixes without air entrainment. *Figure 26* shows bullfloating on a floor slab.

Figure 26. Bullfloating A Floor Slab

The floating must be completed before bleed water appears on the surface, so it is important to work quickly. Surface finishing with bleed water on the surface causes crazing, dusting, and scaling.

When using a wooden bullfloat, follow these procedures:

- Start the direction of the floating at right angles to the direction of the screeding.
- For the first pass, push the bullfloat away from you with the handle held low and the front of the blade raised slightly so it will not dig into the surface.
- Pull the blade back with the handle low so the blade of the bullfloat is flat against the surface.
- Move the float with a push-and-pull, forward-and-back motion to fill in holes and smooth ridges. A wider float makes it easier to locate and remove surface irregularities.
- Make two passes, with the second at a 45 degree angle to the first.
- If dishing occurs in the surface, use a darby to level it off.
- If low spots remain, add concrete to the surface from a wheelbarrow and bullfloat again.

When using a magnesium bullfloat, keep the blade flat against the surface at all times. If the blade is tilted, the surface will be sealed. Bullfloats are used to embed mineral-sprinkle hardeners and other surface toppings. Hand floating or power floating follows bullfloating, according to the requirements of the job.

CAUTION: Metal handles on bullfloats conduct electricity. Be careful if you are working near electrical equipment with this tool.

6.3.3 Darbying

Darbying serves the same purpose as bullfloating, and the same rules apply. Since bullfloating and darbying have the same effect on the surface of fresh concrete, the two operations are not performed on the same surface. A darby is advantageous on narrow slabs and in restricted spaces. Long-handled darbies can be used for better leverage and control of level. Metal darbies are usually unsatisfactory for producing surfaces meeting high-tolerance requirements. *Figure 27* shows the use of the darby.

106F27.tif

Figure 27. Darbying

Darbying calls for moving the blade flat to the surface.

- Move the blade back and forth in a rapid sawing motion while also moving in a sweeping arc.
- Be sure to cover the surface in a regular pattern, moving slowly from one edge to another.
- Move the darby at an angle to the path of the screed.
- If you must move onto the surface of the slab, use kneeling boards and knee pads.

After the surface is level, tilt the darby blade at an angle. Move it in a sweeping motion to fill any holes left by the sawing motion.

SUMMARY

Placing concrete covers activities from delivery of the concrete to the site through bullfloating, darbying, or highway straightedging before the bleed water rises. Placing activity starts with checklists that cover the project as a whole and list preparations for concrete delivery. The equipment used to deliver the concrete is selected as well as any additional equipment needed to convey the concrete while it is on the job site. More detailed checklists are drawn up. Subgrades, forms, and reinforcements are inspected one last time. Defined traffic patterns and runways are marked out as needed. Placing sequences are set up.

The concrete arrives, and the batch ticket is checked. Placing begins. The concrete may be placed into the form using a chute, conveyor, pump, buggy, or other equipment. As it is placed, the concrete is spread with shovels and come-alongs. Then it is consolidated by spading or mechanical vibration.

Strike off follows consolidation with mechanical or motorized screeds. Some screeds include surface vibrators. Straightedges are used for fine tolerances. Bullfloating and highway straightedging follows strike off, and must be complete before bleed water shows. After the bleed water evaporates, finishing operations begin.

References

For advanced study of the topics covered in this task module, the following reference materials are suggested:

Concrete Construction and Estimating, Craig Avery, Craftsman Book Company, Carlsbad, CA, 1991.

Concrete Manual, International Conference of Building Officials, Whittier, CA, 1992.

Concrete Parking Lots: Eight Steps to Quality Construction, Videotape, Portland Cement Association, Skokie, IL.

Concrete Technology, Third Edition, George R. White, Delmar Publishers, Inc., Albany, NY, 1991.

Design and Control of Concrete Mixtures, Thirteenth Edition, Portland Cement Association, Skokie, IL, 1988.

Finishing Concrete Flatwork, Videotape 32:00 minutes, Portland Cement Association, Skokie, IL, 1991.

Guide for Concrete Floor and Slab Construction, ACI 302.1R-96, American Concrete Institute, Farmington Hills, MI, 1997.

The Contractor's Guide to Quality Construction, American Concrete Institute, Farmington Hills, MI, 1997.

ACKNOWLEDGMENTS

Figures 1, 3, 4, 9, 10, 18, and 20 courtesy of Roy Jorgensen Associates, Inc.
Figure 2 courtesy of The Aberdeen Group
Figures 5-7 and 19 courtesy of American Concrete Institute
Figure 8 courtesy of Associated General Contractors
Figures 11, 13, 14, 22, 26, and 27 courtesy of Portland Cement Association
Figures 12, 15, 24, and 25 courtesy of Baker Concrete
Figure 16 courtesy of U.S. Department of Transportation
Figure 21 courtesy of Alberta Ready Mixed Concrete Association
Figure 23 courtesy of SOMERO Enterprises, Inc.

Pre-Construction Meeting Checklist

1. **Scheduling:** Schedule at least two weeks prior to placing floor slabs with a special finish. If shrinkage compensating concrete is involved, are all necessary tests completed and sufficient data available to ensure predictable results? See special checklist.

2. **Location:** A conference shall be held at the project site.

3. **Attendance:** A responsible representative of the following shall attend:
   a. Owner's representative
   b. Architect/engineer
   c. Construction manager
   d. General contractor
   e. Concrete contractor and a superintendent
   f. Rebar or reinforcing contractor (if separate company)
   g. Ready-mix supplier
   h. Rebar or reinforcing supplier
   i. Other subs affected by slab on grade industrial floors
   j. Dry shake floor hardener representative
   k. Testing laboratory
   l. If expansive concrete is involved, Type K or component manufacturer

4. **Purpose:**
   a. To establish close coordination of all parties by not only reviewing in detail all plans, specifications, bid documents, and understanding, but by determining and specifically agreeing on the procedures for installing the slab under the apparent job conditions.

5. **Meeting Agenda:** (Our specific scope of work)
   a. Sequence of construction and milestone dates.

   b. Basis for acceptance of base/subgrade grading (completely resolved by all parties prior to placing), compaction, and elevation (specifically, who determines when we pour and what the minimum square feet available is per day, per week?).

   c. Protection of area from the elements:
      - roof and walls
      - maintain access roads for trucks
      - available storage areas and security
      - available space for pumping operations

   d. Lighting, power, water, water pressure, and ventilation during placing and finishing.

   e. Acceptable pour sequence with joint layout. (If expansive concrete is being used, give special attention to the sequence based on the results of the ASTM C-878 expansion test. Insist on the manufacturer of the expansive product to approve the layout or give specific suggestions.)

   f. Forms, bulkheads, column enclosures.

   g. Placing reinforcement: (if steel fibers are being used, see special details below)
      - procedure
      - location
      - support

h. Necessary prepour clearance procedure.

i. Setting reinforcement:
- chairing reinforcement, method, and timing
- is reinforcement discontinued at all joints, alignment to permit sawcuts to open
- if required, dowel and dowel baskets (chip wires)

j. Concrete:
- mix design(s), specific details on all ingredients and how the combination may be expected to work under the anticipated field conditions, yardage, and starting time
- delivery, mix, slump, provision for one (1) on-site adjustment
- test reports and quality control
- where and when tests are made
- who will receive copies of test reports (and when)
- number of trucks
- type of truck
- turn around time
- wash out area
- if general contractor is buying concrete, do we have all the facts and do we have approval of the mix at the slump we will use

k. Placing and screed methods:
- vibrator
- review entire finishing procedure and necessary tools and equipment, including scraping straightedge

l. Special finishing materials:
- material
- application procedure
- guidelines for quality control

m. Procedure for obtaining surface tolerance:
- when and how tested

n. Type of finish required, edging.

o. Curing:
- type, procedure, and length of time
- water tank if necessary (temperature of water)
- maintain temperature, if required, through the entire curing process

p. Sawcutting:
- timing
- equipment

q. Filling joints:
- timing (review options and predictable results)
- material
- procedure (flush or a slightly crowned joint surface)
- contractor's obligation to further touch up

r. Floor protection where floor can be subjected to various types of traffic, including construction traffic:
- when equipment will be set or when racks will be erected

s. Ventilation during entire placement and finishing procedure.

t. Basis and timing for approving completed slab.

u. Proper process if it is felt there should be a change considered in the plans and specifications during the construction period.

ADDITIONAL ITEMS TO BE CHECKED IF SHRINKAGE COMPENSATING CONCRETE IS TO BE USED:

a. If Type K cement is used, are the detailed results of ASTM C-878 available and do they indicate a successful installation is feasible?

b. If a component expansion mixture will be used with a locally available cement, results from ASTM C-845 (compatibility with two products), plus results from ASTM C-878, must be available.

c. In either case, has the manufacturer of Type K or the manufacturer of the component approved the following:
   - specific mix design and proposed slump at point of placement
   - percent reinforcement details and location
   - pour sequence, time of placing
   - abutting slabs
   - type and location of isolation material
   - finishing and curing procedure

d. Make pours as square as practical, check location of reinforcement.

e. Cure, cure, cure.

AREAS OF POTENTIAL PROBLEMS

- NOT HAVING COMPLETE KNOWLEDGE OF ALL PLANS AND SPECIFICATIONS, INCLUDING ADDENDA AND CONTRACTUAL OBLIGATIONS BEFORE GOING ON-SITE.
- SCOPE OF WORK – Complete agreement with all parties before starting work.
- CONCRETE – Do we have full information on the mix design and supplier as indicated on our quotation form (see "Concrete" section in this manual), summer, winter, truck dump, pump, and access?
- SUBGRADE OR BASE – Has the surface on which we will place concrete been accepted? By whom? In writing? Has it been tested? How and by whom? Proof rolled? Is geotechnical report available?
- VAPOR RETARDER – Are we required to place concrete directly on a vapor barrier? Is the architect/engineer aware of the recommendation in ACI 302 if a vapor retarder is actually required? Don't use sand as fill – not compatible at any depth.
- REINFORCEMENT – Is it discontinued at all joints? Is owner aware of the potential cracks if steel goes through joint? Are dowel baskets in high traffic areas?
- JOINT SPACING – Does spacing appear excessive? Question is often the choice of cracks or joints.
- TREATMENT OF CONSTRUCTION JOINTS – Edged, later sawn, later filled?
- SOFT-CUT – Is it appropriate to use for depth of slab, will later chasing be required?
- JOINT FILLING – Are all parties, including the owner, aware of potential problems with early filling and associated costs of necessary repairs?
- CURING METHOD – Although a method of water moist cure is most effective, everyone should be aware that its use may result in the concrete having a "dull, not shiny" surface. The use of synthetic fiber may result in some staining of the surface.
- PROTECTION AND ACCEPTANCE OF COMPLETED SLAB – When will slabs be "accepted", what is the basis of acceptance, and who has the responsibility of protection?
- RESTRAINT, WHICH CAN BE DIRECTLY RESPONSIBLE FOR CRACKING – IT IS IMPERATIVE THAT THE SLAB SHOULD NOT BE RESTRAINED ("LOCKED IN OR TIED IN") TO ANY OTHER STRUCTURE ELEMENTS OF THE BUILDING:
  - Often occurs at the docks, next to tilt-up panels, around columns, or where there is any penetration in the floor slab.
  - Abrupt change in the thickness of the concrete.
  - Early rack installations.
  - Thermal shock. If the required schedule and job conditions will result in the completed slab being subjected to severe temperature change during or after the slabs are placed.
- REQUIRED FLOOR TOLERANCE (if you have a 1/8" in 10' spec, contact the Corporate Office!) – Remember, if there is only one F number in the specification; this is your minimum number. Can you achieve this under the job conditions? When will the floor be tested, method used, and by whom? ASTM E1155 does not require testing tolerances within 2' of any penetration of the slab or wall.

SI* (MODERN METRIC) CONVERSION FACTORS

APPROXIMATE CONVERSIONS TO SI UNITS

Symbol	When You Know	Multiply By	To Find	Symbol
		LENGTH		
in	inches	25.4	millimeters	mm
ft	feet	0.305	meters	m
yd	yards	0.914	meters	m
mi	miles	1.61	kilometers	km
		AREA		
in^2	square inches	645.2	square millimeters	mm^2
ft^2	square feet	0.093	square meters	m^2
yd^2	square yards	0.836	square meters	m^2
ac	acres	0.405	hectares	ha
mi^2	square miles	2.59	square kilometers	km^2
		VOLUME		
fl oz	fluid ounces	29.57	milliliters	mL
gal	gallons	3.785	liters	L
ft^3	cubic feet	0.028	cubic meters	m^3
yd^3	cubic yards	0.765	cubic meters	m^3
NOTE: Volumes greater than 1000 l should be shown in m^3.				
		MASS		
oz	ounces	28.35	grams	g
lb	pounds	0.454	kilograms	kg
T	short tons (2000 lb)	0.907	megagrams (or "metric ton")	Mg (or "t")
		TEMPERATURE (exact)		
°F	Fahrenheit temperature	5(F-32)/9 or (F-32)/1.8	Celsius temperature	°C
		FORCE AND PRESSURE OR STRESS		
lbf	poundforce	4.45	newtons	N
lbf/in^2	poundforce per square inch	6.89	kilopascals	kPa

APPROXIMATE CONVERSIONS FROM SI UNITS

Symbol	When You Know	Multiply By	To Find	Symbol
		LENGTH		
mm	millimeters	0.039	inches	in
m	meters	3.28	feet	ft
m	meters	1.09	yards	yd
km	kilometers	0.621	miles	mi
		AREA		
mm^2	square millimeters	0.0016	square inches	in^2
m^2	square meters	10.764	square feet	ft^2
m^2	square meters	1.195	square yards	yd^2
ha	hectares	2.47	acres	ac
km^2	square kilometers	0.386	square miles	mi^2
		VOLUME		
mL	milliliters	0.034	fluid ounces	fl oz
L	liters	0.264	gallons	gal
m^3	cubic meters	35.71	cubic feet	ft^3
m^3	cubic meters	1.307	cubic yards	yd^3
		MASS		
g	grams	0.035	ounces	oz
kg	kilograms	2.202	pounds	lb
Mg (or "t")	megagrams (or "metric ton")	1.103	short tons (2000 lb)	T
		TEMPERATURE (exact)		
°C	Celsius temperature	1.8C + 32	Fahrenheit temperature	°F
		FORCE AND PRESSURE OR STRESS		
N	newtons	0.225	poundforce	lbf
kPa	kilopascals	0.145	poundforce per square inch	lbf/in^2

* SI is the symbol for the International System of Units. Appropriate rounding should be made to comply with Section ASTM E380.

REVIEW/PRACTICE QUESTIONS

1. If the temperature of steel forms or reinforcements exceeds 120°F, _____.
   a. they must be sprayed with water before placing concrete
   b. it is time to shut down the project until it cools
   c. the concrete will not adhere to them
   d. the concrete will be damaged by them

2. Straightedges are not used for _____.
   a. precise tolerances on floors and roads
   b. cambered highway surfaces
   c. construction joints at the end of a day's work
   d. restraightening after floating

3. Loose concrete with a high slump _____.
   a. does not move well on a steep conveyor
   b. is difficult to level
   c. places easily under reinforcement
   d. is difficult to discharge from a chute

4. Discharging pumps can _____.
   a. operate under pressures ranging from 150 to 500 psi
   b. pass aggregates larger than ½ the pipe diameter
   c. operate at only one speed
   d. damage form surfaces

5. Combinations of delivery and deposit equipment depend on _____.
   a. slump characteristics of the concrete
   b. budget, scheduling, and availability
   c. cost, availability, and mix characteristics
   d. crew size, slump, and aggregate size

6. If a slab is on a slope, start placement _____.
   a. at the lower end
   b. at the upper end
   c. at the middle
   d. the shortest way across the slab

7. Forms bulging or groaning signal _____.
   a. cement dermatitis
   b. the need to revibrate
   c. compacting has begun
   d. possible form collapse

8. Buggies need a firm runway because _____.
   a. they tend to tip
   b. bouncing will cause segregation of the mix
   c. they can deposit directly in the form
   d. they have the right-of-way

9. Only one craftworker giving signals to equipment operators _____.
   a. slows movement of heavy equipment
   b. avoids confusion and accidents
   c. controls discharge gates on ready-mix trucks
   d. avoids cold joints and bridging

10. When air bubbles rise to the surface _____.
    a. vibration is complete
    b. screeding is complete
    c. floating is complete
    d. pumping is complete

11. In moving concrete from the truck to the point of placement, you must prevent _____.
    a. bridging
    b. honeycombs and bugholes
    c. contamination and segregation
    d. bleeding and cold joints

12. Conveyors are most efficient for concrete with a slump of _____.
    a. 2 to 4½ inches
    b. 1 to 3½ inches
    c. 3½ to 5 inches
    d. 4½ to 6 inches

13. Rodding, spading, and tamping _____.
    a. leave honeycombing, bugholes and lift lines
    b. get rid of the voids and bring aggregates closer together
    c. even out the ridges and hollows left by depositing
    d. cause cement dermatitis

14. Hoppers collect concrete from the truck or portable mixer and _____.
    a. transfer it to handling equipment
    b. drop it vertically through the discharge gate
    c. handle extremely low-slump discharge
    d. discharge it directly into the form

15. By placing lifts in slab forms you can _____.

    a. eliminate bridging around the reinforcement
    b. save concrete
    c. make cold joints horizontally instead of vertically
    d. keep bleed water to a minimum

16. The purpose of spreading is to _____.

    a. use square-end shovels, rakes, and come-alongs
    b. even out the ridges and hollows left by depositing
    c. push the concrete downward into the form
    d. get rid of the voids in concrete

17. Planning and checklists directly influence _____.

    a. how the craftworkers do their jobs
    b. inspection and testing signoffs
    c. precautions for weather conditions
    d. the mix design

18. Working with a bucket, you always _____.

    a. use a cableway to move it
    b. work beside and underneath it
    c. fill it with a hopper
    d. use the gate to control the speed of discharge

19. The batch or delivery ticket does not identify _____.

    a. the time of first mixing the concrete
    b. the amount of water that can be added
    c. the starting number on the revolution counter
    d. the amount of concrete in pounds

20. Lifts are _____ layers in placed concrete.

    a. horizontal
    b. vertical
    c. diagonal
    d. bridging

notes

ANSWERS TO REVIEW/PRACTICE QUESTIONS

Answer	Section Reference
1. a	2.3.0
2. c	6.3.0
3. a	2.2.0
4. d	4.6.0
5. c	2.2.0
6. a	2.6.2
7. d	1.2.3
8. b	2.2.0/Figure 4
9. b	3.1.0
10. a	5.3.3
11. c	4.0.0
12. a	4.4.0
13. b	5.3.1
14. a	4.8.0
15. a	5.1.1/5.1.2
16. b	5.2.0
17. a	2.1.0
18. d	4.5.0
19. d	3.3.0
20. a	5.1.1

NCCER CRAFT TRAINING USER UPDATES

The NCCER makes every effort to keep these manuals up-to-date and free of technical errors. We appreciate your help in this process. If you have an idea for improving this manual, or if you find an error, a typographical mistake, or an inaccuracy in the NCCER's Craft Training Manuals, please write us, using this form or a photocopy. Be sure to include the exact module number, page number, a description of the problem, and the correction, if possible. Your input will be brought to the attention of the Technical Review Committee. Thank you for your assistance.

Instructors – If you found that additional materials were necessary in order to teach this module effectively, please let us know so that we may include them in the Equipment/Materials list in the Instructor's Guide.

Write: Curriculum and Revision Department
National Center for Construction Education and Research
P.O. Box 141104
Gainesville, FL 32614-1104

Fax: 352-334-0932

Craft ______________________ Module Name ______________________

Module Number ______________________ Page Number(s) ______________________

Description of Problem ______________________

____

____

____

(Optional) Correction of Problem ______________________

____

____

(Optional) Your Name and Address ______________________

____

____

notes

Finishing, Part I

Module 23107

FINISHING, PART ONE

NATIONAL
CENTER FOR
CONSTRUCTION
EDUCATION AND
RESEARCH

OBJECTIVES

After completing this module, the trainee will be able to:

1. Describe the basic finishing process.
2. Use the following finishing hand tools: float, edger, groover, and trowel.
3. Mark and cut joints with a saw.
4. Apply a broom finish.
5. Apply a rubbing finish.
6. Associate trade terms with the appropriate processes and equipment.

Prerequisites

Successful completion of the following Task Modules is recommended before beginning study of this Task Module: Core Curricula; Concrete Finishing Level 1, Modules 23101 through 23106.

Required Trainee Materials

1. Trainee Task Module
2. Pencil and paper
3. Appropriate Personal Protective Equipment

COURSE MAP

This course map shows all of the task modules in the first level of the Concrete Finishing curricula. The suggested training order begins at the bottom and proceeds up. Skill levels increase as a trainee advances on the course map. The training order may be adjusted by the local Training Program Sponsor.

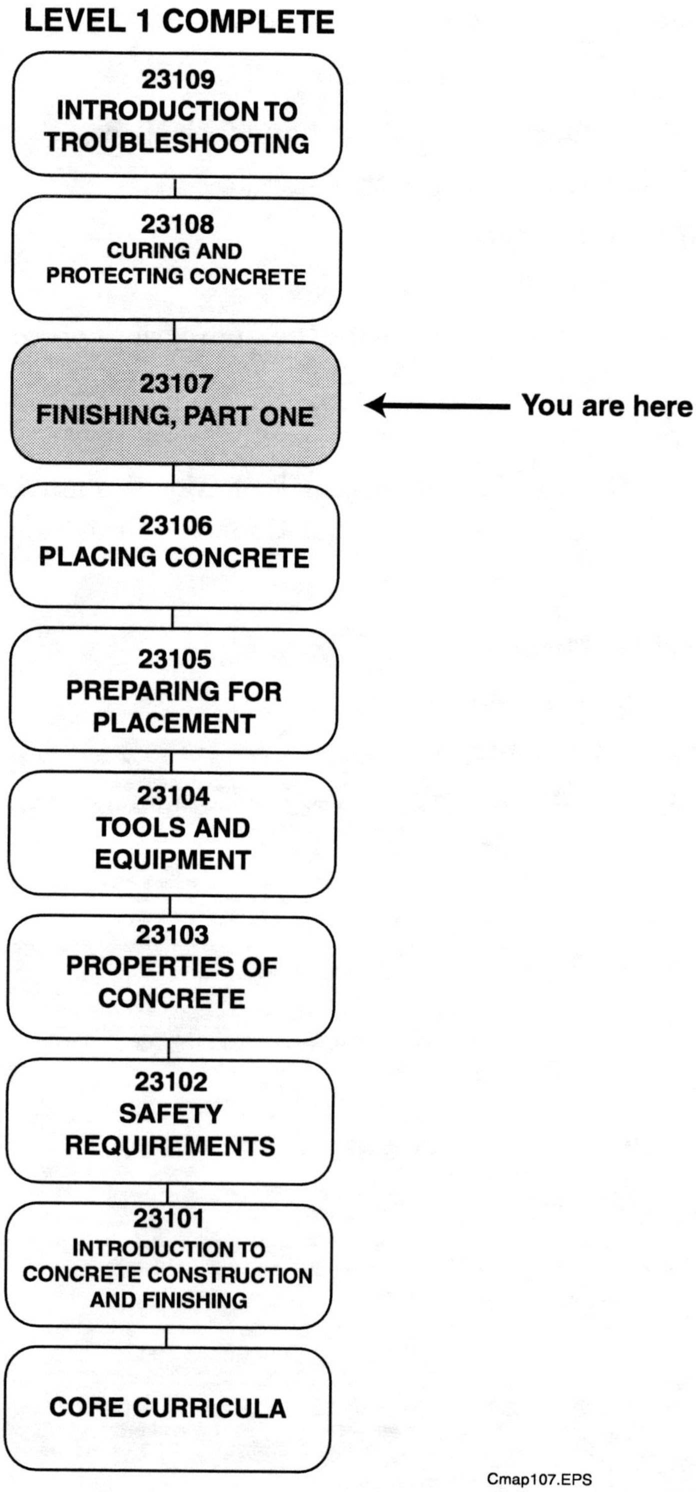

Cmap107.EPS

TABLE OF CONTENTS

Trade Terms Introduced In This Module

Bleed water: Water that rises to the surface of concrete after it is placed; caused by the settlement of cement paste.

Dry cutting: Cutting contraction joints in concrete without the use of water to cool the saw blade.

Dry shake hardener: Mineral or metallic material embedded in the surface of concrete to improve the abrasion and impact resistance.

F-number: A measure of flatness (F_F) and levelness (F_L) of a concrete slab derived through a combination of measurements and calculations.

Spall: A chip, flake, or piece of concrete that has been broken or separated from the concrete surface; usually caused by expansion within the concrete.

Wet cutting: Cutting contraction joints in concrete with the use of a water-irrigated saw.

1.0.0 INTRODUCTION

Concrete finishing is a set of activities that takes newly placed concrete from a rough, open surface to its final smooth state. The finishing process gives the concrete craftworker a chance to affect the appearance of the concrete. If the finishing is poorly done, all the best mixing and placing efforts that preceded it, and all the best curing that follows it, will not improve the final product.

As the concrete finisher, you have a great impact on how the surface will look and feel. It is this phase of the work where skills and ability really pay off. Remember that the finish is the most visible part of the project.

1.1.0 PURPOSE

This module covers the operations and techniques needed to finish flat concrete surfaces. Concrete finishing begins at the point where the concrete has been placed, compacted, and screeded level. Bullfloating, or first-pass floating, is technically part of finishing; however, it has been covered in the previous module, *Placing Concrete*, because it is done before **bleed water** rises. The finishing steps in this module take place after bleed water has evaporated or has been removed. The module ends at the point where the curing process begins. By the end of this module, you will know the basic procedures to finish the surface of freshly placed concrete.

1.2.0 SAFETY

When working with freshly mixed concrete, take care to avoid skin irritation or chemical burns. These can occur through prolonged contact with the concrete or concrete-soaked clothing. There are three reasons fresh concrete causes skin irritation or chemical burns:

- Portland cement is alkaline and therefore caustic.
- Portland cement is hygroscopic, so it absorbs moisture from the skin.
- The sand in fresh concrete is abrasive to bare skin.

Do not allow your clothing to become saturated with moisture from concrete because it can carry alkaline or hygroscopic elements to your skin. Waterproof gloves, rubber boots, a long-sleeve shirt, safety glasses or goggles, and full-length pants are essential.

When handling and placing concrete you need to:

- Protect your eyes by wearing safety glasses with side shields or safety goggles
- Protect your skin by wearing long-sleeved shirts, full-length pants, rubber boots, rubber gloves, and kneepads
- Avoid prolonged contact with fresh mix
- Protect your clothing by using knee boards when on the surface of the concrete

Protection is important because finishing calls for close contact with the fresh concrete. If you stand in freshly mixed concrete, wear rubber boots high enough to prevent concrete from flowing into them. Use waterproof pads and/or kneeler boards between concrete surfaces and knees, elbows, and shoes. Promptly rinse in clean water any clothing that gets saturated with fresh concrete or bleed water, as it can hold irritants.

If you get any concrete on your skin, wash thoroughly with soap and clean water as soon as possible. Mild skin irritation can be relieved by applying a lanolin cream after washing. Persistent or severe discomfort should be seen by a doctor. If your eyes are affected, wash them with clean water and consult a doctor immediately. Do not rub your eyes, and do not use any other treatment without consulting a doctor.

2.0.0 BASIC FINISHING PROCESSES

Concrete needs some basic finishing, regardless of its final use or shape. These basic finishing activities are discussed here in order to give you a sense of sequence and an overall picture of the finishing process. They are described in detail in the sections following this overview. These basic finishing activities are:

- Bullfloating, darbying, or highway straightedging
- Bleed water removal
- Edging

- Jointing
- Floating and troweling
- Surface treatment

Most of these activities are performed to some degree on each job. The sequence can vary, and some steps can be completed once or repeated several times, depending on the particular project. The following sections address the work steps involved with each activity.

2.1.0 BULLFLOATING, DARBYING, AND HIGHWAY STRAIGHTEDGING

As discussed in *Placing Concrete*, a bullfloat or darby is used immediately after the surface has been screeded. On some jobs, a highway straightedge is used. These tools eliminate any high or low spots on the surface, embed large aggregate particles at or near the surface, and bring up a slight amount of cement paste to the surface.

This initial floating needs to be done before any bleed water accumulates on the surface. Floating with water present will weaken the surface of the finished concrete. During this step, you must use care not to overwork the surface of the concrete, as this would cause too much paste to rise, resulting in a less durable surface. This part of the process was presented in *Placing Concrete*. Bullfloating may be repeated after bleed water dissipates, depending on the requirements of the job.

2.2.0 BLEED WATER REMOVAL

Once the surface has its initial float, it is necessary to let some time pass before the remaining finishing steps can be done. This allows the bleed water to rise and evaporate or be removed mechanically.

In fresh concrete, aggregate particles and cement are partly suspended in water. Because they are heavier, they tend to settle. As they settle, water is displaced and moves up toward the surface of the concrete. This water is called bleed water, and the process is called bleeding. This water may appear as a sheen on the surface; it quickly evaporates on a warm or windy day.

Bleed water can also accumulate in greater amounts. Finishing work cannot continue with bleed water on the surface, so sometimes it must be removed mechanically. One method is to set up blowers to speed the evaporation process. Another method is to push off the water by using a squeegee or heavy rubber hose dragged across the surface. This method is used only when the concrete is stiff enough so that the water, not paste, is removed. Sometimes special vacuum mats are used to reduce water content at the surface. This can also improve the strength and wear resistance of the concrete.

2.3.0 EDGING

Edging is the next finishing step in the process. The concrete must be firm enough that a person standing on the surface will not make an imprint more than ¼ of an inch deep. Before the bleed water evaporates, edging starts by inserting the point of a mason's trowel between the form and the concrete to a depth of about one inch along the perimeter of the slab. The trowel is run along the form to cut the concrete away from the form.

After the bleed water evaporates, place the edger with the curved lip in the gap between the concrete and the form, with the plane surface on the surface of the concrete. Move the edging tool along the edge of the concrete, around the form.

This process forms a rounded corner and increases the density of the concrete at the edge. As a result of the edging process, the corner will be less vulnerable to chipping, raveling, or **spalling**. Edging is typically done on sidewalks, driveways, and other outside projects. The remaining finishing work is done after the bleed water is gone.

2.4.0 JOINTING

As concrete hardens, shrinkage cracks may develop at random places across the surface. To minimize these random and unsightly cracks, contraction joints are placed in the surface. These provide planes of weakness where the cracking will more likely occur. Contraction joints are usually formed by using a hand groover or by inserting strips of plastic, wood, metal, or preformed joint material into the unhardened concrete. Another method is to wait until the concrete is hard enough to support foot traffic, then to use a power saw to cut contraction joints in the semi-hardened surface.

Isolation joints separate the concrete slab from other structural members. These are set in the forms before the concrete is placed. Construction joints are made at the end of a day's deposit of concrete. Sometimes they are also used as contraction or isolation joints.

2.5.0 FLOATING AND TROWELING

After edging and jointing, the concrete surface needs to be floated and (if desired) troweled. This final activity removes slight surface imperfections such as bumps and voids, and compacts the mortar at the surface. Hand floating is done using a small wooden or magnesium hand float. Power floating is usually done for large areas.

If a smooth, dense surface is desired, the contractor will use steel hand trowels or power trowels after the floating operation. Troweling is done in several passes, with the blade of the trowel held at an increasing angle for each pass. Troweling seals the surface and gives a polish to the concrete. Troweling is not needed on every project. Large outside projects such as parking lots, sidewalks, and road surfaces are generally not troweled. Other surface finishes, such as brooming, may be applied.

2.6.0 SURFACE TREATMENT

After troweling, or sometimes after floating, the final step is a surface treatment. If a slip-resistant surface texture is desired, the surface can be scored with a fiber or wire finishing broom. For a finer surface, a soft bristled brush can be used. Additional surface treatment can include safety grooving, and stone or sack rubbing. In any case, the concrete must be hard enough to retain the textured finish.

The following sections describe details of the finishing process.

3.0.0 EDGING CONCRETE

Edging forms a radius or rounds the slab edge with a hand tool designed for that purpose. The first step in edging takes place before the bleed water appears, and the remainder occurs after the bleed water has disappeared.

3.1.0 PURPOSE OF EDGING

Edging is used to accomplish the following:

- It loosens the concrete from the forms.
- It rounds over the edges, making them more durable and less vulnerable to spalling and chipping.
- It floats the surface immediately next to the edge where normal floating and troweling are less effective.

Not all slabs are edged, so review the plans to determine which ones require edging. Usually floors are not edged, but sidewalks, driveways, steps, and patios are. Heavy-duty and industrial floors do not require edging. Slabs that will receive a covering of tile or similar materials are not edged. Sometimes edging is used on a construction joint to make it readily visible for later saw cutting.

3.2.0 EDGING PROCEDURES AND TECHNIQUES

To make edging easier, the first step is to cut away or loosen the concrete from the form or adjacent slab. This preliminary step is done right after the bullfloating and before the bleed water begins to appear.

To perform this process, use a pointing or margin trowel. Insert the tip between the form and the concrete to a depth of one inch while keeping the trowel in contact with the form. Move the trowel around the perimeter of the form in order to create space between the side of the form and the side of the slab. Use a cutting or sawing option. It is not necessary to do any prying at this point; inserting the trowel will be sufficient to create this gap. *Figure 1* shows this technique being performed between fresh concrete and an adjacent slab.

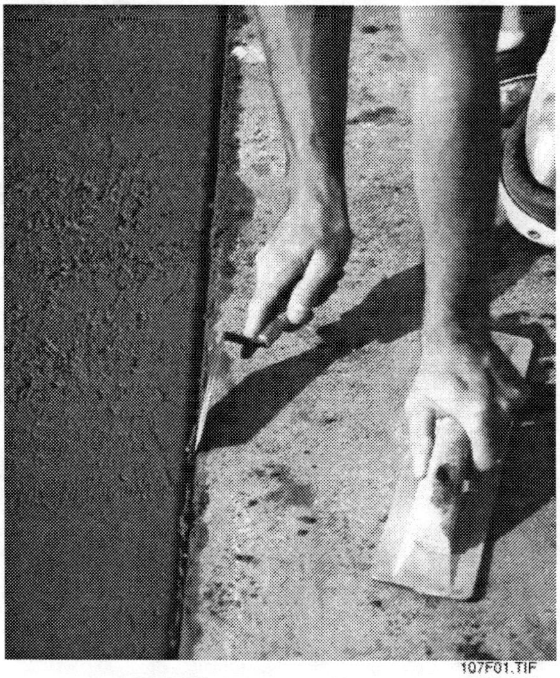

107F01.TIF

Figure 1. Margin Trowel Cutting

After this preliminary step is done, you will have to wait until the bleed water is gone. The surface must be set enough so that a person standing on it will not leave footprints deeper than ¼ of an inch. The actual edging process can then begin.

The edger is a flat metal tool with a curved-down tip on one side. This curve is what shapes the edge of the concrete, as shown in *Figure 2*. The radius of the edge varies from ⅛ of an inch for floors to ½ an inch for sidewalks. Edgers are usually made of steel but can also be made of bronze or malleable iron. Ordinarily, an edger is designed to be used by the craftworker in the kneeling position. It can, however, be fitted with a pole handle to permit the finisher to work in the standing position. These walking edgers should not be used for the first pass, as they give inconsistent results.

107F02.TIF

Figure 2. Edger Rounds The Edge

Follow these basic steps to perform edging:

Step 1 Insert the edging tool into the slot between the concrete and the form. This is the slot made earlier with the trowel.

Step 2 Move the edger forward along the edge of the slab to an arm's length. Keep the leading edge of the lip raked, or tilted slightly up, and the curved radius edge against the form. This will keep the edger from digging into the surface.

Step 3 At the end of your reach, lift the edger up slightly. To avoid pulling up concrete, lift it while you are still moving it forward.

Step 4 Now draw the edger backward to its starting point. Keep the new leading edge slightly raked, as shown in *Figure 2*.

The result will be an edge that is rounded, dense, and smooth. Edging is redone after floating and repeated again if required.

4.0.0 JOINTING CONCRETE

Joints control the location of cracks in concrete by isolating or weakening the slab at preselected locations. They also allow for movement of two surfaces without damage to each other. Three types of joints require some type of finishing activity depending on how they are used.

- *Contraction joints* weaken the slab by deep scoring.
- *Isolation joints* limit movement by separating structural members.
- *Construction joints* mark the ending and beginning of separate placement activity and can be used as contraction or isolation joints.

The procedures for making the joints or finishing around them are described in the following sections.

4.1.0 RESETTING THE CONSTRUCTION JOINT EDGE

During placing, the edge of a construction joint may lose its proper elevation. This can be fixed by reworking the edge with a 3- to 4-foot wood or magnesium straightedge. When done carefully, this gives the tie-in to the new slab a higher flatness tolerance.

Resetting the joint edge is done prior to and after machine floating. Run the straightedge back and forth as you hold it perpendicular to the construction joint form edge. Use the magnesium straightedge on air-entrained concrete. This reflattening is sometimes called "boarding the joint" (*Figure 3*).

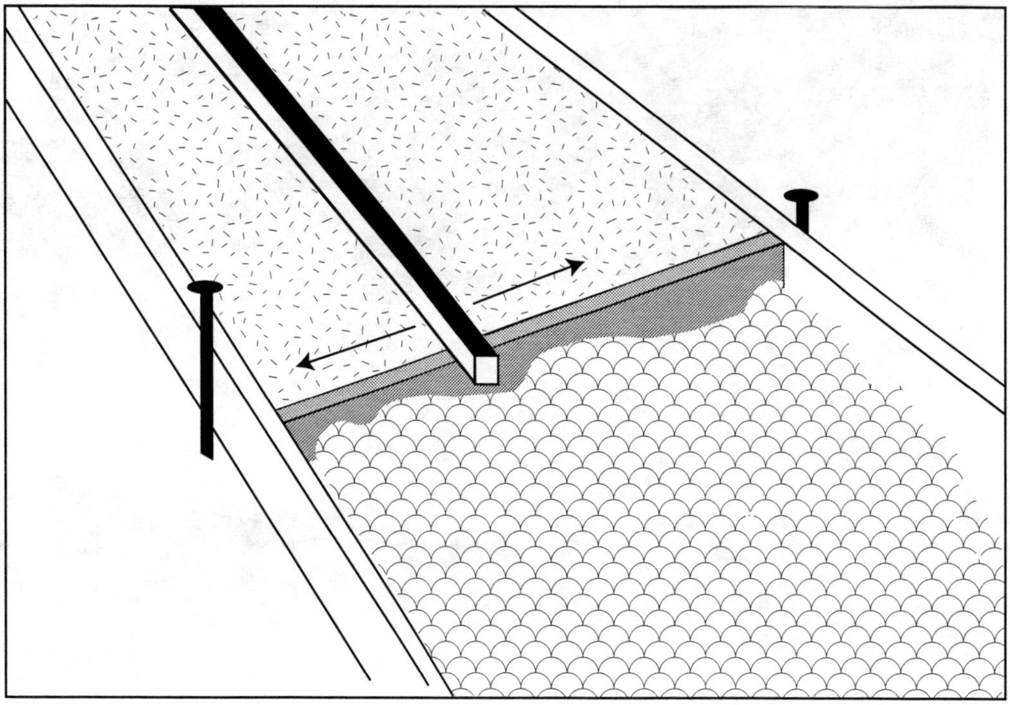

107F03.EPS

Figure 3. Boarding The Construction Joint

4.2.0 JOINTING FRESH CONCRETE

Isolation and construction joints are made when concrete is deposited. Contraction joints can be made in fresh concrete or shortly after the concrete is placed. This is done by hand with a groover or jointer, or by using a power saw.

Jointing follows bullfloating but only after the bleed water disappears. It is usually done after edging but before hand floating. Jointing is done with a hand tool called a groover or jointer. Hand jointing is not practical for slabs deeper than 8 inches because of limits on bit length. The groover has a hand grip or a walking handle. First passes should be done with a hand-held groover to ensure a good cut. Jointing may be regrooved several times to ensure proper depth.

After bleed water dries, mark the location of each joint. Sometimes marks on the form edges indicate where the contraction joints go. Use a string or chalk line stretched between the side forms and across the concrete surface to mark the cut line. Remember that joint spacing in feet is two to three times the slab thickness in inches.

You can use a straight-sided board at least 6 inches wide as a cutting guide, as shown in *Figure 4*. The groover bit should have a depth equal to ¼ the depth of the slab. Remember to lean on the cutting guide, not the slab, as you draw the groover through the surface.

107F04.TIF

Figure 4. Grooving A Contraction Joint

Mark the location of each joint on the side of the forms before starting the concrete placing. It is easiest if they are marked before concrete is placed. For deeper slabs, move the cutting guide next to the mark. Using a pointing trowel, make a preliminary score on the cutting line against the cutting guide. Measure the distance from the bit to the plane edge on the jointer. Then reposition the cutting guide so that the groover bit will fit into the score with its plane edge against the guide. If concrete has hardened to the point where the jointer will not make a cut, use a hatchet or hammer and chisel to make the first cut, then use a jointer to work and smooth the joint. The following three steps complete the jointing process:

Step 1 Start by pushing the groover into the concrete and moving it forward while applying pressure to the back of the tool. This slightly rakes, or tilts, the leading edge of the blade upwards and makes it easier to start.

Step 2 Keep the edge of the groover against the edge of the cutting guide as you move forward.

Step 3 After the joint is cut, turn the tool around. Clean the bit if needed, and pull the groover back over the groove in the opposite direction to get a smooth finish. Again, rake the leading edge slightly.

Use the same groover when the joint is recut later after finish floating and/or troweling. As shown in *Figure 5*, regrooving widens and deepens the joint. Sometimes, power saws are used on large projects to cut contraction joints in fresh concrete.

107F05.TIF

Figure 5. Regrooving A Joint

4.3.0 SAWING JOINTS

On large slabs, it may be more convenient to cut joints with a power saw. Saw cutting should be done before the concrete starts to cool or before random drying cracks can form in the slab, and as soon as the surface is firm enough so that it will not be damaged by the saw blade. Concrete saws use either water-lubricated diamond blades or non-lubricated carborundum blades. The three types of sawing are conventional **wet cutting**, conventional **dry cutting**, and early entry dry cutting.

4.3.1 Types Of Cutting

Conventional wet cut and dry cut sawing can be performed on a finished slab after a waiting period of approximately 4 hours in hot weather and 12 hours in cold weather. The waiting period is longer for concrete with steel fiber reinforcement or embedded mineral aggregate hardeners. The depth of the cut is at least ¼ the depth of the slab or a minimum of 1 inch, whichever is greater. For steel fiber reinforced slabs, the depth of the cut should be ⅓ of the slab depth.

Dry cut saws use carborundum blades with no cooling system. These are small rotary saws like the one shown in *Figure 6*. They are used on small jobs because they tend to overheat after long use. Wet cut saws use diamond blades with a water cooling system. These are large rotary saws used on large jobs. These types of saws are also used to cut concrete after it has hardened completely.

107F06.TIF

Figure 6. Hydraulic Powered Saw

Early entry dry cut sawing is used when early sawing is needed to prevent tensile cracking. This form of cutting uses a lightweight saw with diamond blades without water lubrication, as shown in *Figure 7*. The saw cuts here are not as deep as those made with conventional wet cutting. The maximum size for an early entry dry cut joint is 1¼ inches, and the minimum depth should be 1 inch. These cuts are done after a waiting period of about 1 hour in hot weather or 4 hours in cold weather.

107F07.TIF

Figure 7. Early Entry Dry Cut Saw

Whether using conventional or early entry sawing, it is important to protect yourself by following manufacturer's guidelines for using power equipment.

WARNING! Wear rubber boots and gloves when using electric power equipment on a wet surface. Also, wear eye protection, a respiratory device, and ear protectors when using any power saw.

4.3.2 Jointing With A Saw

Whether using a wet cut or dry cut saw, here are some helpful guidelines for jointing with a saw:

- Always wear appropriate personal protective equipment.
- Start only when a person standing on the surface leaves an impression of ¼ of an inch or less.
- Check that joint locations are clearly marked with string, score, or chalk. If they are not marked clearly, mark them.
- Wear rubber gloves and boots when using an electric-powered tool with water.
- Use a cutting guide for hand-held saws, if possible.
- Move slowly and carefully, observing joint markings.
- Check the edges of the cut for feathering or raveling. Raveling indicates the concrete is too soft for sawing.
- Practice power tool safety by keeping yourself out of the saw path.

4.4.0 SEALING JOINTS

Sawed joints may be sealed to facilitate cleaning and to help support the joint edges under traffic. Project specifications will describe the type of sealant to use. Sealants can be either field-molded materials that are poured, gunned, or troweled into position or preformed material that is rigid or flexible. They are designed to keep dirt and moisture out of wide joints.

The joint surfaces should be cleaned with a dry brush or pressurized air to clear any dust or debris before they are sealed. It is important that the surface of the joint be clean to insure a good bond for the sealant. Read the manufacturer's instructions and follow proper safety procedures when working with chemicals.

Preformed joint fillers of plastic and/or asphalt compounds are popular for wider joints. This type of sealant, however, is not recommended for heavy traffic locations.

Joint filling should be delayed as long as possible to allow concrete to shrink. Early filling can result in separation of the concrete and the sealant. This can be repaired with epoxy. If construction traffic requires early sealing of joints, the contractor may need to return later to complete the work.

5.0.0 FINISHING TECHNIQUES

Several techniques can be used to finish the surface of a slab. This section describes the basic operations that you will need to know to produce a well finished concrete surface. They include floating, darbying, and troweling. These finishing techniques are used after the bleed water is gone and the edging and jointing have been completed. The concrete must be firm enough so that a person standing on the surface will not make an imprint more than ¼ of an inch deep.

Remember that the edges of a slab will set up and harden before the rest of the slab. Float and use the edger around the border before floating the rest of the slab.

5.1.0 HAND FLOATING

Floating compacts and consolidates the concrete surface. Floating operations occur at two separate stages in the finishing process. The first floating activity, generally bullfloating, occurs immediately after screeding and before the bleed water rises. The second floating occurs after bleed water evaporates and is done by hand or power tools. This second floating operation follows edging and jointing. This procedure helps to:

- Embed the large aggregates beneath the surface mortar layer
- Remove slight humps and other minor surface imperfections to produce a level surface
- Further consolidate the surface in preparation for the remaining finishing steps
- Remove laitance

Hand floating prepares the surface for other finishing operations. Hand floats are available in a variety of materials. Magnesium floats are lightweight and so require less effort to use. They are used for air-entrained and lightweight concrete. Wooden floats are best for non-air-entrained concrete and for high-slump concrete. A wooden float does not seal the surface, so air and water trapped in surface pores can escape. Composition floats with resin-impregnated surfaces are smoother than wood and only slightly rougher than magnesium. They are sometimes used for finish floating where troweling is not necessary.

5.1.1 Using A Darby

Darbying serves the same purpose as bullfloating and the same rules apply. A darby is advantageous on narrow slabs and in restricted spaces.

Hand floating can begin by using a darby in a semicircular motion until the surface is relatively free of high and low spots. *Figure 8* shows a darby being used on a small slab.

107F08.TIF

Figure 8. Working A Small Slab With A Darby

Darbying calls for holding the blade flat to the surface while moving in a semicircular area. The main points for using a darby are:

- Move the blade back and forth in a rapid sawing motion while also moving in a sweeping arc.
- Be sure to cover the surface in a regular pattern, moving slowly from one edge to another.
- Move the darby at an angle to the path of the screed.
- If you must move onto the surface of the slab, use kneeling boards and knee pads.

After the surface is level, tilt the darby blade at an angle. Move it in a sweeping motion to fill any holes left by the sawing motion.

5.1.2 Using A Hand Float

Concrete is generally ready for hand floating when the bleed water and water sheen has disappeared and a finisher on knee boards makes less than ⅛ of an inch indentation. Hand floating around the edges of a slab, as shown in *Figure 9*, can start earlier because the edges set faster and the finisher can stay off the slab while performing the work. Hand floating is also used to give a smoother finish around the edges of slabs that have been bullfloated or power floated.

107F09.TIF

Figure 9. Hand Floating Around Edges

Hand floating on the surface of a slab requires moving out onto the slab, as shown in *Figure 10*. This is done using knee boards, a hand trowel, and a hand float.

107F10.TIF

Figure 10. Hand Floating On A Slab

To work in this manner without damaging the slab and being saturated with wet concrete, follow these basic steps:

Step 1 Support yourself by leaning with one hand on a trowel while working the surface with the float in the other hand.

Step 2 Keep the float flat on the surface to avoid sealing the concrete. Keep the float moving in an arc at right angles to the direction of the previous bullfloating or darbying. If there was no previous floating, move at right angles to the direction of the screeding.

Step 3 Lift the float from the surface only while the float is moving, otherwise you may pull up part of the concrete on the surface.

Hand floating can provide the final finish for slabs where a fine textured surface is desired, such as sidewalks, patios, and other outdoor paving. *Figure 11* is a good example of such a slab. Note the semicircular pattern left by the action of the float.

107F11.TIF

Figure 11. Surface After Floating

5.1.3 Using Knee Boards

Knee boards allow finishers to move across the surface of fresh concrete without displacing the concrete. This is useful when finishing slabs with hand tools.

Use a pair of knee boards if you need to go out onto the slab's surface. With wooden knee boards, use one board under your knees and the other under your boots, as shown in *Figure 12*. This protects the surface from damage and protects you from exposure to the wet concrete. To avoid picking up concrete, lift the knee boards with a sliding motion.

To move forward with wooden knee boards, reach for the one under your boots and bring it around to the front. To move backward with the knee boards, back your knees onto the one under your boots. Keep your feet in the air as you move the board that is in front of you to behind you, under your boots.

107F12.TIF

Figure 12. Using Wooden Knee Boards

Newer types of kicker knee boards have cushions for your knees. Instead of fitting under your knees and feet, one board fits under one leg. They have stirrup attachments to fit over your boots, as shown in *Figure 13*. They are easy to slide forward or back by kicking your foot forward or backward.

107F13.TIF

Figure 13. Kicker Knee Boards

5.2.0 POWER FLOATING

Power floating can begin when the concrete can support a finisher on foot with an indentation of less than ¼ of an inch. The float machine should be the lightest available and should not dig into or disrupt the levelness of the surface.

As a general rule, power floating should be started as late as possible. Areas to be floated first generally include surfaces next to edges, columns, and walls, as well as surfaces exposed to sun and wind.

There are two types of power floats.

- Heavy rotating disk floats that can also vibrate
- Troweling machines equipped with float shoes

Rotating disk floats, or Kelly floats, have a rotating disk about 2 feet in diameter and are used for heavy-duty floors made with very low-slump concrete. This type of machine is shown in *Figure 14* . These machines perform the same operation as hand floating but cover larger areas and, of course, are much faster.

107F14.TIF

Figure 14. Revolving Disk Kelly Float

As shown in *Figure 15*, most troweling machines have four blades mounted in a ring attached to the base, with a ring diameter between 36 and 48 inches. Floating shoes slip over the blades, or a different float rotor is attached for use in floating the slab surface. The float blades are wider than troweling blades and are turned up on the edges to prevent them from digging into the surface.

107F15.TIF

Figure 15. Power Floating Machine

Power floating tends to create troughs under the center of the machine and ridges or strings of mortar just outside the edges of the blades. The power float will tend to push the strings up against projections such as columns. This excess mortar must be removed by hand, or the surface will be at a higher elevation. Grade marks and templates are useful to check that the surface is at the designated level around any projections. Power floating also removes the marks left by edgers and groovers. If these marks are intended for decoration, rerun the edger or groover after power floating is complete.

CAUTION: Wear rubber gloves and rubber boots when using electric power tools on a wet surface.

Follow these steps when floating a surface with a power float:

Step 1 Check that the slab is firm enough for machine floating (footprints of ¼ of an inch or less).

Step 2 Check that the blades are in good condition, and check fuel levels if the equipment is gasoline powered.

Step 3 Start the machine at the hardest point of the slab where concrete was first placed.

Step 4 Adjust the blades for floating so they are flat to the surface.

Step 5 Move the machine across the slab at right angles to the direction of bullfloating, floating out footprints and imperfections.

Step 6 Direct the movement of the machine by raising and lowering the handle. To move the machine to the left, raise the handle; to move to the right, lower the handle.

Step 7 Do not operate the machine in any one spot for a length of time, as this will dish a low place on the surface.

Step 8 Continue across the slab until the opposite edge is reached.

Step 9 Step back and maneuver the machine into position to pass again in the opposite direction. Overlap the previous path by about half the width of the rotor, as shown in *Figure 16*.

Step 10 Continue until the slab has been completely floated.

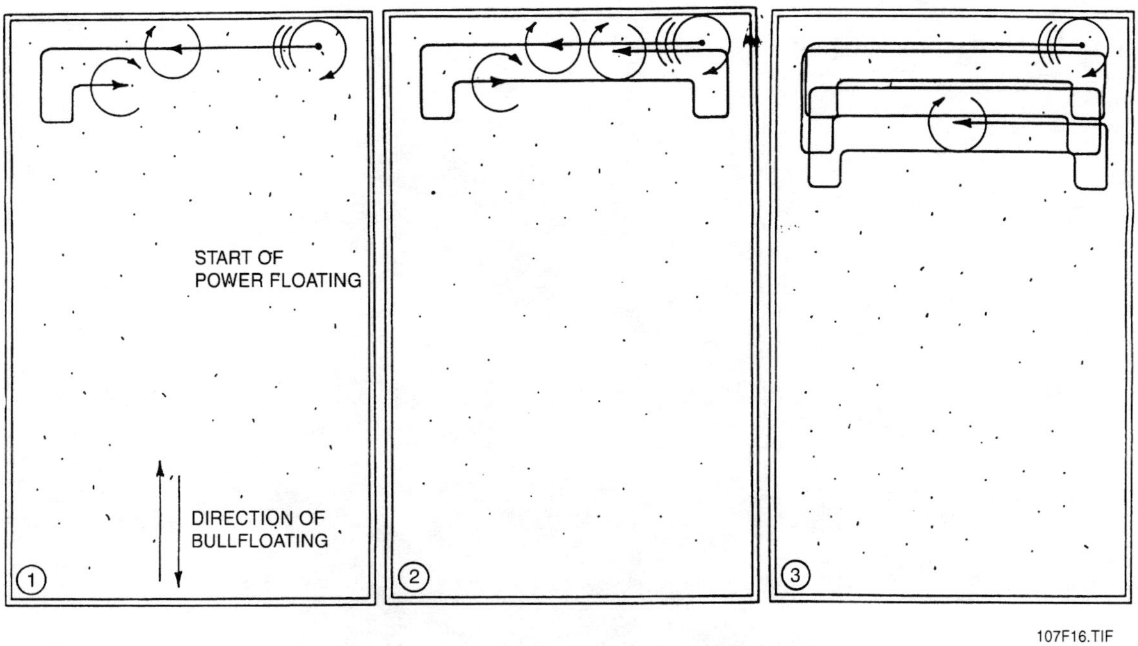

Figure 16. Power Floating Path

With practice, the movement of the machine across the surface will become almost effortless. You may come across a hole in the surface that the machine cannot fill. Carry a hand trowel in your pocket so you can scrape up some mortar from the surface and mend the finish. Cleaning excess paste from the float blades with the trowel may provide enough material to fill the hole.

5.3.0 HAND TROWELING

No troweling should be done on a surface that has not been floated. Initial bullfloating or darbying will not refine the surface enough for troweling without additional floating. Large outside slabs such as parking lots are not troweled but are usually textured with a broom after floating. On jobs where troweling is to be done, it should follow immediately after floating.

Troweling makes the surface hard and dense as it seals and smoothes the surface to its final condition. A steel-troweled surface is quite smooth and becomes slippery when wet. Hand trowels must be made of spring steel or of a special, hard stainless steel and are usually from 3 to 5 inches wide and 10 to 20 inches long.

Hand trowels include the pole-handled fresno trowel, which looks somewhat like a bullfloat. The fresno trowel has a locking adapter for the pole handle so that the angle of the blade can be adjusted and kept constant. *Figure 17* shows the adapter of a fresno trowel. Fresno trowels are useful for troweling large slabs such as sidewalks, where speed of finishing, rather than hardness, is critical.

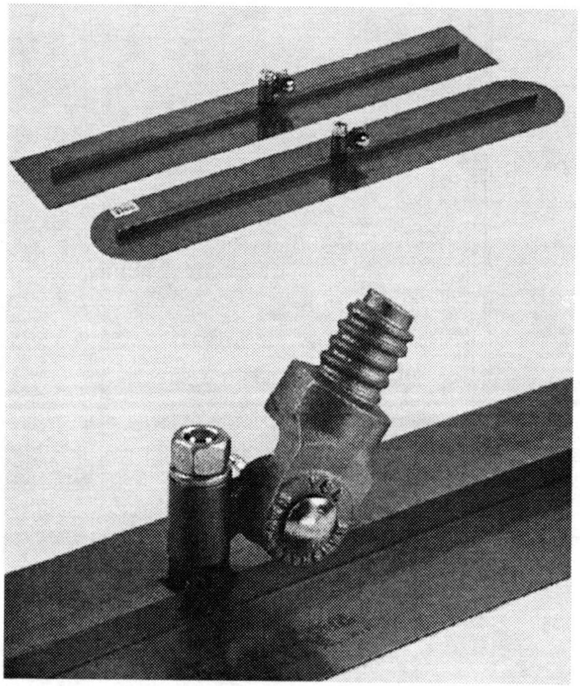

107F17.TIF

Figure 17. Fresno Trowel Adapter

Hand-grip trowels are used in descending size. Larger trowels, held at a slight angle, are used for the first troweling to spread the force over a larger area. As the surface becomes harder, smaller trowels, held at an increased angle, are used to increase the pressure on the surface. The movement of the trowel is in an arc at right angles to the direction of the preceding pass.

The action of a trowel is somewhat different from that of a float. Considerably more pressure is applied to the trowel, so care must be taken to ensure that the leading edge does not dig into the surface. As the trowel is moved across the surface, the leading edge must be kept at a slightly raised angle just enough to keep the edge out of the concrete. Tilting a trowel too much or too soon will result in chatter or washboard marks on the surface. An old trowel is preferred for the first pass because it can be worked almost flat without the edges digging into the concrete. *Figure 18* shows a craftworker doing finish troweling. The finisher is kneeling on knee boards and leaning on a float.

107F18.TIF

Figure 18. Hand Troweling

When hand floating is to be followed by hand troweling, the craftworker can float and trowel in the same operation. This is done by following these steps:

Step 1 Place the knee boards and yourself on the part of the slab that has been bullfloated.

Step 2 Float the area you can comfortably reach, leaning your left hand on the trowel to keep your balance.

Step 3 Change tools and lean on the float.

Step 4 Trowel the area you can comfortably reach.

Step 5 Back the knee boards and yourself away from the troweled area, stopping a comfortable arm's length away from the previous area.

Step 6 Continue with floating, then troweling.

5.4.0 POWER TROWELING

Power troweling produces a dense, smooth, hard surface. It covers a large area in a short time and can bring a floor to almost mirror brightness after several passes, as shown in *Figure 19*.

107F19.TIF

Figure 19. Burnished Power Troweled Slab

Troweling machines come in several sizes, from the walk-behind models shown in *Figure 20* to the larger double-riding power trowel shown in *Figure 21*. Triple ganged riding trowels are also used on large jobs.

107F20.TIF

Figure 20. Walk-Behind Power Trowel

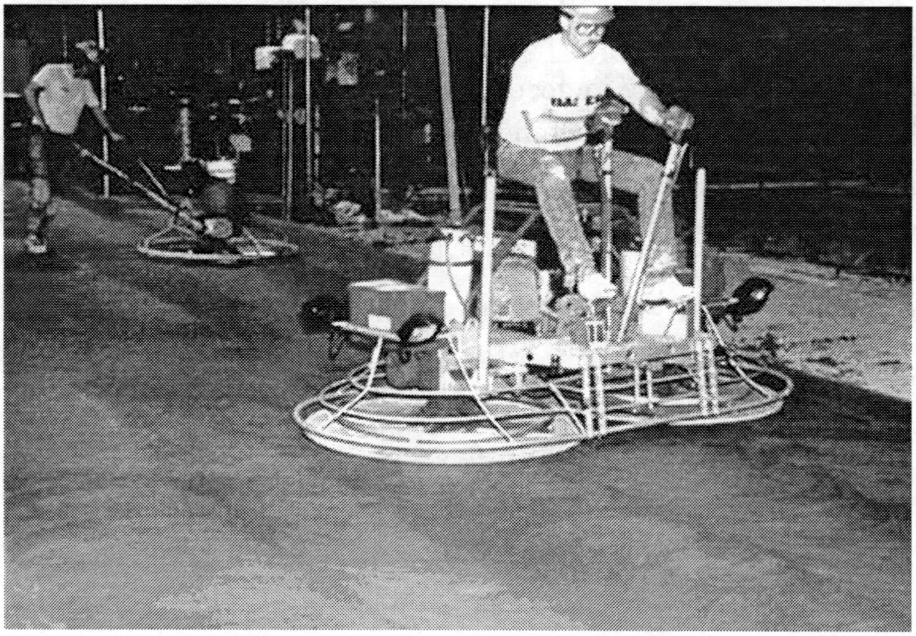

107F21.TIF

Figure 21. Double-Rotor Riding Trowel

Before using a trowel machine, inspect the blades to ensure that all are in the same condition. If one is bent, replace all the blades on the rotor rather than just the damaged one. During the first pass, the blades are tilted very slightly. For each additional pass, the blade tilt is increased. The troweling path goes at a right angle to the previous path, as shown in *Figure 22*.

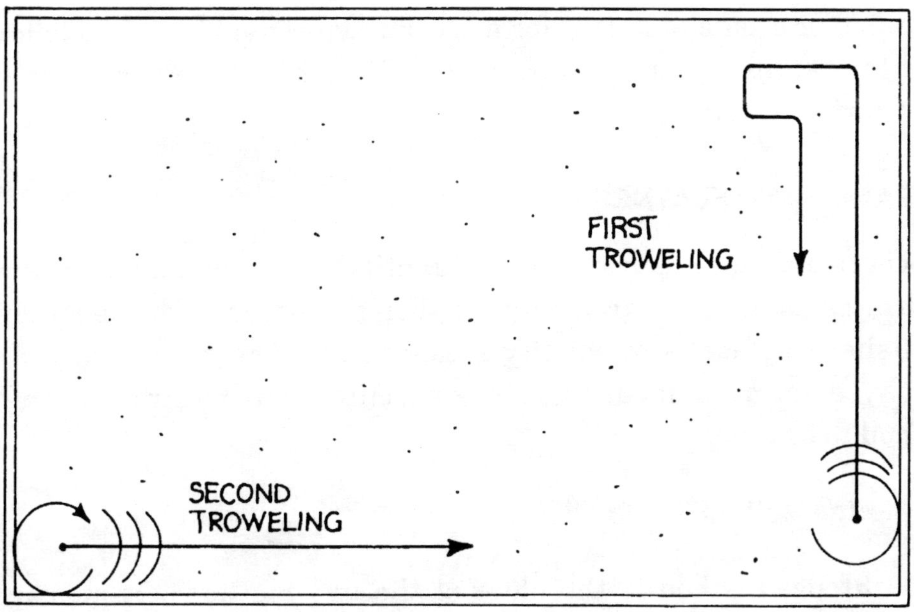

107F23.TIF

Figure 22. Trowel Path

To use a hand-guided troweling machine, follow these steps:

Step 1 Check that the blades are in good condition. Check the safety devices and controls. Check the fuel level if the machine is gasoline powered.

Step 2 Adjust blades for troweling so they are not flat to the surface but at a slight angle.

Step 3 Move the machine across the slab at right angles to the direction of bullfloating.

Step 4 Guide the machine in the direction you want by raising and lowering the handle. To move the machine to the left, raise the handle; to move to the right, lower the handle.

Step 5 Do not operate the machine in any one spot for a length of time. This will scrape a low place on the surface.

Step 6 Continue across the slab until the opposite edge is reached.

Step 7 Step back and maneuver the machine into position to pass again in the opposite direction. Overlap the previous path by about half the width of the rotor.

Step 8 Continue until the slab has been completely troweled.

Step 9 For additional passes, slightly increase the angle of the trowel blades for each pass.

6.0.0 SURFACE TREATMENTS

Job specifications usually provide information on the type of surface treatment needed for each concrete element in a large project. Interior and exterior placements go through similar finishing processes and can get similar surface treatments. There are, however, some steps that may be modified or even eliminated for a specific type of surface. The following sections discuss flatness tolerances, special requirements for exterior and interior surfaces, and different types of surface treatments.

6.1.0 TOLERANCE AND FLATNESS

Inspectors perform tolerance measurements to quantify the flatness and the levelness of the final finished surfaces. These measurements are then compared with the specifications. The quality of a finished surface is generally measured in one of two ways: by a 10-foot straightedge or by **F-number** measures. The specifications controlling the project will identify the method to be used.

6.1.1 Straightedge Method

The 10-foot straightedge method is the older of the two methods. It measures any gaps between the straightedge and the surface. The specifications will provide guidance on the maximum gap that is acceptable. The highest class of surface finish allows a gap of ⅛ of an inch or less. Lower classes of finish allow larger gaps. For sidewalks, driveways, and other projects where extremely flat surfaces are not needed, this method is satisfactory.

6.1.2 F-Number Measuring System

The newer method is the F-number (for "flatness-number") measuring system. This system is preferred over the older straightedge method because it gives more consistent results. It is also more reliable for large areas and for surfaces where flatness and levelness are very important.

The F-number is made of two measures: F_F for flatness and F_L for levelness. Statistical sampling procedures are used to determine the flatness number F_F and the levelness number F_L. The F_F relates to surface irregularities such as waviness, and the F_L relates to the slant or inclination of the surface. *Figure 23* shows an F-number use guide. The dots mark the intersections of F_F on the vertical scale and F_L on the horizontal scale. Note that vehicular use slabs call for higher F_F and F_L levels than carpeting does.

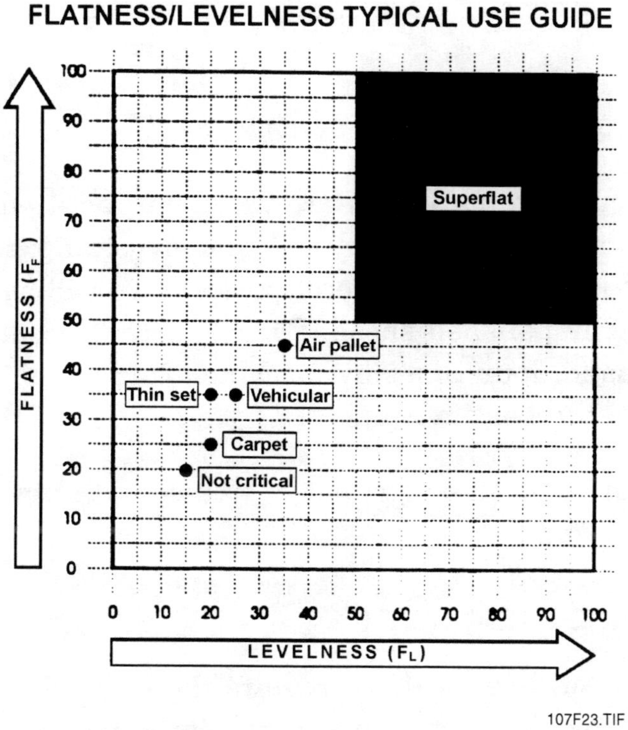

Figure 23. F-Number Use Guide

Measurements are taken with any device capable of measuring point elevation at 12-inch centers. These measurements are entered into a mathematical formula with the results expressed in the form F_F/F_L.

Flatter and more level surfaces will have higher F_F and F_L numbers. A mechanical room or a parking structure slab, where flatness is not critical, may specify values such as 20/15. A critical surface requiring a high degree of flatness and levelness will have values such as 50/30 or even higher.

6.2.0 INTERIOR SURFACE TREATMENTS

Interior concrete surface finishing is done in the same manner and using the same general techniques described in the previous sections. Interior concrete generally requires more extensive finishing and a finer surface treatment. Normally, flatness and levelness tolerances will be higher than for exterior surfaces. The job specifications will provide information on the types of surface treatment needed and some criteria for measuring the flatness tolerances required.

6.2.1 Plain Surface Treatment

A plain surface treatment for interior floors includes the following work. After jointing, floating, and troweling, a second troweling is done 1½ to 2 hours later. Repetitive troweling is continued until the surface meets specifications. This treatment provides a good surface for attaching carpet or tile. For a nonslip surface, texture is applied by brooming with a soft-bristled finishing broom.

6.2.2 Heavy Duty Or Industrial Surface Treatment

A slab for industrial use requires a harder surface treatment, as it will be subject to heavy abrasion. Specifications for industrial floors may call for a low-slump, stiff mixture. For these floors, jointing and floating is followed by several passes of troweling. Troweling continues until the surface has a highly polished sheen. Some sources recommend a minimum of troweling three times by hand, four times by machine. This gives a harder, more wear-resistant surface that is less absorbent to liquids.

For skid resistance, these surfaces can be grooved, or treated with **dry shake hardeners** or other surface or architectural treatments.

6.3.0 EXTERIOR SURFACE TREATMENTS

Generally, exterior concrete surfaces will not require the extensive troweling needed on interior surfaces. They may require edging, which usually is not needed on interior floors. Typical exterior applications will be sidewalks, patios, driveways, parking areas, and roadways. The job specifications will provide information on the types of finishing needed and some criteria for measuring the flatness tolerances required.

Weather impacts exterior surface finishing as well as the surface itself. Areas of the surface exposed to sun will set up more quickly than those protected by shade. Surfaces exposed to wind also require attention sooner than those protected from the wind. You should deal first with those areas that will set faster than the overall placement.

6.3.1 Air Entrainment

Almost all exterior and some interior placed concrete is air entrained because of its high resistance to damage caused by freeze-thaw temperature extremes. Air entrainment also makes concrete somewhat easier to finish, since there is less water in the mix. This is why slightly different finishing procedures are used.

In any hand floating, use a magnesium float. Wood floats drag and take more effort on air-entrained surfaces. Floating by machine can be started sooner than on non-air-entrained concrete. However, too early power floating can result in entrained air becoming lost from the surface of the slab.

6.3.2 Plain Exterior Treatment

A plain exterior surface treatment is typical for driveways, sidewalks, and other small jobs. The sequence of surface treatment steps will include edging, jointing, and floating. The surface will usually be treated with a broom for a nonslip texture.

6.4.0 BROOM SURFACE TREATMENTS

The most common treatment for exterior concrete surfaces is a broomed finish. This is used for projects such as driveways, walks, feeding floors, and lots and ramps for vehicles. This slip-resistant treatment is produced by passing a broom over the surface before the concrete is thoroughly hardened. A rough texture can be made in the concrete surface by running the broom crosswise (perpendicular) to the direction of travel. Almost any degree of coarseness or scoring can be obtained using various types of brooms and amounts of pressure. Since this is the last step in this particular type of finishing process, be sure that all other steps are completed prior to beginning the brooming.

Coarse brooming is usually done using a broom made of a stiff, coarse fiber or fine wire. Coarse brooming can follow the floating step with no troweling. If a finer texture is needed, the concrete must be sealed and floated; then use a soft-bristled broom to brush the surface.

Whether a coarse or fine texture is desired, the steps are the same once brooming begins. The broom is placed at the opposite edge of the surface and pulled toward the finisher. *Figure 24* shows a broom being pulled back across the slab surface. Note the joint in the middle of the pass.

107F24.TIF

Figure 24. Brooming For Texture

When the broom is close to the near edge, lift it up and return to the far edge. Let the next pass slightly overlap the previous pass. Always pull the broom, never push it. This allows more of the broom face to make contact with the concrete surface and gives you better control. Pushing the broom may cause surface paste to build up on the broom which would damage the surface.

Figure 25 shows a closeup of a broomed surface. By moving the broom in regular arcs or waves, you can build an attractive pattern. Skillful brooming can add decoration to an otherwise plain surface.

107F25.TIF

Figure 25. Completed Coarse-Broomed Treatment

6.5.0 GROOVED SURFACE TREATMENTS

This surface is applied mainly to highways and runways where hydroplaning is a potential problem. Grooves are cut with a gang saw across the entire width of the surface. Sometimes the grooves are cut longitudinally for long sections of the road or runway to reduce tire noise.

Outdoor stairs present a potential hazard for slipping. To gain additional traction, grooves are cut by hand using a safety-step edger or a shallow-bitted jointer called a cheater. This safety grooving is done at the same time as edging, using the edging technique described earlier.

6.6.0 RUBBING

Another way to correct surface defects in concrete is to rub them out. Rubbing involves going over the surface of the hardened concrete to fill small surface voids, clean the surface, or smooth irregularities. It can be used to produce a surface of uniform color or sandy texture, or to remove stains from hardened concrete. Rubbing does not improve or change the concrete structurally, it only alters its appearance.

Several different techniques for rubbing are used depending on the method specified or the results desired. Generally, this will be stated in the job specifications. Normally, a good bond between the rubbing grout and the concrete can be achieved after the concrete is hard enough to dry grind without loosening or pulling out any aggregate.

6.6.1 Sack Rubbing

Sack rubbing calls for rubbing concrete surfaces with coarse fabric such as burlap. It can be used to fill minor voids and pits in the surface. It is most effective when done while the concrete is still fresh, although it may be done at any later time up until the surface has been coated with curing compound.

The following procedure can be used to sack rub concrete:

Step 1 Chip off heavy lines and projections above the surface, using a grinder if necessary, until the surface is free of voids and projections.

Step 2 Moisten the surface by thoroughly wetting the concrete then allow it to dry to a slightly moist condition.

Step 3 Mix a mortar to a barely damp consistency. Mortar should consist of 1 part cement to 2-2½ parts of sand, with just enough water so the material will stick together in a tight ball when squeezed in the hand. Sufficient white cement should be blended with common grey cement so that the patches, when dry, will match the color of the existing concrete.

Step 4 Spread the mortar over the concrete surface with a piece of burlap wrapped around a rubber float or similar tool. Make sure the voids are well filled with a slight excess of mortar.

Step 5 After the mortar takes its initial set, rub the surface a second time, removing all mortar from the surface but leaving the voids full. The mortar in the voids should not be raised above or depressed below the concrete.

Step 6 Allow the surface to dry before continuing any other work.

6.6.2 Stone Rubbing

Stone rubbing involves rubbing the concrete surface with a hand tool containing a flat stone, as shown in *Figure 26*. This surface finish can be applied to both vertical and horizontal surfaces. It can be used to produce a surface of uniform color and sandy texture. However, much rubbing or stoning is necessary to blend the grout into the concrete to produce a well-bonded sandy finish. For large areas, the job is best done with power tools because workers using hand tools frequently fail to do a thorough job of rubbing. This results in the grout scaling and peeling later.

Figure 26. Rubbing

Stone rubbing should start as early as possible after the forms are stripped. If rubbing is delayed, the concrete should be water cured. The following procedures can be used to stone rub concrete.

Step 1 Wet the concrete surface thoroughly.

Step 2 Prepare a slurry consisting of 1 part cement to 2-2½ parts of fine sand. The cement should be a mixture of common gray cement and white cement blended to match the existing concrete. The slurry should be of a thick, creamy consistency.

Step 3 Spread the slurry over the surface with a rubber float or similar tool, taking care to fill the voids.

Step 4 Rub vigorously with a carborundum stone. This will bring up a lather of fines that will mix with the slurry on the surface.

Step 5 Spread the resulting mixture uniformly over the concrete until it is approximately $\frac{1}{32}$ of an inch thick.

Step 6 Continue rubbing with the stone until the surface skin on the entire treated area has been removed and replaced by the fines ground out of the concrete, mixed with the slurry, and uniformly spread over the surface.

Step 7 Allow the slurry to harden.

Step 8 Moisten the surface slightly and grind it with a finer carborundum stone until the surface is uniform and smooth.

Step 9 Begin normal curing as soon as the surface can withstand wetting without streaking or other damage.

SUMMARY

Finishing includes the activities of bullfloating or darbying, bleed water removal, edging, jointing, floating, troweling, and surface treating. Edging rounds off and strengthens the corners of outdoor slabs, sidewalks, and driveways and is done both before and during the finishing process, depending on project requirements.

Joints are set in concrete to control the location of cracks (contraction joints), to limit stresses from movement (isolation joints), and to mark the end of a day's work (construction joints). Joints can be set by hand or saw. Sawing techniques include early entry dry cutting, conventional dry cutting, and conventional wet cutting.

Finishing by floating and/or troweling the surface of a slab should not begin until all the bleed water has evaporated and the water sheen is gone. The technique used to finish a surface will depend upon how it will be used. For slabs that are outside and exposed to the elements, a broom finish after floating may be all that is necessary. For slabs that serve as floors for certain industrial or commercial applications, a very smooth, flat surface may be required. In these cases, the finishing technique requires additional passes with a trowel as well as constant checking with a straightedge.

References

For advanced study of the topics covered in this task module, the following reference materials are suggested:

Concrete Construction and Estimating, Craig Avery, Craftsman Book Company, Carlsbad, CA, 1991.

Concrete Manual, International Conference of Building Officials, Whittier, CA, 1992.

Concrete Parking Lots: Eight Steps to Quality Construction, Videotape, Portland Cement Association, Skokie, IL.

Concrete Technology, Third Edition, George R. White, Delmar Publishers, Inc., Albany, NY, 1991.

Design and Control of Concrete Mixtures, Thirteenth Edition, Portland Cement Association, Skokie, IL, 1988.

Finishing Concrete Flatwork, Videotape 32:00 minutes, Portland Cement Association, Skokie, IL, 1991.

Guide for Concrete Floor and Slab Construction, ACI 302.1R-96, American Concrete Institute, Farmington Hills, MI, 1997.

The Contractor's Guide to Quality Construction, American Concrete Institute, Farmington Hills, MI, 1997.

ACKNOWLEDGMENTS

Figures 1, 2, 5, 9, 13, 15, 18-21, and 24 courtesy of Baker Concrete Construction

Figure 3 courtesy of Roy Jorgensen Associates, Inc.

Figures 4, 6, 14, 17, and 23 courtesy of American Concrete Institute

Figures 7 and 8 courtesy of Portland Cement Association

Figures 10-12, 16, 22, 25, and 26 courtesy of Associated General Contractors

SI* (MODERN METRIC) CONVERSION FACTORS

APPROXIMATE CONVERSIONS TO SI UNITS

Symbol	When You Know	Multiply By	To Find	Symbol
		LENGTH		
in	inches	25.4	millimeters	mm
ft	feet	0.305	meters	m
yd	yards	0.914	meters	m
mi	miles	1.61	kilometers	km
		AREA		
in^2	square inches	645.2	square millimeters	mm^2
ft^2	square feet	0.093	square meters	m^2
yd^2	square yards	0.836	square meters	m^2
ac	acres	0.405	hectares	ha
mi^2	square miles	2.59	square kilometers	km^2
		VOLUME		
fl oz	fluid ounces	29.57	milliliters	mL
gal	gallons	3.785	liters	L
ft^3	cubic feet	0.028	cubic meters	m^3
yd^3	cubic yards	0.765	cubic meters	m^3
NOTE: Volumes greater than 1000 l should be shown in m^3.				
		MASS		
oz	ounces	28.35	grams	g
lb	pounds	0.454	kilograms	kg
T	short tons (2000 lb)	0.907	megagrams (or "metric ton")	Mg (or "t")
		TEMPERATURE (exact)		
°F	Fahrenheit temperature	5(F-32)/9 or (F-32)/1.8	Celsius temperature	°C
		FORCE AND PRESSURE OR STRESS		
lbf	poundforce	4.45	newtons	N
lbf/in^2	poundforce per square inch	6.89	kilopascals	kPa

APPROXIMATE CONVERSIONS FROM SI UNITS

Symbol	When You Know	Multiply By	To Find	Symbol
		LENGTH		
mm	millimeters	0.039	inches	in
m	meters	3.28	feet	ft
m	meters	1.09	yards	yd
km	kilometers	0.621	miles	mi
		AREA		
mm^2	square millimeters	0.0016	square inches	in^2
m^2	square meters	10.764	square feet	ft^2
m^2	square meters	1.195	square yards	yd^2
ha	hectares	2.47	acres	ac
km^2	square kilometers	0.386	square miles	mi^2
		VOLUME		
mL	milliliters	0.034	fluid ounces	fl oz
L	liters	0.264	gallons	gal
m^3	cubic meters	35.71	cubic feet	ft^3
m^3	cubic meters	1.307	cubic yards	yd^3
		MASS		
g	grams	0.035	ounces	oz
kg	kilograms	2.202	pounds	lb
Mg (or "t")	megagrams (or "metric ton")	1.103	short tons (2000 lb)	T
		TEMPERATURE (exact)		
°C	Celsius temperature	1.8C + 32	Fahrenheit temperature	°F
		FORCE AND PRESSURE OR STRESS		
N	newtons	0.225	poundforce	lbf
kPa	kilopascals	0.145	poundforce per square inch	lbf/in^2

* SI is the symbol for the International System of Units. Appropriate rounding should be made to comply with Section ASTM E380.

REVIEW/PRACTICE QUESTIONS

1. Basic finishing sequence includes B.
   a. bullfloating, compacting, jointing, troweling, and surface treatment
   b. screeding, bleed water removal, edging, jointing, troweling, and surface treatment
   c. darbying, bleed water removal, vibrating, jointing, and troweling
   d. edging, jointing, consolidating, floating, and troweling

2. Finishing work cannot continue B.
   a. until consolidation is over
   b. with bleed water on the surface
   c. until aggregate settles out
   d. until the concrete separates from the form

3. Raking the leading side of the edger will C.
   a. cut the radius finer
   b. clear the concrete from the edge of the form
   c. keep it from digging into the surface
   d. compact the lip of the slab

4. Contraction joints control cracking by A.
   a. weakening the slab
   b. allowing concrete to shift vertically
   c. cutting the bond of two placements
   d. marking the end of a day's work

5. Isolation joints allow a slab to A.
   a. move horizontally and vertically
   b. avoid plastic shrinkage
   c. be sawn by a wet-cut saw
   d. move the long side 1½ times as much as the short side

6. Contraction joint spacing in feet is equal to B.
   a. the distance between columns
   b. two to three times the slab thickness in inches
   c. two to three times the slab width
   d. the distance between re-entrant corners

7. Joints are not sealed to __D__.
   a. help support joint edges
   b. keep dirt and moisture out
   c. make cleaning easier
   d. prevent raveling

8. Jointing can be done with a __D__.
   a. groover, conventional entry saw, or edger
   b. kelly float, edger, or carborundum saw
   c. kelly float, dowel basket, or edger
   d. conventional entry saw, early entry saw, or groover

9. Cutting joints in dry concrete requires a __B__.
   a. carborundum-blade saw lubricated and cooled by oil
   b. diamond-blade saw lubricated and cooled by water
   c. key-blade saw lubricated and cooled by oil
   d. tungsten-blade saw lubricated and cooled by water

10. Joints are not sawed __B__.
    a. in the same order as the slab was cast
    b. before the bleed water evaporates
    c. before any surface treatment
    d. after the concrete hardens

11. Raveling while sawing indicates __D__.
    a. the concrete is not drying properly
    b. the saw blade needs to be replaced
    c. the saw blade should be carborundum
    d. the concrete is too wet for jointing with a saw

12. Float around the borders of the form first because __A__.
    a. the edges of a slab harden before the rest of it
    b. the bleed water evaporates there first
    c. that is where you start and work forward
    d. the surface needs to be sealed there first

13. By using one tool in each hand, the skilled craftworker can __B__.
    a. double the amount of surface covered
    b. first float, then immediately trowel the same area
    c. first trowel, then immediately joint the same area
    d. first float, then immediately edge the same area

14. When using a power floating machine, do not __A__.
   a. keep the machine in one place
   b. move at a right angle to the direction of the previous floating
   c. start at the point where the concrete was first placed
   d. overlap the previous path by about half the width of the rotor

15. Working with a hand float on a large slab, you can __C__.
   a. lean on the edge of the form
   b. hold onto a kneeboard placed on the slab
   c. support yourself by leaning with one hand on a trowel
   d. kneel at right angles to the direction of the bullfloat

16. Tilting a trowel at too sharp an angle will result in __A__.
   a. washboard marks
   b. dowel baskets
   c. monolithic treatment
   d. raveling

17. The F-number system measures __D__.
   a. isolation joint stresses
   b. contraction joint spacing
   c. irregularities and edging
   d. waviness and slant

18. A plain interior surface treatment for carpet or tile overlay usually requires __A__.
   a. two troweling passes
   b. edging if a nonslip surface is desired
   c. jointing
   d. many troweling passes to harden the surface

19. Specifications for a heavy duty or industrial floor finish usually do not include __D__.
   a. many troweling passes
   b. low-slump concrete
   c. surface treatments for skid resistance
   d. edging

20. Exterior surface areas exposed to sun and wind will __A__.
   a. set up faster than those protected and should be finished first
   b. will have little or no bleeding and should be finished last
   c. not meet flatness tolerances and should be finished first
   d. will have high F-number readings

ANSWERS TO REVIEW/PRACTICE QUESTIONS

	Answer	Section Reference
1.	b	2.0.0
2.	b	2.2.0
3.	c	3.2.0
4.	a	4.0.0
5.	a	4.0.0
6.	b	4.2.0
7.	d	4.4.0
8.	d	4.2.0
9.	b	4.3.1
10.	b	4.2.0
11.	d	4.3.2
12.	a	5.0.0
13.	b	5.3.0
14.	a	5.2.0
15.	c	5.1.2
16.	a	5.3.0
17.	d	6.1.2
18.	a	6.2.1
19.	d	3.1.0/6.2.2
20.	a	6.3.0

NCCER CRAFT TRAINING USER UPDATES

The NCCER makes every effort to keep these manuals up-to-date and free of technical errors. We appreciate your help in this process. If you have an idea for improving this manual, or if you find an error, a typographical mistake, or an inaccuracy in the NCCER's Craft Training Manuals, please write us, using this form or a photocopy. Be sure to include the exact module number, page number, a description of the problem, and the correction, if possible. Your input will be brought to the attention of the Technical Review Committee. Thank you for your assistance.

Instructors – If you found that additional materials were necessary in order to teach this module effectively, please let us know so that we may include them in the Equipment/Materials list in the Instructor's Guide.

Write: Curriculum and Revision Department
National Center for Construction Education and Research
P.O. Box 141104
Gainesville, FL 32614-1104

Fax: 352-334-0932

Craft ____________ Module Name ____________

Module Number ____________ Page Number(s) ____________

Description of Problem

____

____

____

(Optional) Correction of Problem

____

____

(Optional) Your Name and Address

____

____

notes

Curing and Protecting Concrete

Module 23108

Concrete Finishing Trainee Task Module 23108

CURING AND PROTECTING CONCRETE

NATIONAL
CENTER FOR
CONSTRUCTION
EDUCATION AND
RESEARCH

OBJECTIVES

Upon completion of this module, the trainee will be able to:

1. Describe the process of curing concrete.
2. Identify methods of curing concrete.
3. Describe how each method is applied.
4. Identify when each method is used.
5. Associate trade terms with the appropriate processes and equipment.

Prerequisites

Successful completion of the following Task Modules is recommended before beginning study of this Task Module: Core Curricula; Concrete Finishing Level 1, Modules 23101 through 23107.

Required Trainee Materials

1. Trainee Task Module
2. Pencil and paper
3. Appropriate Personal Protective Equipment

COURSE MAP

This course map shows all of the task modules in the first level of the Concrete Finishing curricula. The suggested training order begins at the bottom and proceeds up. Skill levels increase as a trainee advances on the course map. The training order may be adjusted by the local Training Program Sponsor.

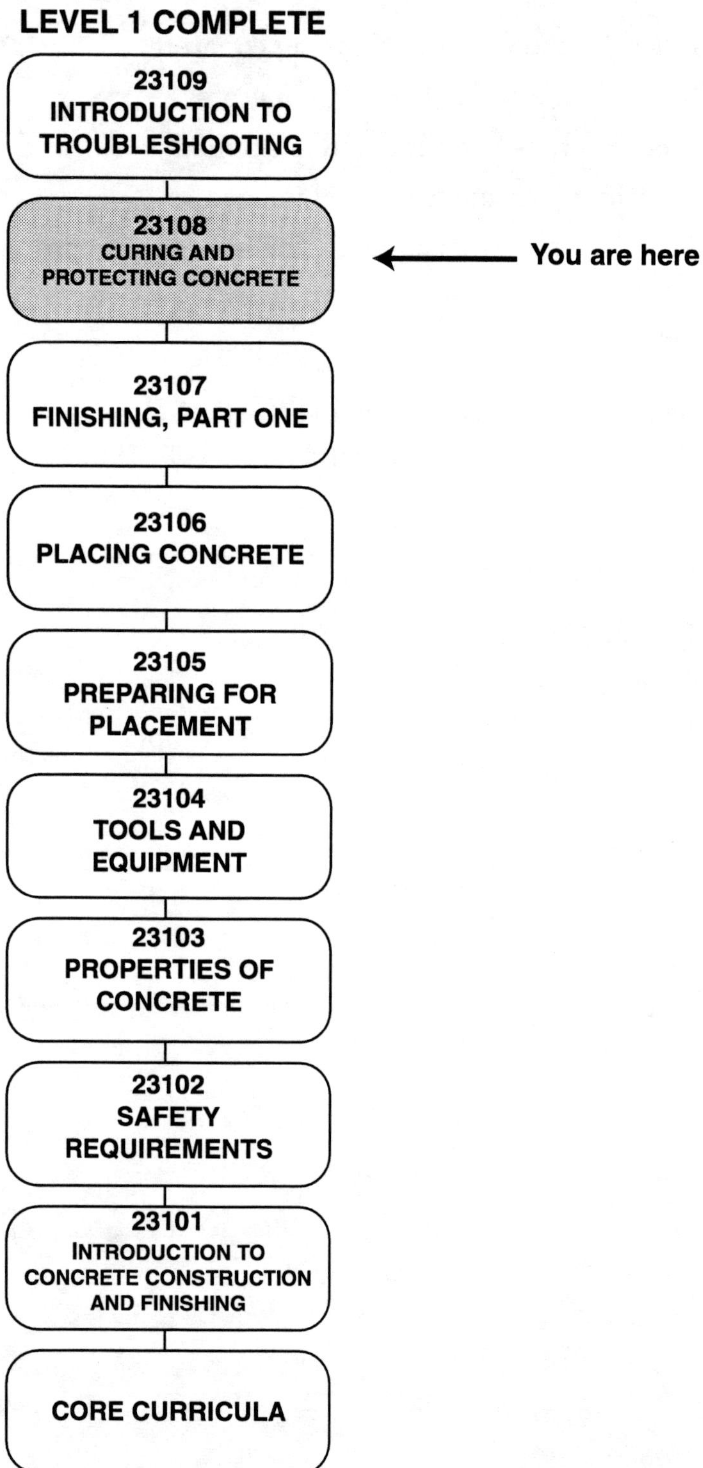

Cmap108.EPS

TABLE OF CONTENTS

Trade Terms Introduced In This Module

Carbonation: The reaction of carbon dioxide with fresh concrete which may cause a weak and dusting surface after the concrete hardens.

Crazing: A network of fine, shallow random cracks or fissures on the surface of the concrete, similar in appearance to the more severe map pattern cracking or pattern cracking. Crazing is not structurally serious nor does it generally indicate the start of future deterioration.

Curing: Keeping fresh concrete at proper conditions of moisture and temperature to allow the cement to hydrate.

Hydration: The chemical reaction that occurs between water and cement or other substances to form hardened concrete.

1.0.0 INTRODUCTION

The **curing** process is the last but very important step in finishing concrete. After concrete has been mixed, placed, consolidated, screeded, floated, and troweled, curing begins.

This module discusses **hydration** and curing, then introduces the processes and equipment used in curing concrete.

1.1.0 PURPOSE

These pages give a review of hydration and an introduction to curing processes. Hydration is the chemical reaction of cement with water that changes the plastic, semi-liquid concrete into a rock-like material. Curing is the set of activities needed on the job site so that hydration can occur under controlled conditions. Curing improves the final product. This information gives you a framework to understand the procedures and equipment discussed later.

1.2.0 SAFETY

Curing involves moisture, which calls for caution at the work site. Wetted or water-covered concrete is slippery. Run-off water makes the ground muddy or slick, and can turn into ice in cold weather. Be aware of your footing and of ground conditions as you work.

Curing may involve the spraying of chemical compounds on the fresh concrete. Read the labels of all chemicals to be sprayed, and take appropriate safety precautions. Read and follow the directions on the Materials Safety Data Sheet (MSDS) furnished with each compound. Wear Appropriate Personal Protective Equipment as required. Curing compounds may also cause surfaces to become slippery.

2.0.0 IMPORTANCE OF CURING

Before beginning the study of curing techniques, read this quick review of hydration. Since curing supports and controls hydration, understanding curing depends on understanding hydration.

2.1.0 HYDRATION

Concrete hardens and gains strength by hydration, a chemical reaction between cement and water. This reaction generates heat. After cement and water mix, the cement remains dormant for a short period of time (approximately one hour), which can be made shorter or longer by the use of admixtures. This dormant period allows concrete to be transported to the job and placed in forms before it starts to harden. As the cement begins to hydrate, the concrete looses plasticity/workability and starts to set up. Hydration of cement continues after concreting is finished and the job is over. The longer cement hydrates, the stronger it becomes. Hydration is quick during the first 3 to 7 days, then continues slowly indefinitely.

2.1.1 Moisture

As cement hydrates, concrete increases in durability, watertightness, wear resistance, and strength. The chemical reaction of hydration uses the water originally mixed into the concrete. Hydration continues as long as water is available. Curing prevents the loss of moisture and adds more moisture, so hydration can continue.

Figure 1 shows the difference moisture makes and how concrete continues to gain strength over time. The bottom of the chart shows elapsed time in days. The vertical axis shows the tested compressive strength of the concrete.

The differences in strength made by curing or by simply letting the concrete harden in the air can be substantial. The bottom line shows that concrete drying without additional moisture will never reach its moist-cured strength. The middle lines show how concrete tests after 7 and after 3 days of curing. With 7 days of moist curing, it will take about 90 days for concrete to reach the same strength that fully moist-cured concrete reaches at 28 days. The curve of the top line shows that concrete moist-cured for 28 days continues to gain strength, even beyond 180 days. This curve suggests that the concrete will continue to gain strength for a long time.

2.1.2 Temperature

Hydration is affected by temperature as well as moisture content. The ideal range of concrete temperature is warmer than 40°F and cooler than 80°F. Concrete hydrating at temperatures below 40°F develops strength slowly; concrete hydrating at temperatures above 80°F develops strength quickly.

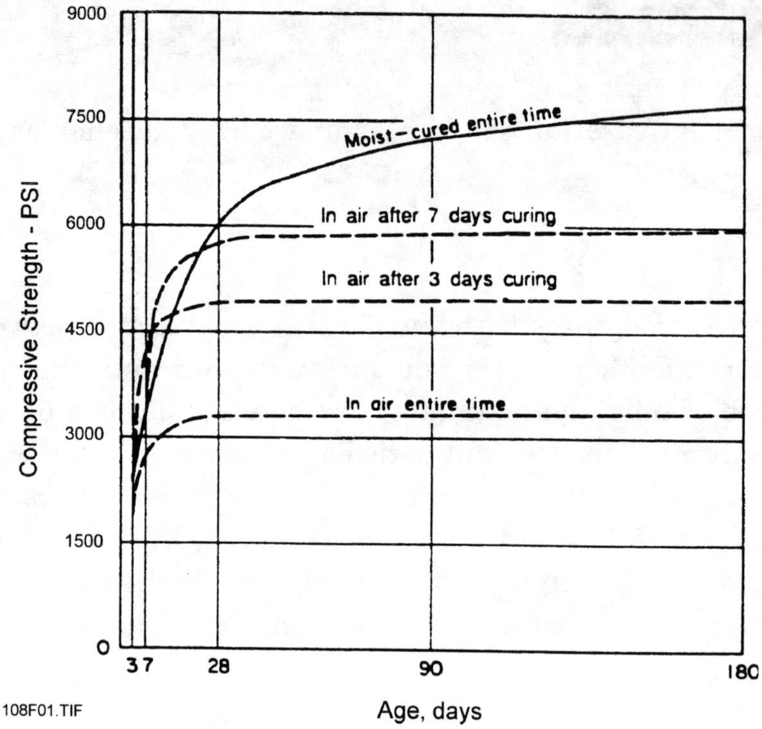

108F01.TIF

Figure 1. The Effect Of Moist-Curing On Concrete Strength With Age

2.2.0 CURING

The purpose of curing concrete is to maintain moisture and adequate temperature in the concrete. Curing is critical for slabs on ground because they have such a large surface for a comparatively small volume. Evaporation from the surface can be so great that a slab will crack as it sets. Variations in temperature across the face of the slab can cause stress cracks. Curing provides preventive measures to stop evaporation and maintain temperatures.

2.2.1 Temperature And Strength

Both hot and cold weather affect curing. Hot weather can cause problems of lower strength, shorter setting time, and rapid drying of fresh concrete surfaces. Sometimes, hot weather may dry out concrete so quickly it cannot be finished.

Hot weather can raise internal concrete temperatures. Concrete cured at 120°F will have lower final strength than concrete cured at 73°F. *Figure 2* shows the effect on strength of curing at high temperatures. It is easy to see that concrete cured at temperatures higher than 80°F ends up with lower strength.

Cold weather can lower curing temperatures inside concrete to a point where hydration almost stops. Concrete setting time is longer and concrete gains strength slowly at cold temperatures. Concrete cured below 40°F will have lower final strength than concrete cured at 40°F to 80°F. Generally, concrete that freezes during setting must be replaced.

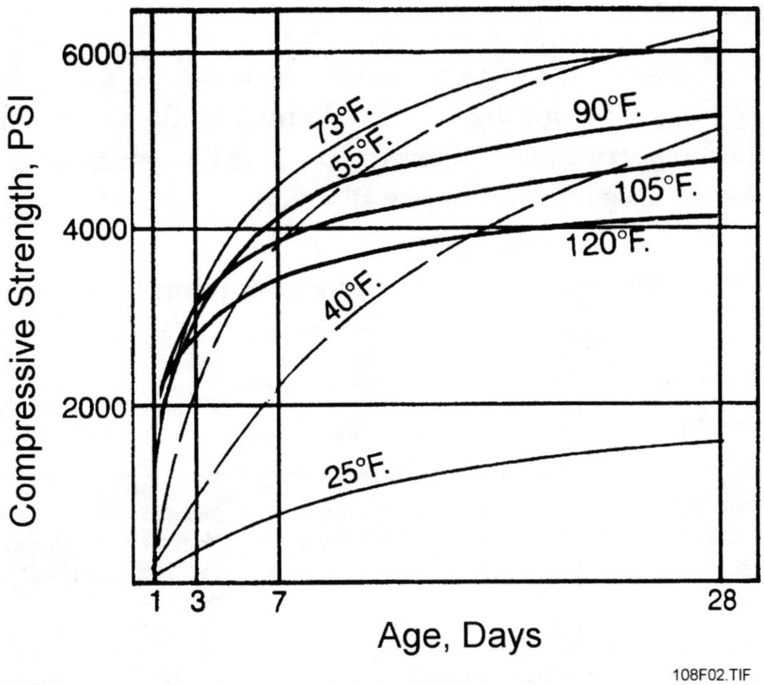

Figure 2. Concrete Strength Gain Changes With Temperature

Specifications generally call for protective measures when placing outdoor concrete while the temperature is less than 40°F.

Curing procedures protect setting concrete from the effects of extreme air temperatures. Curing procedures are designed to allow concrete to harden at internal temperatures between 40°F and 80°F.

2.2.2 Curing Effects On Surface Conditions

It is important to start curing as soon as possible after finishing is ended. Curing is as important as finishing for slabs. Poor or no curing results in a weakened concrete surface. This surface will not resist abrasion and will wear away quickly. It can be discolored and may easily scale or spall.

Tests show 7 days of curing give the surface twice the abrasion resistance as compared to 3 days of curing. Abrasion resistance is reduced by about 65 percent if curing is delayed by only 24 hours. Delaying the start of curing can cause cracks and **crazing** as well.

If the curing surface dries out then gets rewetted, it will craze and may scale easily. To prevent crazing, start curing as soon as possible and keep the surface continuously moist for 3 days or more.

3.0.0 CURING METHODS

Curing methods fall into two main groups: those that add water to the surface of the concrete, and those that prevent water loss from the surface. A third method of curing speeds strength gain by adding heat and moisture to the concrete. Different techniques can reach the same goal. Any of the methods discussed here could be the best for any job. The contractor usually makes the choice, depending on the characteristics of the concrete, the schedule, and other limitations. *Figure 3* lists curing methods used most frequently and the advantages and limitations of each.

Type of Curing	Where to Use	Advantages	Limitations
Spraying/ fogging	Horizontal or vertical work (slabs or walls)	Thorough moist curing, very clean surface for bonding	Needs water source, relies on equipment (timers) or workers to spray at appropriate intervals, drying out can occur, prevents other work from occurring on slab, affected by wind
Wet coverings	Horizontal or vertical work (slabs or walls)	Thorough moist curing, very clean surface for bonding, can help control temperature	Coverings may be messy or difficult to use/remove, must be continually wetted, drying out of covering can draw moisture away from slab, prevents other work from occurring on slab, can discolor surface, potential fire hazard
Impervious paper	Horizontal work (slabs)	Very clean surface for bonding, allows using slab as work surface, inexpensive, will not stain	Must be anchored and sealed at edges and joints
Plastic sheeting	Horizontal work (slabs)	Very clean surface for bonding, allows using slab as work surface, inexpensive, can help control temperature, reusable	Must be anchored and sealed at edges and joints, may discolor concrete, can tear under traffic, not for use where appearance is important
Membrane-forming compounds	Horizontal or vertical work (slabs or walls)	Fast and easy application, inexpensive, can help control surface temperature, may allow traffic on horizontal work after short period	Can interfere with surface bondings of other materials (paints, coatings, tile), may contain solvents
Insulating blankets	Anywhere (slabs, walls, or other formed surfaces)	Retains both heat and moisture, flexible	Must be shaped to member and anchored or held in place
Live steam	Anywhere (slabs, walls, or other formed surfaces)	Supplies heat and moisture, increases the rate of strength gain	Requires enclosed area, requires special equipment, time-consuming setup, expensive, can be dangerous

108F03.EPS

Figure 3. Curing Methods, Advantages, And Limitations

3.1.0 SPRAYING/FOGGING

Fogging, sprinkling, or spraying all cure by maintaining water on the surface. The water must be delivered to all exposed surfaces frequently enough to keep the concrete continuously moist. If the concrete surface dries between sprinklings, it will reduce the effectiveness of curing. A fog nozzle is used for flat surfaces, soaker hoses on vertical or sloping surfaces. Water pressure and flow must be low enough that they do not wash away the new surface.

A mechanical arrangement of pipes and hoses is stretched across or around large slabs of fresh concrete. For small slabs, a garden sprinkler can be set, as shown in *Figure 4*. The system may be set in any way that will continuously maintain water on the concrete surface.

108F04.TIF

Figure 4. Sprinkling System

The advantage of this technique is that it will not discolor the surface; it will leave a clean surface ready for painting or other coating. Difficulty in maintaining the equipment can be one disadvantage. Also, the wind can shift and move the spray or fog off part of the concrete. Another disadvantage is that activity on the slab must stop during curing.

3.2.0 WET COVERINGS

Wet coverings cure by holding water on either horizontal or vertical surfaces, and include burlap, cotton rugs and mats, or other absorbent materials. The material used must be washed and free of anything that can stain the concrete.

The wet covering is thoroughly soaked before being applied to all surfaces of the concrete and it must be kept wet throughout the curing period. *Figure 5* shows a worker spraying a wet covering to replace water lost to evaporation.

108F05.TIF

Figure 5. Rewetting A Wet Covering

Sometimes, burlap is coated with plastic to reflect light and heat. Cotton mats and rugs hold more water than burlap and have less risk of drying out.

Sometimes, wet coverings are used in combination with sprinklers or foggers to make sure the covering stays wet. If this type of covering dries before the curing period is complete, it will actually draw water from the concrete.

The advantages of wet cloth covering are the clean surface it leaves and the low cost of the coverings, as they can be reused. The disadvantages come from the need to keep the coverings wet. Sprinkler systems can break down, or workers can come late. If the covering dries, the slab surface can be seriously damaged. Activity on the slab usually stops when wet coverings are used.

3.3.0 IMPERVIOUS PAPER

Impervious paper placed on the concrete surface will retain moisture by reducing evaporation. The main advantages of impervious paper are that the slab is accessible during curing and it is not necessary to provide water continuously. Impervious paper has two layers of brown kraft paper reinforced and bonded with asphalt. It is vaporproof, strong, nonstaining, and tear resistant. It is commonly used on slabs.

The concrete must be wetted down before the paper is placed. It is unrolled flat and wrinkle-free, as shown in *Figure 6*. The paper must cover all the exposed slab surfaces, sides, and edges, as well as the face.

108F06.TIF

Figure 6. Placing Waterproof Paper

The paper lap joints must be sealed with nonstaining glues or tapes. The edges of the lapped paper must reach to the ground and be covered with sand or lumber to seal in moisture. The paper must be weighted down at regular intervals to make sure it stays in contact with the surface.

Another advantage of impervious paper is that it leaves the floor clean and ready to bond. Because it wrinkles less than plastic sheets, it is less likely to discolor the concrete.

3.4.0 PLASTIC SHEETING

Plastic sheeting cures by stopping evaporation. The film should be at least 0.004 inches thick, and can be clear, black, or white. Black film absorbs heat during the winter, and white film reflects heat during the summer.

Figure 7 shows a worker laying a roll of plastic sheeting commonly used for slabs. The surface of the concrete is wetted before the plastic is laid. The sheeting is unrolled flat and overlapped 6 inches. The plastic is weighted so it keeps in contact with the concrete surface.

108F07.TIF

Figure 7. Placing Plastic Sheets

It is important to place the plastic flat because wrinkles can cause surface discoloration. *Figure 8* shows how water vapor collects under the wrinkles, through condensation. The pooling water can discolor the surface of the freshly cured concrete; however, this generally fades with time.

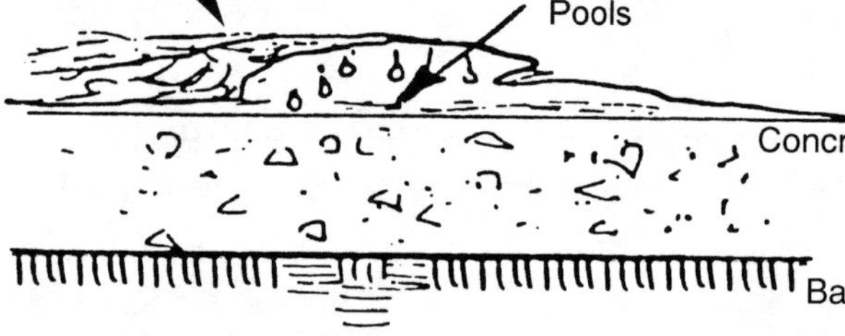

108F08.TIF

Figure 8. Surface Discoloration Under Plastic

The plastic sheeting must be wrapped around the exposed surfaces and sealed to the ground with lumber or sand. Sealing the plastic around the edges of the concrete stops moisture leaks and evaporation. Notice that the plastic is sealed around the starting edge with braced lumber in *Figure 7*.

The advantages of plastic sheeting include low cost and ease of placing, except in a high wind. It can be reused. Its disadvantage is that it easily wrinkles, and, thus, discolors the surface. Plastic is not recommended where concrete is colored or where appearance is important. Plastic can tear easily if there is traffic over the fresh slab.

3.5.0 MEMBRANE-FORMING COMPOUNDS

Membrane-forming compounds cure by reducing evaporation. These liquids contain waxes, natural and synthetic resins, and solvents. Properly applied to concrete, they form a moisture-retaining film.

These compounds are applied by brush, roller, or pressurized spray, as shown in *Figure 9*. They are applied on damp, not wet, concrete immediately after final finishing. When spraying, all exposed surfaces need to be covered. Walking on the wet spray will break the film, so it is important to keep off the surface.

108F09.TIF

Figure 9. Spraying A Membrane-Forming Compound

These compounds can be pigmented to indicate the area of coverage. Application should be in a criss-crossing pattern, as shown in *Figure 10*, to make sure of coverage. The best coverage is two coats at right angles to each other.

There are several common types of compounds.

- Clear wax-resin base compound is inexpensive and takes from 2 to 9 months to disintegrate. It easily takes a vanishing dye to show coverage. It should not be used when the slab will be covered by tiling, painting, or other surface treatment.

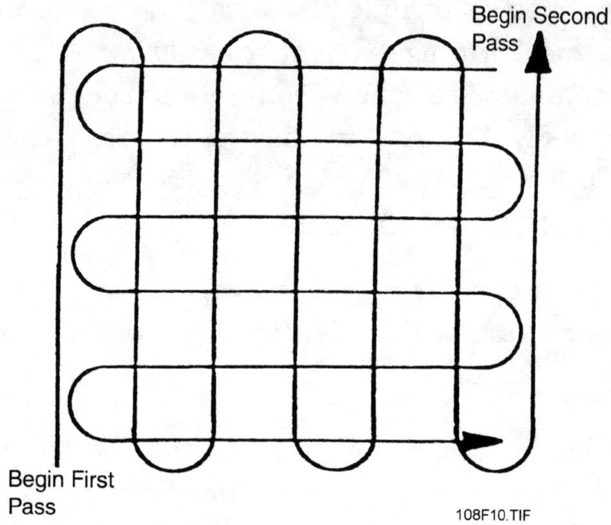

Figure 10. Compound Spraying Pattern

- Clear resin compound costs more but dries and disintegrates in 30 to 60 days. This gives a harder abrasion-resistant surface less likely to wear off under traffic. It allows painting or other surface treatment and takes a vanishing dye.
- White-pigmented compounds have wax or resin bases. The pigment reflects light and can reduce heat buildup, a benefit in hot weather. These compounds have the same characteristics as their base.
- Black and gray-pigmented compounds usually have an asphalt base, and are very inexpensive. They absorb heat, a benefit in cold weather. They are good for waterproofing and provide a good adhesion base for linoleum or tile. They do not provide a good adhesion base for other types of coverings or finishes.
- Newer membrane-forming compounds include chlorinated rubber, acrylics, and epoxies. These can have different limits on their use, their results, and their residues. It is important to read the instructions or guidelines for these products.

WARNING! The Environmental Protection Agency regulates the use of some solvents used in newer membrane compounds. Be sure to read the instructions for any chemicals used at the job site. Use Appropriate Personal Protective Equipment.

The advantages of compounds are that they are simple to apply, inexpensive, and allow activity soon after application. The disadvantages are few. Because compounds leave a residue that interferes with some types of bonding, sealing, or painting, these final surface treatments and coverings should be considered when selecting curing methods. The compounds will be damaged if they are stepped on before they dry. They need to be covered with kraft paper or other material to protect them from traffic until curing is complete.

3.6.0 INSULATING BLANKETS OR LAYERS

Insulating blankets or layers cure by holding in both heat and moisture. In cold weather, these blankets are wrapped around forms containing fresh concrete to retain the heat of hydration and protect them from the cold weather.

Figure 11 shows insulating blankets wrapped around column forms and tied in place. Although these blankets were designed for use on slabs, they are easily adapted to other shapes. Their flexibility makes it easy to cover all surfaces of regular or irregular shapes.

108F11.TIF

Figure 11. Insulating Blankets Wrapped Around Columns

Insulating materials used in these blankets are of several types. Sheets of polystyrene or polyurethane are cut to shape and fitted around forms. Pliable vinyl foam blankets with extruded vinyl backing can be made with embedded electrical heating elements. Fiberglass or cellulose fibers can be bound within waterproof casings to form large mats, batts, or rolls. For some jobs, urethane foam is sprayed on the outside of the forms to make a custom insulating blanket.

For small slabs, an insulating blanket can be made at the job site, as shown in *Figure 12*. This hand-made insulating blanket is a thick layer of straw spread between two polyethylene sheets, or tarpaulins.

Figure 12. Hand-Made Insulating Blanket

The sheet on the bottom allows easy pickup of the loose insulating material. Bricks or lumber must hold down the edges of this hand-made blanket to retain heat and moisture, as shown in *Figure 12*. Other waterproofing and insulating materials may be used to build this type of blanket.

3.7.0 FORMS

Formwork left in place after concreting cures by limiting evaporation. If forms are used for curing, or for partial curing, it is important to loosen them, wet them, and keep them wet. Any surface exposed by the open top of the form must be wetted enough so that water runs down the inside of the loosened form. The exposed surface can be wetted by a covering layer, as shown in *Figure 13*, or by sprinkling. Plastic should not be used where appearance is critical.

Figure 13. Wet Coverings Over Loosened Forms

If forms are removed before curing is complete, the concrete needs to be covered completely by the chosen curing agent. Again, this can be a wet covering or a membrane-forming compound, as they are easy to apply. In some cold-weather areas, this type of curing may be finished by using steam, as described below.

3.8.0 INCREASING THE RATE OF STRENGTH GAIN

Two types of curing rapidly increase the rate of strength gain by adding heat and moisture to the concrete. Quickening the curing process can be done with steam or by adding heat through other means, such as heating coils or electrically heated forms. The latter is typically used in precast or prestressed concrete operations.

3.8.1 Steam Curing

Two kinds of steam curing are used with concrete: live steam and high-pressure steam.

Live steam curing takes place in a temporary housing, a tent, or a structure of some kind to enclose the steam. Steam can be applied 3 to 5 hours after concrete has been placed. The process has four stages: delay, feed, hold, and cool, as shown in *Figure 14.*

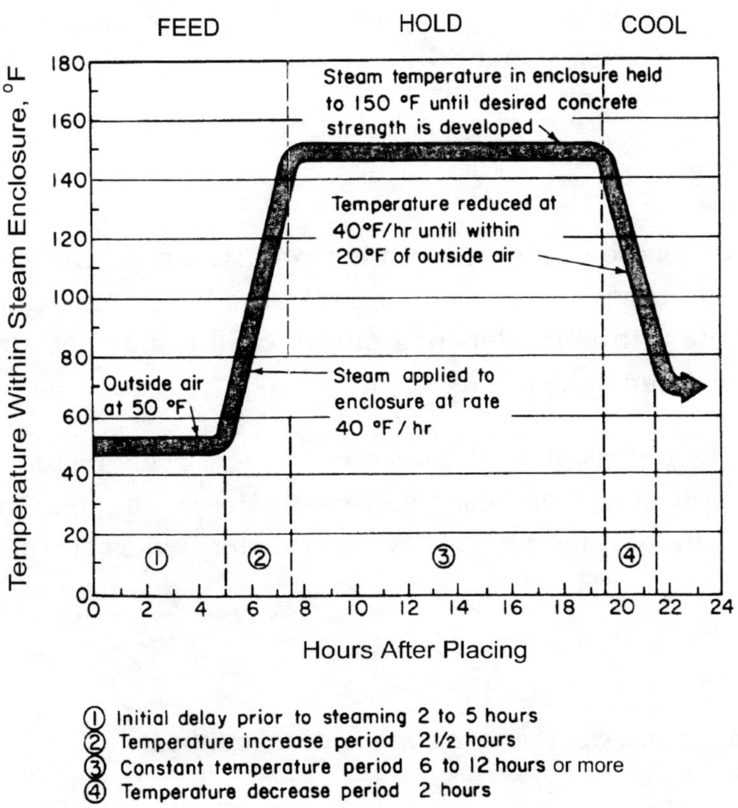

Figure 14. Live Steam Cycle

The initial delay is the time between the completion of placing and the beginning of steaming. The steam is fed into the enclosure slowly, so the temperature does not rise more than 40°F per hour. Holding starts when the temperature reaches 150°F. The temperature should not be allowed above 170°F, or strength will be reduced. The holding time, from 6 to 12 hours, is set by the mix specifications. During cooling time, the steam vents slowly, so the temperature does not fall more than 40°F per hour.

High-pressure steam follows the same cycling but uses an autoclave for a housing. Curing temperatures can range from 325°F to 375°F, under pressures of 80 to 170 pounds per square inch. Hydration is very fast under these conditions. High-pressure steam is used on small manufactured castings at a precast plant. This technique is limited by the size of the high-pressure housing needed. High-temperature curing allows chemical reactions that do not occur at lower temperatures. It is best to avoid curing temperatures very much higher than the concrete will experience during its service life.

3.8.2 Other Heat Curing

Precast concrete can be cured at quickened rates by electrical and hot oil heat.

In electrical curing, current is passed through the concrete or through the reinforcing steel to generate heat. Steel forms may be electrically heated, or hot oil may be circulated through hollow-walled steel forms. Electrical and oil heating are rarely used on job sites because of the expense.

4.0.0 PROTECTION

As discussed before, concrete cures best under conditions of favorable moisture and temperature for a period of time. The moisture is controlled by the curing methods mentioned above. The time past the minimum depends on job conditions. The heat is provided by hydration, but the weather can quickly affect that.

Fresh concrete needs to be protected against its environment. This includes extreme weather conditions and traffic. Different techniques can reach the same goal. Any of the methods discussed here could be the best for any job. The contractor makes the choice, depending on job limits.

4.1.0 HEAT

The two curing problems in hot weather are evaporation and heat.

Hot weather can increase the rate of evaporation and cause cracking. An increase of 20°F in concrete temperature can mean a reduction in finishing time of 50 percent. Hot weather curing techniques must slow evaporation.

The second problem is the heat itself. Concrete temperatures can rise significantly. When the internal temperature of the hydrating concrete rises above 80°F, ultimate strength is reduced, as shown in *Figure 2* earlier. The heat needs to be dissipated to keep the mix from getting too hot. Hot weather curing techniques must dissipate heat.

On-site in hot weather:

- Monitor fresh concrete temperature.
- Work and schedule work rapidly, so concrete is finished before it sets.
- Cool mix ingredients, especially water. Use ice as part of the mix water, or add shaved ice directly to the mixer. In extreme weather, liquid nitrogen can be added into the mixing water to turn it to slush.
- Use a retarding admixture to slow setting time and lengthen curing time, if specifications permit.
- Protect curing concrete against heat and evaporation by paying special attention to maintaining or retaining moisture.
- Be sure the subgrade is thoroughly moistened in advance, so it will not absorb water from the concrete. Be sure there is no standing water on the subgrade.
- Wet the forms in advance, or use a release agent on them. Spray them periodically after filling to cool them down. Shade them from the sun with temporary reflective (white) covers.
- Use continuous spraying.
- Use white-pigmented membrane-forming compounds, white plastic sheeting, or white-coated wet coverings to reflect heat and hold in water.
- Taper off curing gradually, so the concrete will not be shocked by sudden drying.
- Place concrete during cooler parts of the day: evening, morning, or night.
- Use mineral admixtures to reduce the heat of hydration.

4.2.0 COLD

At cold temperatures, fresh concrete hardens very slowly. When temperatures drop below 40°F, concrete must be protected against freezing. Moisture usually does not need to be used for curing in cold weather, unless dry heat is applied.

On the job site, take these precautions against freezing:

- Prepare in advance by having heaters on hand.
- Heat the water and the materials added to the concrete to raise the temperature of the concrete.
- Use an accelerating admixture to speed hydration and shorten finishing time, if specifications permit.

- Use air-entrained concrete.
- Use insulating blankets overnight to protect the subgrade before placing concrete. Do not place concrete on a frozen subgrade.
- Preheat form surfaces.
- Wrap and seal insulating blankets around forms.
- Avoid placing concrete during extreme cold weather.
- Use black-pigmented membrane-forming compounds, black plastic sheeting, or kraft paper, as appropriate, to absorb heat and retain moisture.
- Taper off curing gradually, so the concrete will not be shocked by sudden cold.
- Monitor fresh concrete temperatures.

Concrete should be heated to between 50°F and 70°F at the time of placing. The temperature should be maintained at 50°F for 5 days, or at 70°F for 3 days. To keep temperatures in this range, temporary winter enclosures can be used.

4.2.1 Temporary Winter Enclosures

During ongoing severe cold, a temporary housing can be built around the placement site from wood, plastic, or prefab panels. *Figure 15* shows a temporary housing of tarpaulins that provides protection for a bridge during severe and prolonged winter weather.

108F15.TIF

Figure 15. Temporary Winter Enclosure

Housing needs to be weathertight and may need heat to keep the temperature suitable for curing. When heat is used, hot, dry air can cause plastic shrinkage problems. To avoid this, water must be added to the curing process in winter when a heated housing is used.

Space heaters for temporary housings must be vented, or they can damage concrete and people. Carbon dioxide from unvented space heaters will bond with wet concrete surfaces. This process, called **carbonation**, causes a weak, dusting surface when the concrete dries.

WARNING! Be sure any space heater is vented to the outside. Carbon monoxide poisoning is hard to detect.

4.2.2 Applying Steam

The advantage of steam is that it shortens curing time to 24 hours or less. Steam can be applied as soon as 3 hours after placement is complete. Under severe conditions, live steam may be used at a job site, inside a housing. As discussed above, live steam is applied in a four-stage process. The temperature is slowly increased, held, and then slowly dropped.

Live steam on the job site can create a safety hazard. It is important to be aware of steam and to protect yourself by staying clear.

***CAUTION*:** Live steam can scald lungs. Stay out of steam enclosures and stay away from venting steam.

4.3.0 WIND

Winds affect curing by speeding the evaporation of water. The combination of hot, dry weather and wind can dry water from the concrete surface faster than bleed water can replace it. Surprisingly, strong wind can speed evaporation in cold weather as well.

In strong wind, surfaces dry faster, so scheduled waterings need to occur more frequently. Wind shifts water from sprayers, sprinklers, and foggers away from the concrete. Wind unseals lap joints and edge seals of plastic sheeting or impervious paper. Strong wind lifts and tears unsecured plastic sheeting or paper.

Windbreaks can be built out of tarpaulins to protect fresh concrete from evaporation.

Work can start late in the afternoon to take advantage of lower wind speed at sunset.

4.4.0 RAIN

Rain creates problems by adding too much moisture. Rain on freshly placed concrete can erode the surface and dilute the cement paste at and near the surface. Rain can wash away fines and cause voids in concrete being placed. Finishing of unformed concrete cannot be completed in the rain because it will be damaged.

Freshly placed concrete must be sheltered completely from rain, or work stopped until the rain is over. Temporary rain shelters can be built from tarpaulins or plastic sheeting to protect concrete surfaces. It is important to check that the rain will not drip off the edges of the shelter onto the slab.

4.5.0 TRAFFIC OVER SLABS

As concrete cures, it gains strength over time. During this time, construction traffic and loads driven over new slabs can damage the surface, can exceed the strength of the slab, and can lead to cracked or damaged slabs. The best protection is to keep traffic off the slab for the first 7 days.

If activity must continue, it should start with smaller equipment, foot traffic, and light material loads. This traffic is acceptable after 3 to 7 days, the initial cure time. The loads must be light, and the concrete surface must be protected. Kraft paper, plywood, or fiber sheets can be used to provide traffic paths over a slab. Care should be taken to prevent damage to edges by routing paths away from edges and by reinforcing path material at the edges of the slab.

Concrete without reinforcement is strong in compression (crushing force) but not in tension (pulling force) or flexure (bending force). Loading or heavy traffic can bend the concrete, especially at the edges of slabs, and cause cracks. Use common sense to protect new concrete from loads.

Residential slabs, driveways, and walks are not designed to carry heavy loads such as ready-mix trucks or heavy materials. These types of loads should not be allowed on residential concrete.

SUMMARY

This module reviews hydration and discusses curing. Hydration is the chemical reaction between cement and water that binds together concrete ingredients. Curing controls the moisture condition of fresh concrete by adding water or maintaining water. When the water is gone, hydration stops. Curing controls the temperature of the mix by adding or removing heat. The temperature controls the speed of hydration.

Curing methods concentrate on adding moisture or preventing evaporation. A few methods cure by speeding hydration. Curing methods can overcome problems caused by heat, cold, wind, or rain. This module discusses how to cure by spraying, wet covering, and using impervious paper, plastic sheeting, membrane-forming compounds, insulating blankets, live steam, and other methods of adding heat.

References

For advanced study of topics in this task module, the following books and videos are suggested:

Concrete Construction and Estimating, Craig Avery, Craftsman Book Company, Carlsbad, CA, 1991.

Concrete Manual, International Conference of Building Officials, Whittier, CA, 1992.

Concrete Parking Lots: Eight Steps to Quality Construction, Videotape, Portland Cement Association, Skokie, IL.

Concrete Technology, Third Edition, George R. White, Delmar Publishers, Inc., Albany, NY, 1991.

Design and Control of Concrete Mixtures, Thirteenth Edition, Portland Cement Association, Skokie, IL, 1988.

Finishing Concrete Flatwork, Videotape 32:00 minutes, Portland Cement Association, Skokie, IL, 1991.

Guide for Concrete Floor and Slab Construction, ACI 302.1R-96, American Concrete Institute, Farmington Hills, MI, 1997.

The Contractor's Guide to Quality Construction, American Concrete Institute, Farmington Hills, MI, 1997.

ACKNOWLEDGMENTS

Figures 1 and 14 courtesy of Portland Cement Association
Figure 2 courtesy of International Conference of Building Officials
Figures 3-6, 11, 13, and 15 courtesy of American Concrete Institute
Figures 7 and 9 courtesy of Associated General Contractors
Figures 8 and 12 courtesy of National Association of Homebuilders
Figure 10 courtesy of Roy Jorgensen Associates, Inc.

SI* (MODERN METRIC) CONVERSION FACTORS

APPROXIMATE CONVERSIONS TO SI UNITS

Symbol	When You Know	Multiply By	To Find	Symbol
		LENGTH		
in	inches	25.4	millimeters	mm
ft	feet	0.305	meters	m
yd	yards	0.914	meters	m
mi	miles	1.61	kilometers	km
		AREA		
in^2	square inches	645.2	square millimeters	mm^2
ft^2	square feet	0.093	square meters	m^2
yd^2	square yards	0.836	square meters	m^2
ac	acres	0.405	hectares	ha
mi^2	square miles	2.59	square kilometers	km^2
		VOLUME		
fl oz	fluid ounces	29.57	milliliters	mL
gal	gallons	3.785	liters	L
ft^3	cubic feet	0.028	cubic meters	m^3
yd^3	cubic yards	0.765	cubic meters	m^3
NOTE: Volumes greater than 1000 l should be shown in m^3.				
		MASS		
oz	ounces	28.35	grams	g
lb	pounds	0.454	kilograms	kg
T	short tons (2000 lb)	0.907	megagrams (or "metric ton")	Mg (or "t")
		TEMPERATURE (exact)		
°F	Fahrenheit temperature	5(F-32)/9 or (F-32)/1.8	Celsius temperature	°C
		FORCE AND PRESSURE OR STRESS		
lbf	poundforce	4.45	newtons	N
lbf/in^2	poundforce per square inch	6.89	kilopascals	kPa

APPROXIMATE CONVERSIONS FROM SI UNITS

Symbol	When You Know	Multiply By	To Find	Symbol
		LENGTH		
mm	millimeters	0.039	inches	in
m	meters	3.28	feet	ft
m	meters	1.09	yards	yd
km	kilometers	0.621	miles	mi
		AREA		
mm^2	square millimeters	0.0016	square inches	in^2
m^2	square meters	10.764	square feet	ft^2
m^2	square meters	1.195	square yards	yd^2
ha	hectares	2.47	acres	ac
km^2	square kilometers	0.386	square miles	mi^2
		VOLUME		
mL	milliliters	0.034	fluid ounces	fl oz
L	liters	0.264	gallons	gal
m^3	cubic meters	35.71	cubic feet	ft^3
m^3	cubic meters	1.307	cubic yards	yd^3
		MASS		
g	grams	0.035	ounces	oz
kg	kilograms	2.202	pounds	lb
Mg (or "t")	megagrams (or "metric ton")	1.103	short tons (2000 lb)	T
		TEMPERATURE (exact)		
°C	Celsius temperature	1.8C + 32	Fahrenheit temperature	°F
		FORCE AND PRESSURE OR STRESS		
N	newtons	0.225	poundforce	lbf
kPa	kilopascals	0.145	poundforce per square inch	lbf/in^2

* SI is the symbol for the International System of Units. Appropriate rounding should be made to comply with Section ASTM E380.

REVIEW/PRACTICE QUESTIONS

Select the one correct answer for each question.

1. An increase of 20°F in concrete temperature can ___C___.
   a. extend finishing time by 50 percent
   b. melt away heat quickly
   c. cut finishing time by up to 50 percent
   d. evaporate moisture slowly

2. Winds affect curing by ___B___.
   a. raising the temperature of hydration, in cold weather
   b. speeding the evaporation of water, in all weather
   c. slowing the evaporation of water, in hot weather
   d. lowering the temperature of hydration, in all weather

3. Membrane-forming curing compounds are applied ___B___.
   a. to stop traffic over the slab
   b. immediately after final finishing
   c. to color the finished surface
   d. before bleed water dries off

4. Adding moisture during the curing process will allow the concrete to ___B___.
   a. reach its maximum strength in 7 days
   b. obtain a higher maximum strength
   c. reach its maximum strength in 28 days
   d. remain plastic for 7 days

5. Rain on freshly placed concrete can ___A___.
   a. dilute the cement paste near the surface
   b. be used instead of sprinkling
   c. help slow down hydration
   d. speed up final finishing time

6. Impervious paper is useful for curing because it is ___C___.
   a. strong
   b. bonded by asphalt
   c. vaporproof
   d. lightweight

7. The purpose of curing concrete is to control _C_.
   a. moisture and carbonation, to prolong hydration
   b. carbonation and temperature, to shorten hydration
   c. moisture and temperature, to prolong hydration
   d. moisture loss and temperature, to shorten hydration

8. Traffic paths over fresh slabs _A_.
   a. can be made of plywood or kraft paper
   b. should use the edges of the slabs
   c. can be used by ready-mix trucks
   d. lead to cracks and chips

9. A disadvantage of plastic sheeting is that it can _D_.
   a. hold curing temperatures below 40°F
   b. be reused many times
   c. be laid over fresh concrete
   d. discolor surfaces

10. Concrete cured in cold weather does not _D_.
   a. take longer to set
   b. gain strength slowly
   c. need protection from freezing
   d. harden quickly

11. Unvented space heaters can _B_.
   a. heat temporary housings
   b. carbonate the surface of fresh concrete
   c. avoid rapid hydration
   d. protect workers

12. Spraying, fogging, or sprinkling can prevent _A_.
   a. crazing and plastic cracks
   b. damage from deicers
   c. bonding of carbon dioxide and calcium hydroxide
   d. dissipating heat

13. Insulating blankets can be _C_.
   a. sprayed onto fresh concrete
   b. made of scrap lumber
   c. wrapped around irregularly shaped forms
   d. transparent plastic sheeting

14. Hydration is __A__.
    a. a chemical reaction between cement and water
    b. the management of the curing process
    c. the bonding of carbon dioxide and calcium hydroxide
    d. a placed mix that has not reached its strength and hardness

15. Steam curing does not __C__.
    a. need an enclosure
    b. require gradual start-up and cool-down
    c. slow down hydration
    d. reduce total curing time

16. As they shape concrete, forms can be __D__.
    a. removed after 6 hours
    b. covered completely by the curing agent
    c. used to hold down plastic sheeting
    d. used for partial curing

17. Wet coverings must be __B__.
    a. completely dried before applying
    b. thoroughly soaked before applying
    c. partially dried before removing
    d. partially soaked before applying

18. Fresh, or green, concrete __B__.
    a. has admixtures or sprays to add color
    b. needs protection against heat, cold, wind, rain, and traffic
    c. has an unfinished surface
    d. needs protection against moisture, asphalt, and traffic

19. In cold weather, concrete that is curing needs __B__.
    a. insulating blankets overnight
    b. to be maintained at between 50°F and 70°F
    c. white-pigmented membrane-forming compounds
    d. to be heated to between 60°F and 80°F

20. The advantages of wet curing do not include __C__.
    a. temperature control
    b. clean surfaces for bonding
    c. reusable materials
    d. moist curing to strengthen

notes

ANSWERS TO REVIEW/PRACTICE QUESTIONS

Answer	Section Reference
1. c	4.1.0
2. b	4.3.0
3. b	3.6.0
4. b	2.1.1
5. a	4.4.0
6. c	3.4.0
7. c	2.1.0/2.2.0
8. a	4.5.0
9. d	3.5.0
10. d	2.2.1
11. b	4.2.1
12. a	2.2.2/3.1.0
13. c	3.7.0
14. a	2.1.0
15. c	3.9.1
16. d	3.8.0
17. b	3.2.0
18. b	4.0.0
19. b	4.2.0
20. c	3.0.0/Figure 3

NCCER CRAFT TRAINING USER UPDATES

The NCCER makes every effort to keep these manuals up-to-date and free of technical errors. We appreciate your help in this process. If you have an idea for improving this manual, or if you find an error, a typographical mistake, or an inaccuracy in the NCCER's Craft Training Manuals, please write us, using this form or a photocopy. Be sure to include the exact module number, page number, a description of the problem, and the correction, if possible. Your input will be brought to the attention of the Technical Review Committee. Thank you for your assistance.

Instructors – If you found that additional materials were necessary in order to teach this module effectively, please let us know so that we may include them in the Equipment/Materials list in the Instructor's Guide.

Write: Curriculum and Revision Department
National Center for Construction Education and Research
P.O. Box 141104
Gainesville, FL 32614-1104

Fax: 352-334-0932

Craft ____________________ Module Name ____________________

Module Number ____________________ Page Number(s) ____________________

Description of Problem

____

____

____

(Optional) Correction of Problem

____

____

(Optional) Your Name and Address

____

____

notes

Introduction to Troubleshooting

Module 23109

Concrete Finishing Trainee Task Module 23109

INTRODUCTION TO TROUBLESHOOTING

NATIONAL CENTER FOR CONSTRUCTION EDUCATION AND RESEARCH

OBJECTIVES

Upon completion of this module, the trainee will be able to:

1. Describe a basic troubleshooting methodology that can be used to identify a variety of concrete construction problems and their causes.
2. Identify problems with fresh concrete and describe ways to prevent them.
3. Identify different concrete defects such as crazing, cracking, dusting, scaling, popouts, and efflorescence, and describe ways to prevent them.

Prerequisites

Successful completion of the following Task Modules is recommended before beginning study of this Task Module: Core Curricula; Concrete Finishing Level 1, Modules 23101 through 23108.

Required Trainee Materials

1. Trainee Task Module
2. Pencil and paper
3. Appropriate Personal Protective Equipment

COURSE MAP

This course map shows all of the task modules in the first level of the Concrete Finishing curricula. The suggested training order begins at the bottom and proceeds up. Skill levels increase as a trainee advances on the course map. The training order may be adjusted by the local Training Program Sponsor.

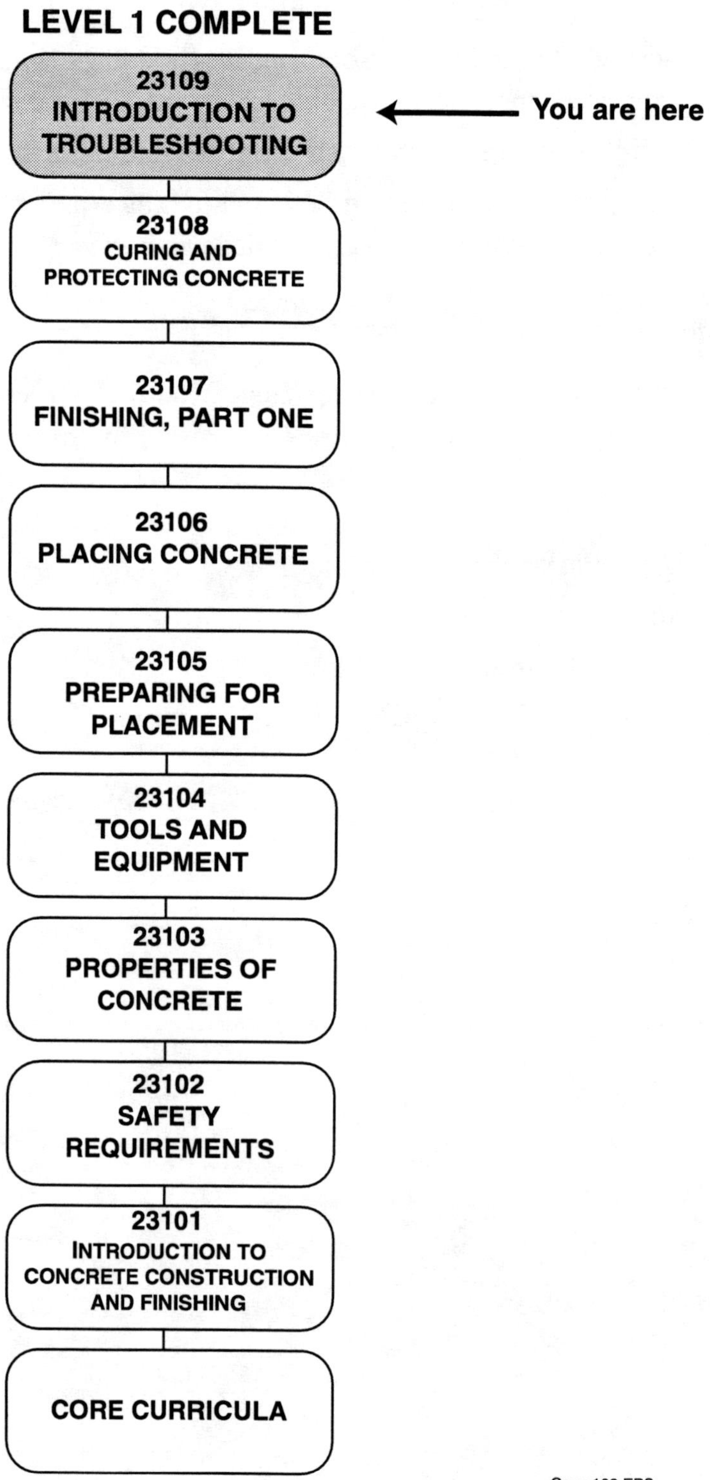

Cmap109.EPS

TABLE OF CONTENTS

Trade Terms Introduced In This Module

Cracking: Irregular surface fractures in the concrete that have a variety of causes.

Crazing: A network of fine, shallow random cracks or fissures on the surface of the concrete, similar in appearance to the more severe map pattern cracking or pattern cracking. Crazing is not structurally serious nor does it generally indicate the start of future deterioration.

Dusting: A fine, powdery substance that appears on the concrete surface, often caused by a mix that has been finished with bleed water on the surface.

Efflorescence: A white deposit on the concrete surface consisting of dissolved minerals deposited after moisture evaporation from the surface.

Honeycomb: Voids left in concrete where cement and sand particles have not filled the spaces between the coarse aggregate particles.

Laitance: A weak, milky layer of cement and fine aggregates that rises to the surface of concrete because of too much mixing water or excessive vibration. It dries to a dusty finish.

Popout: A shallow, conical fragment that breaks out of the concrete surface leaving a conical hole. Usually caused by internal pressure.

Scaling: The flaking or peeling away of the concrete surface caused by hydraulic pressures.

Spall: A chip, flake, or piece of concrete that has been broken or separated from the concrete surface; usually caused by expansion within the concrete.

1.0.0 INTRODUCTION

Versatility, durability, and economy have made concrete the most widely used construction material in the world. Good concrete requires careful planning, the use of quality materials, and a knowledgeable and skilled craftworker. With these ingredients, concrete that is mixed, placed, finished, and cured properly will last for decades. But making this happen requires planning and hard work.

The best way to prevent defects and problems in concrete is to use good placing, finishing, and curing techniques. However, all concrete work will not be problem-free. Some problems are caused by external factors, like site conditions or the weather. Plans can be made to work around these kinds of problems as they arise. Other problems are not so easy to foresee, but they will need to be fixed also. For this reason, it is important that you know how to recognize potential problems and prevent them if possible. Generally, prevention is more economical than correction and is easier to accomplish. Since not all problems can be prevented all the time, you need to know how to identify and correct problems once they have occurred.

1.1.0 PURPOSE OF MODULE

This module presents a basic troubleshooting procedure you can use to help you identify surface imperfections and problems in fresh concrete. It also presents some typical defects as well as their causes, and measures to prevent them. This module concentrates on preventing defects because prevention is the best and easiest fix for any problem.

1.2.0 SAFETY

The safety concerns associated with troubleshooting concrete problems are similar to those for working with concrete in general. It is important to wear the proper protective gear and clothing. Repair of defective concrete may require use of saws, power tools, and chemicals. Fresh concrete is very caustic. Prolonged direct contact between skin surfaces and wet, unhardened concrete can cause skin irritation and severe chemical burns. You need to protect yourself by wearing Appropriate Personal Protective Equipment while performing your task.

Your employer should provide safety equipment, good quality tools, and a safe work environment. You should also be kept informed of new and changing safety regulations. It is your job to act responsibly and safely, and to accept responsibility for your actions and how they affect you, other employees, and the public.

The final responsibility for on-the-job safety rests with each and every worker. It is every workers' responsibility to learn:

- Safety gear and clothing required for each job
- Safety rules and procedures associated with each job
- General work site safety and the hazards associated with each job
- Environmental safety as it affects each job

2.0.0 TROUBLESHOOTING METHOD

When problems do occur, you need a troubleshooting method to help you identify the cause of the problem, to decide on a way to fix it, and to prevent it from recurring. In order to do this, you have to know the proper procedures for mixing, placing, finishing, and curing concrete, and the damage that results from incorrect procedures. You should also follow a logical, step-by-step method for locating and correcting the exact cause of the problem. Successful troubleshooting depends on a logical process of thinking through a problem until the cause and an effective solution become clear.

Some concrete contractors and cement manufacturers provide troubleshooting procedures for common concrete construction problems. When available, they should be followed. If there are no troubleshooting procedures provided, however, a general troubleshooting procedure can be followed. This general procedure can be applied to a variety of problems.

Troubleshooting involves the following three steps, as shown in *Figure 1*:

- Investigate the symptoms.
- Research the problem.
- Apply corrective action.

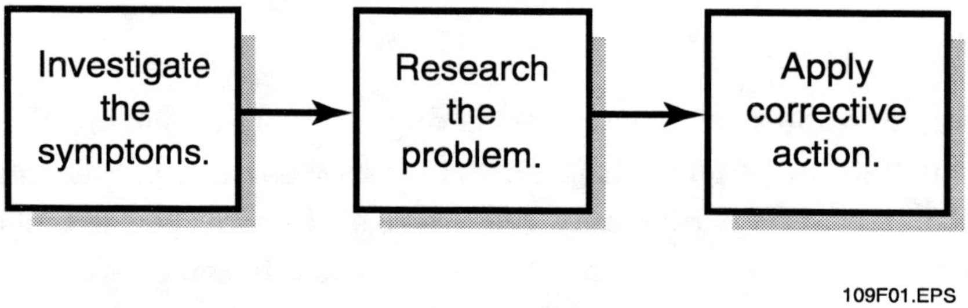

Figure 1. Troubleshooting Methodology

2.1.0 INVESTIGATE

The first step in any troubleshooting procedure is to investigate. In order to fix a problem, you need to know exactly what that problem is. Therefore, it is important to collect as much information on when and how the problem first appeared. You should also check out the problem to see it firsthand.

Generally, when you investigate a problem, you look at the symptoms of the problem, not the cause. A cause is what makes a problem happen. A symptom is a result of the cause, the way the problem appears. In concrete construction, for example, typical symptoms might be white powder, flaking concrete, or shallow cracks. These symptoms are clues that can lead you to the cause of a problem.

Investigating the symptoms can help you locate and identify the cause, to prevent it from recurring. Some common causes are poor quality materials, faulty construction practices, exposure to weathering, and accidental damage.

A crack in concrete, for example, may originate from faulty subgrade, poor jointing, or overloading. In order to repair the crack appropriately, you have to determine the cause of the damage or deterioration and its extent.

2.2.0 RESEARCH

Once you have examined the symptoms, analyze the information to develop an idea about the cause of the problem. It is difficult to determine the cause of a problem. Instead of one single cause, a problem is often a result of a combination of causes, such as weather conditions and incorrect construction procedures.

One way to identify the causes of a problem is to compare the symptoms with normal conditions. Information about normal conditions can be found in the manufacturer's literature, job documents, and your own experience in working with concrete. Knowing the difference between how the concrete looks when it is handled well versus how it looks when it is not handled well can help you recognize the symptoms and find the causes of the problem.

The causes of some problems are not obvious. You may not have enough experience to troubleshoot on your own. Senior workers, supervisors, engineering consultants, testing laboratories, and industry publications are resources for information. The American Concrete Institute, the Portland Cement Association, and state contractor associations have many helpful publications on all aspects of concrete work.

2.3.0 APPLY CORRECTIVE ACTION

As you research, you will find that some problems have many causes. Corrections may be as simple as a change in how you do the work. Other causes may call for changes that are out of your control and must be approved by the project engineer and/or the client.

Once you have identified the cause(s) of the problem, you will need to adjust the procedures that caused the problem to stop it from happening again. Talk to your supervisor about what you find.

If you decide that the mix you are using is too wet, for example, the foreperson will need to request a change in the proportion of water. Or, if you decide that you are not using a finishing tool properly, you need to ask for more training so you will know the right procedures.

If you find an area of possible future trouble, you should discuss preventive actions with your supervisor. This action may involve a change in procedures, a new design, or training of personnel. The project manager or company owner may have to approve some of these changes. This makes it important to thoroughly check your findings before you talk to your supervisor.

Taking these steps should keep the same problem from happening again from the same causes. One last task remains: to repair the symptoms.

There are several different methods of repair. Whichever method is chosen, the repair must be completely bonded to sound concrete, free of cracks, compatible in appearance, impermeable, and resistant to weathering or other attack. The best fix should meet the following guidelines:

- It should fix the cause of the problem as well as its symptoms, so that the problem will not recur.
- It should fix the cause of the problem as quickly and economically as possible.
- It should fix the cause of the problem without creating any new problems.

At this stage of learning concrete work, prevention is your best repair for any problems in fresh concrete. As you learn more, you will be able to prevent problems by recognizing and acting on their causes before the symptoms set in.

3.0.0 FRESH CONCRETE

Concrete is usually selected as the building material of choice because of its strength, durability, long life, low maintenance, and overall serviceability. Unfortunately, there are certain common problems which occur that give a different impression. Most are not highly technical in nature but are common problems relating to mix design, construction practices, testing, and specifications.

The causes and means of prevention are not unsolved mysteries, but prevention lies in applying what is known. Sometimes, it is possible to correct defects after the job is done but prevention is far more economical and more easily done.

The key to prevention is in knowing what causes the problem. Rather than a single cause, it is often a combination of marginal practices and job conditions, no one of which may seem to be important enough to seriously affect the concrete quality.

How well concrete performs is decided by several factors, including mix proportions, aggregate characteristics, and workmanship. These factors can affect concrete workability, and eventually its strength and durability. Working with fresh concrete, you will find these typical problems:

- Excess bleeding
- Segregation
- Rapid or slow set
- Plastic shrinkage **cracking**
- Low spots

As you can see in *Figure 2*, some of these problems are caused by shortcomings in the mix, some are caused by weather, and some are caused by poor craftwork. If you are not familiar with a particular corrective procedure, do not attempt to apply a quick fix. This may produce further problems and result in more costly work. As you study *Figure 2*, decide which corrections you can handle by yourself and which need your supervisor's approval.

PROBLEM	CAUSE	PREVENTION OR CORRECTION
Excess Bleeding	Insufficient fines in mix	Increase percent of fines.
	Excess mix water	Reduce water content.
	Vapor barrier under slab	Compact 3 to 6 inches of sand or other drainable fill on top of vapor barrier.
Segregation	High slump (excess water)	Reduce water content. Use superplasticizer.
	No homogeneity in mix	Use air-entraining admixtures. Reduce coarse aggregate proportion in mix.
	Overvibration	Insert vibrator at closer intervals, use until concrete is flowing. Make sure concrete is thoroughly mixed.
Sticky finish	Air and/or sand too high	Promote bleeding. Reduce air and sand percentages.
	Rapid drying of the surface	Dampen subgrade and forms. Apply fog spray to surface.
Rapid set (Hot weather)	High cement content	Use water-reducing retarder and/or fly ash.
	High concrete temperature, high air temperature	Use maximum retarder dose. Use fly ash. Cool mix water, add ice or liquid nitrogen. Cool aggregate piles by sprinkling. Place concrete during coolest parts of day.
	Trucks waiting in sun	Schedule trucks for shortest wait, in shade if possible. Sprinkle outside of mixer.
Slow set (Cold weather)	Lean mix, especially with fly ash or slag	Increase cement. Use accelerator. Heat aggregates and water.
	Cold or wet subgrade, cold air temperature	Warm subgrade by covering with mats or polyethylene sheeting. Warm forms and reinforcements. Use temporary housing and heat source.
Plastic shrinkage cracking	Rapid evaporation of water from the surface, through wind, heat, low humidity	Reduce mixing water to stiffen mix. Induce water gain on surface by reducing sand and air entrainment. Avoid use of vapor barrier, or cover it. Fog spray on surface when finishing. Provide windbreaks. Reduce concrete temperature by using cold water in mix. Keep embedded materials from absorbing radiant heat.

109F02.EPS

Figure 2. Problems In Fresh Concrete

3.1.0 EXCESS BLEEDING

Bleed water rises during the setting process, when the cement and aggregate particles suspended in the water sink because of their weight. As they move downward, the displaced water moves upward and appears at the surface as bleed water. *Figure 3* shows a surface covered with bleed water.

Some amount of bleeding is acceptable. In fact, it can actually result in a slightly stronger paste. However, if bleeding continues, **laitance** may form on the surface. This laitance will dry to a weak, permeable surface susceptible to abrasion. If finishing operations work the bleed water back into the surface, the dry surface will **craze**, **dust**, and/or **scale**. Sometimes, the foreperson will adjust the mix as it is placed.

Sometimes, there is no time to let bleed water evaporate. Mechanical ways of dealing with bleed water include laying absorbent mats, using vacuum mats, and using a squeegee or rubber hose to push the water off the surface. The important thing is that the water be removed before finishing continues.

109F03.TIF

Figure 3. Bleed Water On Surface

3.2.0 SEGREGATION

Segregation occurs when the coarse aggregate separates from the paste, as shown in *Figure 4*. The main causes of segregation are too much water in the mix, incomplete mixing, poor placing work, or over- or under-vibrating the placed concrete.

109F04.TIF

Figure 4. Segregation

To prevent segregation, check that the concrete has been thoroughly mixed. Place it with a square-end shovel and a come-along. Do not move it with a vibrator. Use a drop chute or elephant trunk as appropriate. Do not over- or under-vibrate it. It pays to avoid segregation because it can be a major repair job in the hardened concrete.

3.3.0 STICKY FINISH

Sticky finish or "sticky mix" is a common finisher's complaint. The worst condition would be the combination of wet, oversanded, air-entrained mix on a hot, dry day. Air entrainment makes the mix more flowable or workable, requiring less water for the same slump and making the concrete workable at a lower slump. Remove the excess water and sand and concrete won't be sticking to the float or knee boards as it sometimes does.

Harsh mixes, the kind that seem to lack enough paste to close the surface, are another common complaint. These are most likely to occur in low strength mixes requiring relatively low cement contents. A coarse sand would aggravate the problem. If air entrainment is not already required for durability, at least 3 to 4 percent air would be helpful for workability. A small addition of fine sand or fly ash would also be beneficial, as would an addition of cement.

Sometimes, hard-to-finish mixes can be overcome by sprinkling water on the surface if there is rapid evaporation; however, this practice can be risky. A fog spray that drifts across the slab without providing excess water to be worked into the surface is much safer.

3.4.0 RAPID SET

Once the concrete is mixed, usually about 90 minutes is allowed for placing and finishing before it sets. Quick work is particularly important in hot weather. High temperatures speed chemical reactions, including hydration. As a result, in hot weather, it is crucial to place and finish concrete on the job as quickly as possible before it can harden.

Rapid set or flash set occurs when the concrete sets rapidly before it can be finished. This happens in hot weather and in windy, dry weather. However, rapid set can also occur in cold weather when very hot water is added directly to the cement to prevent it from freezing.

To minimize the effect of rapid set, be sure that the site is ready for the concrete and work quickly. Rapid set can be prevented by extending the setting time of the concrete. You can do this by:

- Cooling the concrete materials by sprinkling the coarse aggregate piles with water before mixing
- Replacing a portion of the mix water with an equal amount of crushed ice
- Adding hot water to aggregates, not directly to the cement
- Change the mix by adding a retarder or using fly ash

Any of the above actions may require engineering approval depending on the specifications and local codes.

3.5.0 SLOW SET

Since concrete sets more slowly in cold weather and can be damaged by freezing, measures should be taken to keep the concrete temperature above 50°F. Any of the concrete's tendencies toward bleeding will be considerably aggravated by the slower setting, and more work will be required to take care of it properly. Many extra hours of finishing will be required unless acceptable means can be found to shorten the setting time. Often, extra expense is justified to speed up the operation. This usually involves one of the following approaches:

- Adding calcium chloride or a non-corrosive, non-chloride accelerating mixture where available
- Changing to high-early-strength cement (Type III) or using a larger amount of Type I or II cement than usually required
- Increasing the concrete temperature to 70°F
- Applying dewatering techniques early on after first strike-off and bullfloating to provide early access for floating
- Reducing the water-cementitious material ratio and selecting the minimum slump that can be easily handled and placed

Overworking the concrete during the strike-off and bullfloating operations should be avoided.

3.6.0 PLASTIC SHRINKAGE CRACKING

Plastic shrinkage cracks occur while the concrete is still plastic or workable, often before final finishing. They occur in the interior of a slab, are generally parallel to each other, and are several inches to several feet in length (*Figure 5*). They are relatively shallow V-shaped cracks of about 1 to 3 inches in depth, rarely the full depth of the slab. Almost never do they extend to the edge of the slab.

109F05.TIF

Figure 5. Plastic Shrinkage Cracks

The cracks occur because of surface tension created by the rapid evaporation of surface moisture. Low humidity and wind are the primary causes.

Other contributing factors are high concrete temperature in relation to air temperature, placement of concrete directly on a vapor barrier, and high slump mixes. If there is more water than needed in the mix, and the concrete is on a totally non-absorptive subgrade, the excess water will rise to the top. The shrinkage occurring with rapid loss of moisture is then increased. If a vapor barrier must be used, 2 to 3 inches of sand should be compacted on the vapor barrier before placing the concrete.

It is important to note that plastic shrinkage cracking does not usually create any concrete performance problems and does not reflect on the quality of the concrete. The cracks do not enlarge or connect with each other during service. If the area is to be covered, sealing then would not serve any useful purpose.

3.7.0 LOW SPOTS

Low spots are depressions in the concrete surface. They appear as puddles on an outdoor concrete slab after a rain or on a floor after hosing. These spots can affect slab drainage or serviceability if items placed on the slab need to be level. Low spots are often caused by poor lighting during placing and finishing, subgrade settlement, inadequate slope, and poor placing and finishing techniques. They can also be caused by forms and screeds that are set incorrectly or that shift during use.

Low spots can be prevented by following these guidelines:

- Work with adequate lighting. If you cannot see well, tell your supervisor.
- Set the proper grades for forms and screeds, brace them well, and check grades frequently to make sure the settings remain intact.
- Strike off with a screed that is straight and true, with a 1½- to 2-inch surcharge.
- Check surfaces with highway straightedges often, and build up low spots when detected.

Low spots can be corrected by applying additional concrete to fill them. If low spots are filled with extra-wet concrete, this concrete will settle more than the surrounding areas during the interval between strike-off and floating operations. The best way to fill low spots is with the same concrete. To do this, discharge a small amount of concrete in a container during placing operations. Use this material to level off the low spots.

4.0.0 HARDENED CONCRETE

Concrete is capable of providing a highly durable, serviceable, and attractive surface. When it does not do so, there are always reasons. By keeping the cause of certain imperfections in mind, it is possible to reduce the likelihood of unsatisfactory results. Most flaws develop in hardened concrete when it is not placed, finished, and/or cured correctly. If concrete is allowed to dry out too fast after placing, for example, or if it is finished excessively, surface flaws usually develop. The most common flaws are crazing, cracking, dusting, and scaling. These and other flaws are discussed in the following sections.

Another common cause of surface imperfections is the lack of prompt curing. The degree to which this can be accomplished, especially in dry or windy weather, will greatly improve the quality of the flatwork surface. Moist curing is best, provided the slab is kept continuously moist.

Figure 6 is a chart of the types of flaws that appear after the concrete sets up and hardens. Rarely will there be a single cause for a given imperfection; usually, some combination will be responsible. The influence of any cause will vary with the degree of its departure from best practice, with the properties of the materials used, and with the ambient temperature and other weather conditions present during the work. Satisfactory results are more likely to be obtained if the causes mentioned for the various kinds of problems are carefully avoided.

PROBLEM	CAUSE	PREVENTION OR CORRECTION
Hairline surface cracks or crazing	High-slump concrete	Reduce water content.
	Excess and/or early troweling	Do not trowel or overwork a wet surface.
	Rapid drying of wet surface	Provide cure immediately after finishing.
Shrinkage cracks	Inadequate spacing or depth of control joints	Space joints in feet about 2-1/2 times slab thickness in inches. Cut joints 1/4 of slab depth. Place joints at changes in direction or thickness or width of slab.
	Late joint sawing	Saw as soon as possible. Avoid sawing during drop in concrete temperature.
	Bond not broken between slab and other elements	Break bond with expansion joint where slab abuts other elements.
	Excessive shrinking	Reduce water in mix. Cure immediately after finishing.
Dusting	Same as Hairline	Same as Hairline.
	Excess water to surface	Remove water before finishing. Early application of chemical surface hardener.
	Carbonation	Grind and refinish surface.
Scaling and Spalling	Freeze-thaw exposure with application of deicer salts	Keep air content at 5 to 8 percent, more with aggregate less than 3/4 inch. Test air content frequently.
	Excess finishing of overwet surface	Protect against rain; minimize surface finishing, use broom or burlap finish. Use minimum slump, 4 inches maximum.
	Inadequate curing and drying before deicers applied	Pavement needs at least 30 days drying after curing before surface deicers applied. Avoid deicers during first winter.
	Quality of mix	Use minimum slump necessary for placing. Strength should be 3500 to 4000 psi before exposure to deicers.
Popouts	Expansion or splitting or porous aggregate near surface	Use better quality aggregate. Use two lift construction.
Blistering	Surface sealed too soon; rising air and water trapped	Delay start of floating, troweling. Use wood float to open surface. Press or pound blisters down.
Honeycomb	Inadequate vibration for workability of mix	Insert vibrator more frequently, closer together. Vibrate near form surface. For heavily reinforced or hard-to-reach areas, consider superplasticizer. Reduce lift depth. Check coarse aggregate gradation and sand content.
Voids or Bugholes	Air or water trapped against form surface and not worked out during placement	Reduce lift height. Reduce percent of sand in mix. Check form treatment; use form release agent. Insert vibrator more frequently. Vibrate as near to form face as possible. Use spading around slab edges.

109F06.EPS

Figure 6. Flaws In Hardened Concrete

4.1.0 CRAZING

Crazing is a network of fine cracks or fissures that occur randomly on the surface of the concrete, as shown in *Figure 7*. These cracks are very shallow, often resembling a crushed eggshell; they usually become visible once the concrete has dried. Crazing cracks are unsightly and can collect dirt, but crazing is not a serious structural defect. Crazing normally does not indicate the start of future deterioration.

109F07.TIF

Figure 7. Crazing

Crazing is caused by incorrect finishing or curing methods. Crazing occurs during finishing when the surface is troweled too early, with bleed water present. Crazing occurs during curing when the surface can dry too rapidly. This is common in hot weather, when there is low humidity, high air temperature, hot sun, and/or drying wind.

To prevent crazing, follow these guidelines:

- If there is water on the surface, do not sprinkle dry cement over it in an attempt to absorb it.
- Limit troweling and do not trowel too early. Excessive troweling can work excess water back into the concrete and increase the water-cement ratio at the surface.
- Dampen the subgrade before placing in order to prevent it from absorbing too much water from the concrete.
- During dry weather conditions, cure the concrete quickly, right after final troweling. If it is very hot, with full sun, use a water curing process to stop rapid drying and lower the surface temperature.

4.2.0 SHRINKAGE CRACKING

Cracking appears as irregular fractures in the concrete surface. It can be caused by many factors: the settlement, expansion, and contraction of concrete; wind and low humidity; and the weight of applied loads. The most obvious cause is the shrinkage that occurs as the concrete dries. Concrete also expands and contracts as temperature changes. This will also cause shrinkage cracks.

Cracks that occur after hardening, as shown in *Figure 8*, are usually the result of long-term drying shrinkage, or thermal expansion and contraction, such as during freeze-thaw cycles. These drying-shrinkage cracks may take several weeks to appear.

109F08.TIF

Figure 8. Drying-Shrinkage Cracks

It is not possible to avoid cracks totally, but they can be controlled. The major way to control cracks is to relieve the stress caused by shrinkage and temperature changes. This is usually accomplished by placing joints in the concrete. There are three kinds of joints, two of which are expressly for crack control.

- Control or contraction joints are induced cracks that are placed in the concrete so that cracking will occur in these joints rather than in a random manner. These joints are at least one-fourth the thickness of the slab, to form a plane of weakness along which the slab will crack when it contracts.
- Isolation joints are deliberate disconnections that create a barrier between the slab and another structure such as a wall, column, or adjoining slab. The isolation joint prevents the two structures from bonding, effectively isolating the slab from the effect of any difference in movement between it and the other structure.
- Construction joints mark the end of a day's placing. They are often spaced to act as isolation or control joints.

Another way to control cracks is to increase the tensile strength of the concrete. This is done by adding fibers, wire mesh, or reinforcing steel bars to the concrete. The reinforcement reduces the size of the cracks, distributes them more uniformly, and holds them together tightly. Small cracks of up to ⅛ of an inch wide can be repaired by filling them with a stiff paste of cement and water or cement-based caulking material. However, larger cracks require much more work to repair.

Cracking can be reduced by taking the following precautions:

- Make sure the subgrade is firm and uniformly compacted.
- Slow the evaporation time as much as possible. Dampen the subgrade in warm weather, just before placing the concrete. Use sunscreens and windbreaks.
- Prevent moisture loss in warm or windy weather by covering the surface with white plastic sheeting or wet burlap immediately after darbying or bullfloating. Keep the burlap continuously wet until the concrete is ready for finishing.
- Start curing quickly and properly. Good curing helps the concrete gain strength before shrinkage stresses can develop.
- In cold weather, keep concrete warm, but add moisture to the air to retard the evaporation rate.
- Keep the concrete temperature uniform during curing. Protect it from sudden or extreme temperature changes.

4.3.0 DUSTING

Dusting is the development of a fine, powdery material that easily rubs off the surface of concrete, as shown in *Figure 9*. Dusting is caused by a too-wet mix, a mix with too little cement, surface freezing, improper curing methods, or working the surface when it has bleed water on it. Floating and troweling with bleed water present can cause the sand and cement to work to the surface, weakening its strength and durability, and giving rise to dusting.

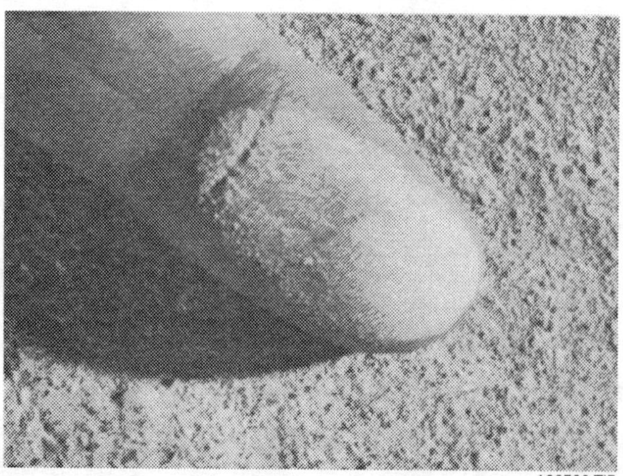

109F09.TIF

Figure 9. Dusting

Dusting is common in cold weather, when the surface freezes before it has a chance to dry and harden. In the winter, heaters are often used to keep concrete warm while it is setting. The carbon monoxide generated from heaters can react chemically with fresh concrete, causing carbonation which produces dusting.

Dusting occurs in spring and fall when the setting concrete becomes colder than the air at night. This temperature change results in condensation forming on the surface of the concrete. If floating and troweling have not been completed yet, this moisture can weaken the cement paste and cause dusting.

Dusting can be prevented by employing the following methods:

- Wait for bleed water to evaporate before finishing the concrete, or remove it by squeegee or vacuum.
- Follow proper curing procedures for weather conditions.
- Protect the concrete from extreme temperatures.
- Protect the concrete from dry heat and from gas buildup by venting all heaters properly to the outside. If you must use heaters, relocate them periodically to prevent any one area from being subjected to a harmful amount of heat. Also, put humidifiers on heaters to increase the relative humidity.

4.4.0 SCALING

Scaling appears as flaking of hardened concrete at its surface, as shown in *Figure 10*. This occurs when the surface cement surrounding the aggregates peels away, exposing the coarse aggregate. One of the most common causes of concrete scaling is early freezing of the surface before the concrete has gained sufficient strength to withstand freezing stresses. Eventually, it will disintegrate to the depth the fresh concrete has frozen. Another cause is repeated cycles of freezing and thawing of non-air-entrained concrete.

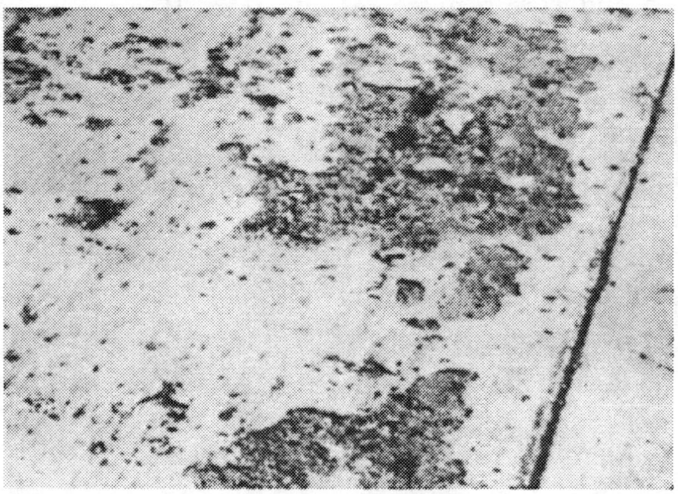

109F10.TIF

Figure 10. Scaling

The more porous the surface, the more water can be absorbed by the concrete, creating the possibility for damage by scaling.

In cold weather, a deicer solution may be sprinkled over concrete to prevent ice buildup. Unfortunately, this deicer can enter the pores in the surface of the concrete and trap moisture, which will then freeze inside the concrete. Deicing salts should not be applied until concrete has dried out and been through one winter.

Faulty workmanship, such as finishing while bleed water is present on the surface, or insufficient curing, can also lead to scaling of concrete surfaces exposed to freezing temperatures. Scaling can be prevented in much the same way other defects can be prevented.

- Follow good finishing practices and do not work the concrete when bleed water is present. Any finishing step done while water is on the surface contributes to scaling. Remixing water into the top of the slab may cause laitance to form, which can scale off.
- Keep the concrete at a temperature well above freezing while it cures.
- Follow good curing practices and cure promptly so that the surface will be as strong and durable as the rest of the concrete.
- Allow sufficient air-drying time before deicers are applied. To be safe, do not use deicing chemicals the first winter. Deicers containing ammonium nitrate, ammonium sulfate, or other harmful chemicals should never be used on any concrete. They will cause disintegration.

4.5.0 SPALLING

Spalling, as shown in *Figure 11*, marks a deeper defect in the concrete than scaling, although the two terms are sometimes used interchangeably. Spalling can appear as oval cavities on the surface or as broken fragments along a joint. When concrete spalls, surface fragments break off. Spalled areas of concrete can be repaired if no more than the top ⅓ of the slab is damaged and the underlying material is sound. If more than ⅓ of the thickness is damaged, or if steel reinforcement is uncovered, a full depth repair is required. Note that the spall in *Figure 11* is deep enough to reveal the reinforcement. The immediate cause of the spalling can be a blow to the surface, freezing and thawing, or rusting of retarders. However, the real cause of spalling could be the corrosion of embedded metals such as rebar, or a poor mix with soft, porous, weak aggregates.

To prevent spalling, the concrete mix should be designed for its environment. The next step is to follow proper finishing and curing practices. Proper curing is one of the most important ways to avoid spalling. Good overall curing can greatly increase the strength of concrete.

109F11.TIF

Figure 11. Spalling

4.6.0 POPOUTS

A **popout**, as shown in *Figure 12*, is a shallow, conical fragment that breaks out of the surface of the concrete, leaving a hole that may vary in size from ¼ of an inch to as much as 1 foot in diameter. There is usually a piece of aggregate at the bottom of the hole and at the point of the popout's cone, as in *Figure 12*. Generally, popouts occur when porous aggregates split or expand near the surface of the concrete. This expansion creates internal pressure that causes a piece of the concrete to pop out of the surface.

109F12.TIF

Figure 12. Popouts

Popouts are fairly common in flatwork, such as sidewalks. They occur in moderate or severe climates and can develop any time after the concrete is placed. Popouts are caused by porous aggregate particles. Because porous rock has a high rate of absorption, it absorbs moisture. When this moisture freezes, the rock expands, creating internal pressures that are sufficient to rupture the concrete surface.

Popouts may also occur from the chemical reaction between the alkalis in the cement and reactive siliceous aggregates. The easiest way to reduce popouts is to use good-quality aggregates. Popouts can also be reduced by using two-lift construction with clean, sound rock and aggregates.

Popouts can be repaired by patching with special repair materials. To make a proper repair, the spalled particle at the bottom of the hole should be drilled out before filling the hole with the appropriate patch material.

4.7.0 BLISTERS

Blisters appear on the surface of a concrete slab during finishing operations as bumps of varying sizes, as shown in *Figure 13*. Blisters range from ¼ of an inch to 4 inches in diameter and approximately ⅛ of an inch deep. They are caused by bubbles of air or water that are trapped just below the surface after it has been sealed.

109F13.TIF

Figure 13. Blisters

Blisters can appear at any time during finishing and without apparent cause. However, there are three principal causes of blistering.

- *Entrapped air* – An excess amount of entrapped air in a sandy mix can result in a sticky concrete that more easily seals when it is floated or finished. Sticky mixes have a tendency to form a top crust under drying winds while the remainder of the concrete remains plastic. As a result, the entrapped air inside will rise to the surface.
- *Finishing* – Finishing when the concrete is still spongy or wet can entrap air toward the surface. During the second or third pass, the air is forced ahead of the trowel blade under the surface skin until enough is concentrated to form the blister.
- *Incorrect vibration* – Too little vibration does not release entrapped air, while too much vibration brings too many fines to the surface. This latter leads to crusting, and trapping air.

Usually, the blisters do not surface until some time after the first troweling. This is because the surface of the concrete shrinks as it dries, but the air trapped under it does not shrink. Instead, this air expands, probably from the heat of hydration, and the pockets appear as surface blisters.

The following general rules can be followed to help prevent blisters:

- Do not work the concrete too much. Overworking it may cause the aggregate to settle and bleed water to rise.
- Do not seal the surface too soon. Sealing the surface too quickly can trap the air inside.
- Use a wood bullfloat on non-air-entrained concrete to avoid sealing the surface.
- Protect the surface from evaporation. In cooler weather, the subgrade may need to be warmed.
- Use flat troweling on the first pass to avoid enfolding air into the surface under the blade action.

If you find that blisters are forming, try to either flatten the trowel blades, or open the surface with a wood float and delay finishing as long as possible. If the entrapped air can escape or subside, the skin of the blister can renew contact with the concrete underneath. Delaying troweling will allow blisters to subside. Blisters can also be pressed or pounded down, but the top skin may still flake off later.

4.8.0 EFFLORESCENCE

Efflorescence, as shown in *Figure 14*, appears as a white deposit on the concrete surface. It occurs naturally over time when the mineral salts contained in cement and brick dissolve in moisture. The moisture carries the mineral salts to the surface. The water then evaporates, leaving the salts on the surface.

Most salts are highly soluble in water, so the first rain will often wash them off the surface. Some minerals may persist. Over time, efflorescence will become lighter and less extensive.

Light-colored surfaces show the deposits much less than darker surfaces. Although the deposits are harmless and will eventually disappear, this may take a couple of years. The easiest way to remove efflorescence is to dry brush, scrub, or sandblast it away.

109F14.TIF

Figure 14. Efflorescence

4.9.0 DISCOLORATION

Surface discoloration of concrete flatwork, as shown in *Figure 15*, may appear as contrasting or gross color changes in large areas of concrete, white patches, or random spotted or mottled, light or dark blotches. Discoloration can be caused by poor mix design. Poorly mixed calcium chloride admixtures also cause discoloration.

109F15.TIF

Figure 15. Discoloration

Uneven curing can cause uneven drying of the concrete, which can also result in discoloration or uneven coloring. Therefore, if even coloration is important, make sure that the curing membrane is applied evenly over the entire surface of the slab. Moisture barriers such as waterproof paper must be thoroughly sealed at joints, must be anchored carefully along edges, and must be as wrinkle-free as possible. Avoid using polyethylene sheeting, as it may wrinkle and cause mottling through condensation.

Discoloration can result from trying to hard-trowel the surface after it has become too stiff to trowel properly. Vigorous troweling can compact a surface to the point where the water-cement ratio is drastically decreased in localized areas. This dense, low-water concrete in the hard-troweled area is almost always darker than adjacent concrete.

The following guidelines can be used to help prevent discoloration and color variation:

- Use consistent ingredients and proportions for all batches.
- Use proper placing and finishing techniques.
- Cure the concrete thoroughly and uniformly, using a wet-cure method that does not require plastic sheeting.
- Apply a clear coating or sealer. This will help prevent discoloration caused by air pollution.
- Avoid calcium chloride admixtures.

4.10.0 HONEYCOMBS

Honeycombs are irregular holes that usually occur in walls, as shown in *Figure 16.* They occur when the cement mortar does not completely fill the voids between the coarse aggregate particles. This becomes evident once forms are removed. Usually, honeycomb results from poor consolidating. It can be eliminated in wet concrete by inserting the vibrator at proper spacing, operating the vibrator long enough, and extracting it slowly.

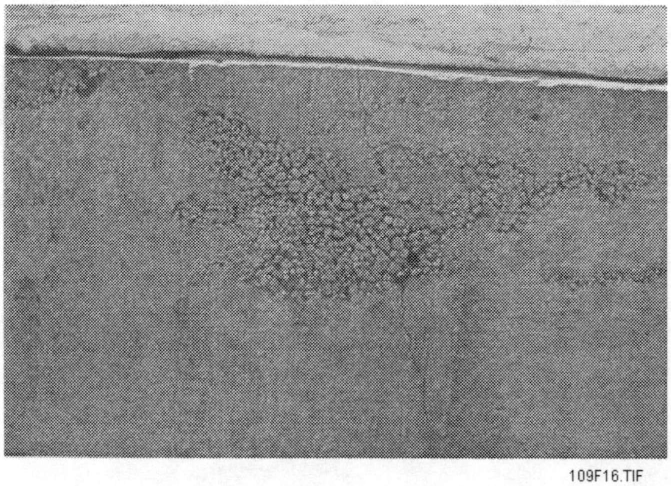

109F16.TIF

Figure 16. Honeycombs

4.11.0 AIR POCKETS (BUGHOLES)

Air pockets occur when water or air becomes trapped against the surface of the form. This creates small pockets or voids, as shown in *Figure 17*. These pockets, or bugholes, occur more often with air-entrained concrete because the air bubbles in the concrete make it more difficult to work the entrapped air or water to the surface.

109F17.TIF

Figure 17. Bugholes

Air pockets can be reduced in wet concrete by reducing the height of the lift, and by increasing vibration. Make sure you move the vibrator as close to the form face as possible and insert the vibrator frequently. Using a form release agent is also helpful.

SUMMARY

Concrete is extremely durable and can last several lifetimes. In order to be durable, concrete must be mixed, placed, finished, and cured properly. The use of improper construction procedures can result in damage and defects to the concrete.

An improper mix or poor agitation can cause defects in fresh concrete, although most defects are caused by improper finishing or curing procedures. Defects in fresh concrete include excess bleeding, segregation, rapid set, and low spots. They usually affect the surface of the concrete.

Improper finishing and curing procedures can also cause defects in hardened concrete. Surface defects in hardened concrete include crazing, cracking, dusting, scaling, popouts, blisters, and efflorescence. Defects that affect deeper levels of the concrete include honeycombs, air pockets, and spalling. These defects tend to make extensive repairs necessary.

The best way to prevent these defects from occurring is to use proper mixing, placing, finishing, and curing procedures.

References

For advanced study of topics in this task module, the following books and videos are suggested.

Concrete Construction and Estimating, Craig Avery, Craftsman Book Company, Carlsbad, CA, 1991.

Concrete Manual, International Conference of Building Officials, Whittier, CA, 1992.

Concrete Parking Lots: Eight Steps to Quality Construction, Videotape, Portland Cement Association, Skokie, IL.

Concrete Technology, Third Edition, George R. White, Delmar Publishers, Inc., Albany, NY, 1991.

Design and Control of Concrete Mixtures, Thirteenth Edition, Portland Cement Association, Skokie, IL, 1988.

Finishing Concrete Flatwork, Videotape 32:00 minutes, Portland Cement Association, Skokie, IL, 1991.

Guide for Concrete Floor and Slab Construction, ACI 302.1R-96, American Concrete Institute, Farmington Hills, MI, 1997.

The Contractor's Guide to Quality Construction, American Concrete Institute, Farmington Hills, MI, 1997.

ACKNOWLEDGMENTS

Figure 1 courtesy of Roy Jorgensen Associates, Inc.
Figures 2-4, 6, 16, and 17 courtesy of American Concrete Institute
Figures 5, 7-13, and 15 courtesy of Portland Cement Association
Figure 14 courtesy of International Conference of Building Officials

SI* (MODERN METRIC) CONVERSION FACTORS

APPROXIMATE CONVERSIONS TO SI UNITS

Symbol	When You Know	Multiply By	To Find	Symbol
		LENGTH		
in	inches	25.4	millimeters	mm
ft	feet	0.305	meters	m
yd	yards	0.914	meters	m
mi	miles	1.61	kilometers	km
		AREA		
in^2	square inches	645.2	square millimeters	mm^2
ft^2	square feet	0.093	square meters	m^2
yd^2	square yards	0.836	square meters	m^2
ac	acres	0.405	hectares	ha
mi^2	square miles	2.59	square kilometers	km^2
		VOLUME		
fl oz	fluid ounces	29.57	milliliters	mL
gal	gallons	3.785	liters	L
ft^3	cubic feet	0.028	cubic meters	m^3
yd^3	cubic yards	0.765	cubic meters	m^3
NOTE: Volumes greater than 1000 l should be shown in m^3.				
		MASS		
oz	ounces	28.35	grams	g
lb	pounds	0.454	kilograms	kg
T	short tons (2000 lb)	0.907	megagrams (or "metric ton")	Mg (or "t")
		TEMPERATURE (exact)		
°F	Fahrenheit temperature	5(F-32)/9 or (F-32)/1.8	Celsius temperature	°C
		FORCE AND PRESSURE OR STRESS		
lbf	poundforce	4.45	newtons	N
lbf/in^2	poundforce per square inch	6.89	kilopascals	kPa

APPROXIMATE CONVERSIONS FROM SI UNITS

Symbol	When You Know	Multiply By	To Find	Symbol
		LENGTH		
mm	millimeters	0.039	inches	in
m	meters	3.28	feet	ft
m	meters	1.09	yards	yd
km	kilometers	0.621	miles	mi
		AREA		
mm^2	square millimeters	0.0016	square inches	in^2
m^2	square meters	10.764	square feet	ft^2
m^2	square meters	1.195	square yards	yd^2
ha	hectares	2.47	acres	ac
km^2	square kilometers	0.386	square miles	mi^2
		VOLUME		
mL	milliliters	0.034	fluid ounces	fl oz
L	liters	0.264	gallons	gal
m^3	cubic meters	35.71	cubic feet	ft^3
m^3	cubic meters	1.307	cubic yards	yd^3
		MASS		
g	grams	0.035	ounces	oz
kg	kilograms	2.202	pounds	lb
Mg (or "t")	megagrams (or "metric ton")	1.103	short tons (2000 lb)	T
		TEMPERATURE (exact)		
°C	Celsius temperature	1.8C + 32	Fahrenheit temperature	°F
		FORCE AND PRESSURE OR STRESS		
N	newtons	0.225	poundforce	lbf
kPa	kilopascals	0.145	poundforce per square inch	lbf/in^2

* SI is the symbol for the International System of Units. Appropriate rounding should be made to comply with Section ASTM E380.

REVIEW/PRACTICE QUESTIONS

Choose one correct answer for each question.

1. Prolonged contact between skin surfaces and wet, unhardened concrete can cause _____.
   a. drying
   b. severe chemical burns
   c. bleeding
   d. cuts

2. A troubleshooting method is a logical procedure for _____.
   a. identifying laitance
   b. researching problems
   c. preventing defects
   d. solving problems

3. The first step in a troubleshooting procedure is to _____.
   a. adjust the procedures
   b. research the problem
   c. investigate the symptoms
   d. fix the problem

4. One way to identify the causes of a concrete problem is to _____.
   a. wait for complete deterioration
   b. compare the symptoms with normal conditions
   c. cut the slab into small pieces
   d. use etching acid on the surface

5. A fine white powder on the surface of a concrete slab would be a _____ of a dusting problem.
   a. cause
   b. simulation
   c. model
   d. symptom

6. Problems in fresh concrete do not include _____.
   a. bleeding
   b. honeycombs
   c. flash set
   d. plastic shrinkage cracks

7. Excess bleed water may cause _____.
   a. symptoms
   b. laitance
   c. popouts
   d. spalling

8. A network of fine cracks that occur randomly on the concrete surface is called _____.
   a. cracking
   b. spalling
   c. crazing
   d. scaling

9. The primary way to control cracks is by _____.
   a. increasing strength
   b. creating furrows
   c. increasing stress
   d. creating joints

10. Dusting is most prevalent _____.
   a. in cold weather
   b. in hot weather
   c. during evaporation
   d. before condensation

11. Flaking of the concrete is known as _____.
   a. cracking
   b. spalling
   c. crazing
   d. scaling

12. Oval cavities on the concrete's surface are known as _____.
   a. cracking
   b. spalling
   c. crazing
   d. scaling

13. Popouts _____.
   a. are fairly common in vertical work
   b. are caused by a very dry mix
   c. always leave huge holes
   d. occur when aggregates expand near the surface

14. What is one cause of blisters in concrete slabs?
    a. Entrapped air created during the finishing process
    b. Freezing temperature during placing
    c. Chemicals in the aggregate reacting with the cement paste
    d. Poor form construction

15. A white powdery substance in the concrete surface that is caused by mineral salts is known as _____.
    a. crazing
    b. spalling
    c. efflorescence
    d. dusting

16. Honeycombs _____.
    a. become evident once forms are removed
    b. are caused by excessive vibration
    c. occur often in sidewalks
    d. are caused by excessive clearances

17. Large air pockets just beneath the slab surface are also referred to as _____.
    a. scaling
    b. blisters
    c. bugholes
    d. popouts

18. What does not cause discoloration?
    a. Poor mix design
    b. Inadequate curing practices
    c. Poor finishing practices
    d. Surface irregularities

19. The last step in troubleshooting is _____.
    a. identifying symptoms
    b. researching problems
    c. applying corrective actions
    d. talking to your supervisor

20. Slow set and rapid set are related to _____.
    a. repair methods
    b. surface defects
    c. finishing techniques
    d. weather problems

notes

ANSWERS TO REVIEW/PRACTICE QUESTIONS

Answer	Section Reference
1. b	1.2.0
2. d	2.0.0
3. c	2.1.0
4. b	2.1.0
5. d	2.2.0
6. b	3.0.0
7. b	3.1.0
8. c	4.1.0
9. d	4.2.0
10. a	4.3.0
11. d	4.4.0
12. b	4.5.0
13. d	4.6.0
14. a	4.7.0
15. c	4.8.0
16. a	4.10.0
17. c	4.11.0
18. d	4.9.0
19. c	2.3.0
20. d	3.0.0/Figure 2

NCCER CRAFT TRAINING USER UPDATES

The NCCER makes every effort to keep these manuals up-to-date and free of technical errors. We appreciate your help in this process. If you have an idea for improving this manual, or if you find an error, a typographical mistake, or an inaccuracy in the NCCER's Craft Training Manuals, please write us, using this form or a photocopy. Be sure to include the exact module number, page number, a description of the problem, and the correction, if possible. Your input will be brought to the attention of the Technical Review Committee. Thank you for your assistance.

Instructors – If you found that additional materials were necessary in order to teach this module effectively, please let us know so that we may include them in the Equipment/Materials list in the Instructor's Guide.

Write: Curriculum and Revision Department
National Center for Construction Education and Research
P.O. Box 141104
Gainesville, FL 32614-1104

Fax: 352-334-0932

Craft ______________________ Module Name ______________________

Module Number ______________________ Page Number(s) ______________________

Description of Problem

____

____

____

____

(Optional) Correction of Problem

____

____

____

(Optional) Your Name and Address

____

____

____

notes